P9-AFS-304

DATE DUE

DEMCO 38-296

Separation Process Technology

Chemical Engineering Books

Separation Process Technology

Jimmy L. Humphrey
J. L. Humphrey & Associates
Austin, Texas

George E. Keller II
Union Carbide Chemicals & Plastics Company
South Charleston, West Virginia

McGraw-Hill

New York   San Francisco   Washington, D.C.   Auckland   Bogotá
Caracas   Lisbon   London   Madrid   Mexico City   Milan
Montreal   New Delhi   San Juan   Singapore
Sydney   Tokyo   Toronto

Riverside Community College
'98   Library
4800 Magnolia Avenue
Riverside, CA 92506

TP 156 .S45 H86 1997

Humphrey, Jimmy L.

Separation process
technology

olication Data

ʝy L. Humphrey, George E. Keller II.

—) and index.

., George E.   II. Title.

TP156.S45H86   1997
660'.2842—dc21

97-12
CIP

McGraw-Hill

*A Division of The **McGraw·Hill** Companies*

Copyright © 1997 by The McGraw-Hill Companies, Inc. All rights reserved. Printed in the United States of America. Except as permitted under the United States Copyright Act of 1976, no part of this publication may be reproduced or distributed in any form or by any means, or stored in a data base or retrieval system, without the prior written permission of the publisher.

1 2 3 4 5 6 7 8 9 0   FGR/FGR   9 0 2 1 0 9 8 7

ISBN 0-07-031173-0

The sponsoring editors for this book were Zoe G. Foundotos and Bob Esposito, the editing supervisor was Bernard Onken, and the production supervisor was Pamela A. Pelton. It was set in Century Schoolbook by Dina E. John of McGraw-Hill's Professional Book Group composition unit.

Printed and bound by Quebecor/Fairfield.

McGraw-Hill books are available at special quantity discounts to use as premiums and sales promotions, or for use in corporate training programs. For more information, please write to the Director of Special Sales, McGraw-Hill, 11 West 19th Street, New York, NY 10011. Or contact your local bookstore.

Information contained in this work has been obtained by The McGraw-Hill Companies, Inc. ("McGraw-Hill") from sources believed to be reliable. However, neither McGraw-Hill nor its authors guarantees the accuracy or completeness of any information published herein and neither McGraw-Hill nor its authors shall be responsible for any errors, omissions, or damages arising out of use of this information. This work is published with the understanding that McGraw-Hill and its authors are supplying information but are not attempting to render engineering or other professional services. If such services are required, the assistance of an appropriate professional should be sought.

This book is printed on recycled, acid-free paper containing a minimum of 50% recycled, de-inked fiber.

To the number one person in each of our lives, our wives, Barbara and Judy. Without their love and encouragement, this book would not have been written.

Jimmy Humphrey and George Keller

Contents

Foreword

Needs for separating mixtures are pervasive in the process industries. With increased attention in recent years to alternative fuels and feedstocks, along with needs for preventing pollution of the environment and recovering new products, the variety of separations to be accomplished has become much wider. As well, the arsenal of available methods of separation and means of implementing particular methods of separation has become more and more extensive. Hence there has been a growing need for practicing engineers and scientists to develop an understanding of the capabilities of different methods of separation and different types of separations equipment on a comparative basis. This book fills that need.

I have had the pleasure and fulfillment of working closely with both of the authors, George Keller and Jimmy Humphrey, in various capacities over the years. They are uniquely qualified in terms of their long and complementary practical experience with a wide variety of separations and applications, as well as the very creative way in which they have approached particular separations needs. Both authors combine strong intuitive understanding of separations with their practical expertise.

C. Judson King
Provost and Senior Vice President—Academic Affairs
University of California
Professor of Chemical Engineering
Berkeley Campus

Preface

This book will help engineers, chemists, managers, and others increase profitability in industrial plants by optimizing performance of separation processes. It focuses on selection and scaleup of separation processes to increase plant capacity, improve product quality, and reduce costs.

Addressing both product purification and environmental applications, it focuses on distillation, extraction (liquid/liquid and supercritical), adsorption, and membrane processes, and their hybrid systems. Professionals throughout the world in the chemical, petroleum, pharmaceutical, food, paper, and textile industries can benefit from this book.

Specific information includes: process descriptions, flow sheets, ranges of performance, process trade-offs, applications examples, advantages and disadvantages of processes, performance data, scaleup information, economic case studies, principles of thermodynamic efficiency, energy consumption data, guidelines for process selection, and coverage of leading-edge technologies. This information is provided to answer the following questions.

- *Which process?* When building a new plant or retrofitting an existing one, the first step is to select the type of process to use. For example, should distillation or adsorption be used? Such decisions can significantly affect plant profitability.

- *Which process configuration?* After the type of process is selected, one must then decide which process configuration to use. For example, if distillation is selected, should the column be equipped with trays or a packing? In adsorption, what regeneration configuration should be selected?

- *What "leading-edge" technologies are available?* The authors have given considerable attention to coverage of "leading-edge" technologies.

- *What follow-up information is available?* Performance data, guidelines, and references are included in each chapter. Equipment suppliers and their products are provided in the Appendix.

The authors would be pleased to receive comments or experiences from readers who wish to provide suggestions or information for possible inclusion in the second edition. Contact information for both authors follows:

Dr. Jimmy L. Humphrey
J. L. Humphrey & Associates
3605 Needles Dr.
Austin, TX 78746
Phone: 512/327-5599
Fax: 512/328-6725
e-mail: jlhsepns@aol.com

Dr. George E. Keller II
Union Carbide Corporation
P.O. Box 8361
So. Charleston, WV 25303
Phone: 304/747-5484
Fax: 304/747-5570
e-mail: agekrd1@peabody.sct.ucarb.com

Acknowledgments

Cecelia Senter had the primary responsibility for completing the various parts of this book. This included typing of the manuscript, and completion of most of the figures, graphs, and tables. Without her many talents and hard work this book would not have been possible. The results of many discussions and projects with Frank Seibert (University of Texas at Austin, and a Research Associate with J.L. Humphrey & Associates) found their way into these pages. Appreciation is given to Hoechst Celanese Corporation, Advanced Technology Group, specifically, Ravi Prasad and H. Wayne Swofford, for allowing the results of an economic analysis for recovery of volatile organic compounds (VOCs) from air to be included in this book.

Zoe Foundotos and Bob Esposito of McGraw-Hill provided key and timely guidance in the preparation of all parts of the book. Liz Clare and Mary Clare provided excellent suggestions on how to improve the format of the manuscript. Max Hoberman used some of his many computer skills to complete most of the more complex flowsheets. Joan Rodberg of Union Carbide was especially helpful in supplying information on membrane technology, as was Jim Williams on waste treatment costs.

During the writing of this book, notes were sent out to industry, academia, and government asking for suggestions on how to make it valuable to readers. More than 50 responses were received. They were very helpful and appreciated.

1

Introduction

It is easy to mix two liquid components like ethanol and water. They are miscible in all proportions and will form a homogeneous solution with only a modest amount of mixing. However, after mixing, suppose it is desired to separate the components back into their pure states.

The first approach might be to boil off one of the components. Unfortunately, an *azeotrope* is formed, and complete separation is not possible. Another approach might be to use a solvent to extract the ethanol. However, an effective solvent may not be commercially available, and one must find a way to later separate the solvent from the ethanol so that the solvent can be recycled. The use of an adsorbent to selectively remove either ethanol or water from the solution is possible, but is such an adsorbent available? If so, where can it be purchased? On the other hand, it may be better to use a membrane separation process, but how could one find out if such a membrane is available, and if so, where can it be purchased? Of these various processes, which one should be used? Of course, for any process selected, one would want to consider using leading-edge technology. It is our goal to provide some of the answers to these types of questions.

Roles of Separation Processes in Industry

Separation processes play critical roles in industry, including the removal of impurities from raw materials, purification of products, separation of recycle streams, and removal of contaminants from air and water effluents (Figure 1.1). Overall, separation processes account for 40–70% of both capital and operating costs in industry and their proper application can significantly reduce costs and increase profits.

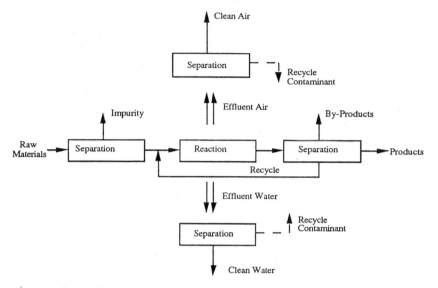

Figure 1.1 Separation processes in manufacturing.

Separating Agents

The heart of the separation process is the *separating agent* which can take the form of energy or mass (King, 1980). In distillation and evaporation, the separating agent is heat, which is a form of energy. In extraction, the solvent causes the separation, and because it is a mass, it is defined as a *mass separating agent,* a subclass of separating agents. In adsorption, the mass separating agent is the adsorbent, and in membrane processes, it is the membrane material.

In evaluating relative economics of processes, one approach is to determine what the performance of the mass separating agents must be to compete economically with the baseline process, which is normally distillation. This approach was used in comparing distillation with a membrane process (Hinchcliffe and Porter, 1995). Some examples of separation processes and their separating agents are given in Table 1.1.

Technological Maturity of Processes

Analyses to determine technological and use maturity of various types of separation processes have been completed (Keller, 1987). As part of these analyses, 16 separation experts were asked to estimate how closely different separation processes approached their "techno-

TABLE 1.1   Examples of Separation Processes and Separating Agents

Process	Separating agent(s)	Application(s)
Absorption	Solvent*	Removal of carbon dioxide and hydrogen sulfide from natural gas with amine solvents
Adsorption and ion exchange	Adsorbent*/resin*	Separation of meta- and paraxylene, air separation, water demineralization
Chromatography	Adsorbent*	Separation of sugars
Crystallization	Heat removal	Production of beverages such as "ice" beer
Distillation	Heat	Propylene/propane separation, production of gasoline from crude oil, air separation
Drying	Heat	Drying of ceramics, plastics, and foods
Electrodialysis	Membrane*	Water desalination
Evaporation	Heat	Water desalination, sugar manufacture
Extraction	Solvent*	Recovery of benzene/toluene/xylenes from gasoline reformate, removal of caffeine from coffee
Membranes	Membrane*	Separation of hydrogen from hydrocarbons, concentration of fruit juices, water desalination
Stripping	Stripping gas*	Removal of benzene from wastewaters

*Because these separating agents are in the form of a mass, they are defined as "mass separating agents," a subclass of separating agents.

logical asymptote" and "use asymptote." At the "technological asymptote" it is assumed that everything is known about the process and no further improvements are possible. If a process is used to the fullest extent possible, it has approached its "use asymptote." Since neither the technological nor use asymptotes have been reached by any process, experts were asked to estimate how closely each process approached their asymptotes. The results of this survey are given in Table 1.2.

Figure 1.2 graphically illustrates how closely each process was judged to approach its asymptote. As expected, distillation, the most frequently used process, is closer to its technological and use asymptotes than other processes. Because distillation is by far the most frequently used separation process, the driving force to improve it is larger than for the other processes.

TABLE 1.2 Technological Maturity—Results of Survey

Process	Technological maturity*	Use maturity*
Distillation	87	87
Extractive/azeotropic distillation	80	65
Solvent extraction	73	61
Gas absorption	81	76
Supercritical gas absorption/extraction	28	25
Adsorption: gas feed	57	52
Adsorption: liquid feed	50	40
Ion exchange	60	60
Membranes: gas feed	39	27
Membranes: liquid feed	37	30
Liquid membranes	13	13
Chromatography: liquid feed	30	22
Affinity separations	15	9
Crystallization	64	62
Electrical and other field-induced separations	24	13

*Values represent the averages of the percentage of approach to the technological or use asymptote, that is, a score of 100 would denote that a given process has reached an asymptote.

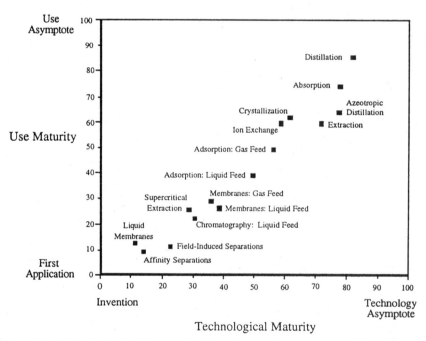

Figure 1.2 Technological and use maturities of separation processes. (*Adapted from Keller, 1987 with permission of the American Institute of Chemical Engineers. Copyright © 1987 AIChE. All rights reserved.*)

Efficiency Versus Capacity

The two key design and operating parameters in any separation process are *efficiency* and *capacity*. Efficiency is related to the mass transfer and product purity. Capacity, on the other hand, is related to the hydraulics and the rate of material which can be processed without a loss in efficiency.

Efficiency and capacity are interdependent. In any separation process, it is often necessary to compromise factors which promote efficiency (and product purity) versus factors which enhance capacity (and product rate). For example, a distillation packing that has a high surface area promotes higher efficiency, but may cause flow restrictions which result in a lower capacity. A second example is in membrane processes where capacity (flux) can be increased by increasing the size of the membrane pores, allowing molecular species to pass more quickly. However, increased pore sizes allow a broader range of molecular species to pass, thereby reducing efficiency. In the practical world, the relationship between efficiency and capacity is often a "give and take" situation as illustrated in Figure 1.3.

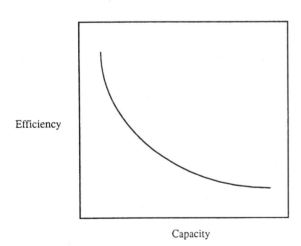

Capacity

Figure 1.3 Characteristic curve—efficiency versus capacity.

Emergence of the Monolith as a Contacting Device

A device used to provide phase contact, which has been found to give both good efficiency and capacity, is the *monolith*. The monolith is a structure composed of individual parts, which together form an organized whole without the presence of joints or seams. In recent years, the monolith has emerged as an important contacting device for a number of separation processes. Initially used in distillation in the form of structured packing, the monolith has spread into reactive distillation, extraction, adsorption, and membrane separation processes (Figure 1.4).

Structured packing used in distillation is an example of a "honeycomb-type" of monolith (Figure 1.5). A similar device is used in reactive distillation where catalyst, held by panels of screen wire, is configured like structured packing used in distillation (Figure 1.6).

In extraction, structured packing has been installed in over 30 commercial extractors (Figure 1.7). In this process, the packing provides ordered channels of flow to promote countercurrent contact of the dispersed and continuous liquid phases.

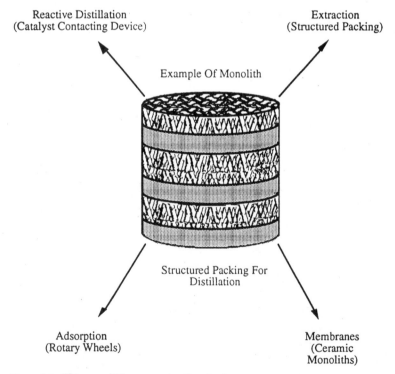

Reactive Distillation
(Catalyst Contacting Device)

Extraction
(Structured Packing)

Example Of Monolith

Structured Packing For
Distillation

Adsorption
(Rotary Wheels)

Membranes
(Ceramic
Monoliths)

Figure 1.4   The monolith as a contacting device.

Figure 1.5 Structured packing used in distillation. (*Provided by Norton Chemical Process Products Corp.*)

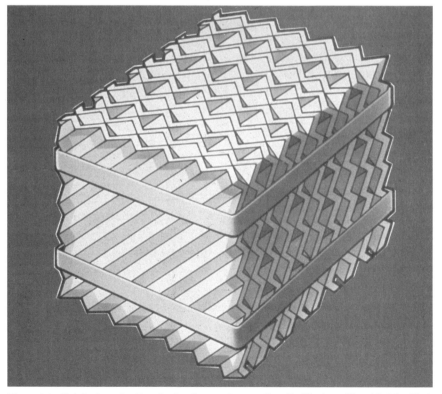

Figure 1.6 Catalyst contacting device for use in reactive distillation. (*Provided by Koch Engineering Co., Inc.*)

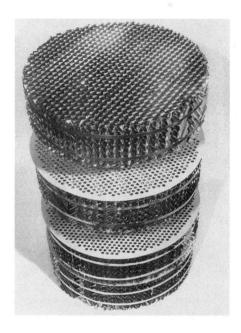

Figure 1.7 Structured packing for extraction. (*Provided by Koch Engineering Co., Inc.*)

In adsorption, the monolith takes the form of rotary wheels with surfaces coated with activated carbon or hydrophobic molecular sieves to remove organics from air. Rotary wheels find application where organic concentrations are low, volumes are high, and removal requirements of organics are not stringent, about 97% or less (Figure 1.8). Details are provided in Chapter 4.

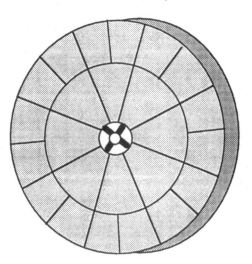

For Typical Large-Scale Application:

Wheel thickness: 3.4 m

Wheel diameter: 0.45 m

Air rate: 60,000 cu m/hr

Figure 1.8 Rotary wheel for adsorption processes.

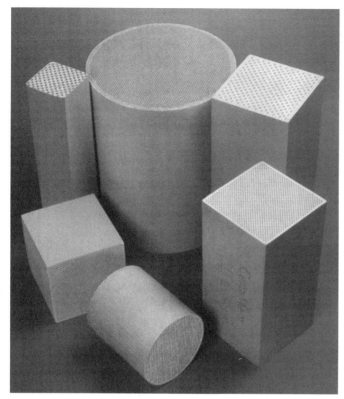

Figure 1.9 Ceramic membrane modules. *(Provided by CeraMem Corp.)*

Ceramic membrane modules, configured in the form of a monolith, are available for ultrafiltration and microfiltration applications and for filtration of solid particles from gas mixtures (Figure 1.9).

Steps Common to Designing All Separation Processes

It is important to realize that the design for any process will not be any better than the accuracy of the data upon which it is based. However, in its haste to complete projects, industry often overlooks the importance of basic data such as phase equilibria and physical properties. The steps below are common to the design of all separation processes and all of them are important.

1. Establish bases
   a. Composition of feed and product(s)
   b. Rate to be processed
   c. Operating conditions (pressure, temperature)
   d. Special conditions (including presence of suspended solids, or excursions of pH, temperature, or pressure)

2. Obtain basic data
   a. Phase equilibria or flux data (for membranes)
   b. Density, viscosity, diffusion coefficients
   c. Efficiency or mass transfer data (use plant data if available)

3. Perform process selection
   a. Critically influenced by bases and basic data
   b. Complete performance and economic evaluations

4. Complete process design
   a. Verify economics
   b. May need to change process selection

References

Hinchcliffe, A. B. and K. E. Porter, "Gas Separation Using Membranes. Part 2: Cost Optimisation of the Separation Process," *Industrial & Engineering Chemistry,* January 1995 (submitted for publication).

Keller, G. E., *Separations: New Directions for an Old Field,* American Institute of Chemical Engineers Monograph Series, New York, **83**(17), 1987.

King, C. J., *Separation Processes,* 2nd ed., McGraw-Hill, 1980.

Supplemental References

Humphrey, J. L. et al., "An Overview of Commercially Available Software for Distillation, Extraction, Adsorption, and Membrane Processes, *CACHE News,* **32,** 21, 1991.

Humphrey, J. L. et al., "Separation Technologies—Advances and Priorities," U.S. Department of Energy Final Report, DOE/ID/12920-1, February 1991.

Meyers, R. A., ed., *The Encyclopedia of Physical Science and Technology,* Academic Press, 1987.

Rousseau, R. W., ed., *Handbook of Separation Process Technology,* Wiley, 1987.

Ruthven, D. M., ed., *Encyclopedia of Separations Technology,* Wiley, 1997.

Schweitzer, P. A., ed., *Handbook of Separation Techniques for Chemical Engineers,* 2nd ed., McGraw-Hill, 1988.

Wankat, P. C., *Equilibrium Staged Separations,* Elsevier, New York, 1988.

2

Distillation

Introduction

With approximately 40,000 columns in operation in the United States, distillation is used to make 90–95% of all separations in the chemical process industry (Humphrey et al., 1991). In making these separations, it consumes the energy equivalent of 1.2 million barrels per day of crude oil. Though not energy-efficient, distillation has a simple flowsheet and is a low-risk process. It is indeed the benchmark with which all newer processes must be compared.

In this chapter, we will characterize the distillation process, present fundamentals, discuss performance, equipment, economics, and introduce new distillation configurations. Detailed design procedures for tray and packed distillation columns, as well as lists of equipment suppliers, are included in the Appendix. Because distillation is the baseline process in a number of industries, we include more information and fundamentals for distillation than for other processes.

Process description

Distillation is a column-type process that separates components of a liquid mixture by their different boiling points. Vapor and liquid phases flow countercurrently within the mass transfer zone of the column where trays or packings are used to maximize interfacial contact between the phases.

Figure 2.1 shows a distillation column equipped with seven sieve trays. Industrial columns typically contain more trays. Tray spacings generally vary from 12 to 24 in. The liquid from a tray flows through a downcomer to the tray below and vapor flows upward through the holes in the sieve tray to the tray above. Intimate contact between the vapor and liquid phases is created as the vapor passes through the

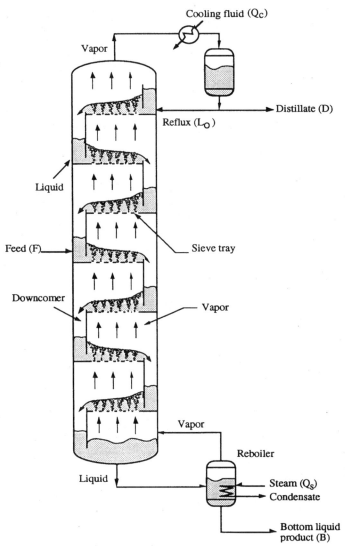

Figure 2.1 Distillation column equipped with sieve trays.

holes in the sieve tray and bubbles through the pool of liquid residing on the tray. The concentration of the more volatile components in the feed increases in the vapor from each tray going upward while the concentration of the least volatile components increase in the liquid from each tray going downward.

The overhead vapor from the column is condensed to obtain a distillate product. Part of the distillate is removed as overhead product (D) while, to increase performance, some distillate is refluxed (L_0) to the top tray. The *reflux ratio* is defined as the ratio of the rate of reflux (L_0) to the rate of product (D) which is removed. The liquid leaving the bottom tray enters a reboiler, where it is heated and partially vaporized. The vapor from the reboiler is sent back to the bottom tray and the liquid which remains is removed as the bottom product. In this way heat, which is normally provided in the form of steam, is added to the column.

Advantages

The advantages of distillation are its simple flowsheet, low capital investment, and low risk. If components to be separated have a relative volatility of 1.2 or more and are thermally stable, distillation is hard to beat. Relative volatility is defined in the section on thermodynamic relationships. A more detailed listing of the advantages of distillation is given in Chapter 7.

Disadvantages

Distillation has a low energy efficiency and requires thermal stability of compounds at their boiling points. It may not be attractive when azeotropes are involved or when it is necessary to separate high boiling components, present in small concentrations, from large volumes of a carrier, such as water.

Factors favoring distillation

The factors which favor distillation are:

- Relative volatility is greater than 1.2.
- Products are thermally stable.
- Rate is 5000–10,000 lb/day or more.
- High corrosion rates/unwanted side reactions/explosive conditions do not exist.

Applications

Distillation applications range widely from the separation of bulk petrochemicals like propylene/propane and ethylbenzene/styrene to the cryogenic separation of air into nitrogen and oxygen. The petroleum refining industry uses distillation to separate large volumes of crude oil into gasoline and other liquid fuels. In spacecraft, distillation can be used on a very small scale to recover water from the urine of astronauts.

New developments

In response to the pressure of international competition, distillation has been improved significantly in the last 5 years. New developments include:

- New trays and packings (examples include cocurrent trays and Optiflow packing)
- New process configurations, such as catalytic distillation
- Hybrid systems, such as distillation/adsorption systems to dehydrate ethanol/water azeotropes
- New design tools, such as the rate-based design method

We will cover each one of these new developments, but first we will address design concepts and fundamentals needed to properly understand and evaluate distillation processes.

Design concepts

The design procedure requires determinations of the column *diameter* and *height*. The diameter is a function of capacity (or rate) whereas column height is more dependent on the degree of separation (or product purity). The design of a distillation column involves eight fundamental steps (Figure 2.2).

Because the hydraulic flow pattern also affects tray or packing efficiency, hydraulic calculations should be made and column diameter determined before column height. In the design procedure, column diameter is determined to achieve a specified *capacity*. To determine column diameter, the *flood point* must be determined based on internal vapor and liquid rates, which in turn is a function of the feed and product rates and the reflux ratio.

Column diameter is determined based on percentages of flow rates at flooding. At flooding the vapor or liquid rate has reached a point where significant decline in efficiency occurs. In tray columns, column diameter is selected so that operating rates are 50–85% of those at flood.

After hydraulic calculations have been made and column diameter determined, the number of equilibrium stages, and an efficiency factor for the tray or packing, are used to determine the height of the column. Detailed procedures for both tray and packed distillation columns are presented in the Appendix on pages 341 through 369. And further details on design procedures and approaches, as well as their limitations, are presented in Kister (1992).

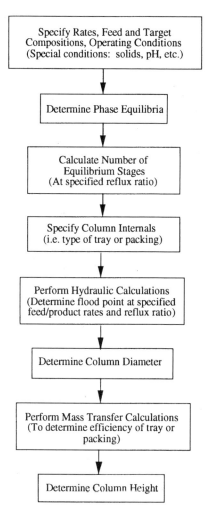

Figure 2.2 Eight fundamental steps required for distillation column design.

Equilibrium

Equilibrium and the equilibrium stage

It is useful to define the concepts of *equilibrium* and the *equilibrium stage*. These concepts can be illustrated by considering the vapor/liquid apparatus shown in Figure 2.3. As heat is supplied by the electrical resistance heater, the liquid mixture contained in the apparatus will begin to boil and generate a vapor. The vapor circulates around the loop and reenters the bottom of the apparatus. At the bottom of the apparatus, vapor is dispersed, with the use of a distributor, into the liquid phase.

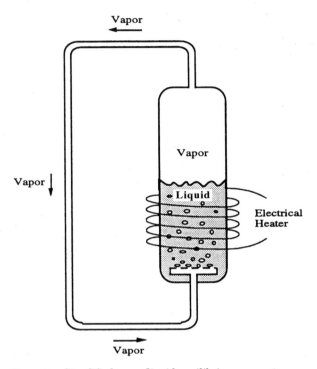

Figure 2.3 Simplified vapor/liquid equilibrium apparatus.

The components present in the original liquid mixture will distribute between the vapor and liquid phases according to their volatilities (or boiling points); the lower-boiling components will tend to concentrate more in the vapor phase whereas the higher-boiling ones will concentrate in the liquid phase. Equilibrium between the liquid and vapor phases is reached when the operation of the apparatus is continued to the point where no further changes in composition, temperature, or pressure occur.

It is not possible to achieve a higher degree of separation than that achieved at equilibrium. At equilibrium the *chemical potential* of the vapor and liquid phases are equal. At this point the apparatus is performing as if it were an "equilibrium," "theoretical," or "ideal" stage and the vapor and liquid compositions are the equilibrium compositions.

Figure 2.3 is simplified to illustrate the basic concepts of vapor/liquid equilibria. It does not include all of the necessary pieces of operating apparatus. For example, it does not include the additional features which are needed to overcome pressure drop to maintain vapor flow in the side loop. Further details of the types of equipment used to experimentally determine vapor/liquid equilibria are presented in Palmer (1987).

Thermodynamic relationships

Equilibrium ratio (_K value_). At equilibrium, the concentration of any component present in the liquid phase mixture may be related to its concentration in the vapor phase mixture by the _equilibrium ratio_, also called the _equilibrium constant_ or "_K value_."

$$K_i = \frac{y_i}{x_i} \tag{2.1}$$

where K_i = equilibrium ratio
  y_i = mol fraction of component i in vapor phase, dimensionless
  x_i = mol fraction of component i in liquid phase, dimensionless

The more volatile components in a mixture will have the higher values of K_i, whereas less volatile components will have lower values of K_i.

Relative volatility. The key _separation factor_ in distillation is the _relative volatility_. As the value of relative volatility increases, the more easily components may be separated by distillation. It is defined as:

$$\alpha_{ij} = \frac{K_i}{K_j} \tag{2.2}$$

where α_{ij} = relative volatility (component i relative to component j)

Ideal systems

Dalton's law. Ideal systems of vapor and liquid mixtures obey Dalton's and Raoult's laws, respectively. Dalton's law relates the concentration of a component present in an ideal gas or vapor mixture to its partial pressure.

$$p_i = Py_i \tag{2.3}$$

where p_i = partial pressure of component i in gas mixture, force/length^2
  P = total pressure, force/length^2

y_i = mol fraction of component i in gas or vapor phase, dimensionless

Raoult's law. Raoult's law relates the partial pressure of a component in the vapor phase to its concentration in the liquid phase.

$$p_i = P_i°x_i \tag{2.4}$$

where p_i = partial pressure of component i in vapor mixture, force/length^2
   $P_i°$ = vapor pressure of pure component i at the system temperature, force/length^2
   x_i = mol fraction of component i in the liquid phase, dimensionless

Combining Equations (2.3) and (2.4) yields

$$Py_i = P_i°x_i \tag{2.5}$$

Combining Equations (2.1) and (2.5)

$$K_i = \frac{P_i°}{P_T} \tag{2.6}$$

Combining Equations (2.2) and (2.6) gives the relative volatility for an ideal mixture at equilibrium.

$$\alpha_{ij} = \frac{P_i°}{P_j°} \tag{2.7}$$

Equation (2.7) shows that for ideal systems, α_{ij} is independent of pressure and composition. Vapor pressures ($P_i°$) for many compounds are published in the literature. Vapor pressures as a function of temperature are correlated by the Antoine equation (Reid et al., 1977):

$$\ln P_i° = A_i - \frac{B_i}{C_i + T} \tag{2.8}$$

where A_i, B_i, C_i are Antoine constants
   T = absolute temperature, °R or °K

Since vapor pressures of components depend on temperature, equilibrium ratios (K_i's) are a function of temperature. Because α_{ij} is proportional to the ratio of the vapor pressures, and vapor pressures increase with increasing temperature, α_{ij} is less sensitive to changes in temperature than is K_i or K_j.

Within a family of similar compounds, as temperature increases, the vapor pressure of the more volatile component tends to increase at a slower rate than the less volatile component. As temperature decreases, the vapor pressure of the more volatile component decreases more slowly than that for the less volatile component. Thus, though not a strong function of temperature, α_{ij} generally decreases with increasing temperature and increases with decreasing temperature. For a binary system, the relative volatility of the two components (i and j) is given by:

$$\alpha_{ij} = \frac{K_i}{K_j} = \frac{y_i(1-x_i)}{x_i(1-y_i)} \tag{2.9}$$

which can be rearranged to give

$$y_i = \frac{\alpha_{ij}\, x_i}{1 + (\alpha_{ij}-1)x_i} \tag{2.10}$$

Equation (2.10) is used to express the concentration of a component in the vapor as a function of its concentration in the liquid and relative volatility. Equation (2.10) is shown plotted in Figure 2.4 for various values of relative volatility. Figure 2.4 shows that when relative volatility increases, the concentration of the most volatile component in the vapor increases. For example, when the value of the relative volatility is five, a liquid containing 50% of more volatile component is in equilibrium with a vapor containing 83% of this component. When the relative volatility is equal to 1, the concentrations of the

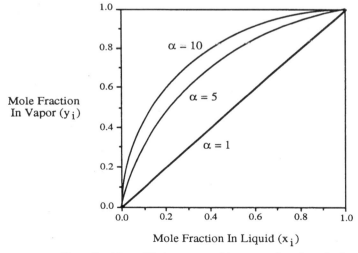

Figure 2.4 Vapor/liquid equilibrium compositions as a function of relative volatility.

most volatile component in the liquid and vapor phases are equal and a vapor/liquid separation is not feasible.

Henry's law. For the special case of low concentrations, the relationship between the concentration of a component in a liquid mixture and its partial pressure in the vapor (or gas) phase can be expressed as a linear relationship. This relationship, which is a modified form of Raoult's law, is known as *Henry's law*.

$$p_i = H_i x_i \qquad (2.11)$$

where H_i = Henry's law constant for component i, force/length^2/mol fraction liquid

If both sides of Equation (2.11) are divided by the total pressure, then

$$y_i = H_i' x_i \qquad (2.12)$$

and

$$H_i' = H_i/P \qquad (2.13)$$

where H_i' = modified Henry's law constant, mol fraction gas/mol fraction liquid

Nonideal systems

Most liquid mixtures are nonideal (i.e., they do not obey Raoult's law). In such cases, Equation (2.4) must be modified to include a correction factor called the liquid-phase activity coefficient.

$$p_i = \gamma_i P_i^\circ x_i \qquad (2.14)$$

where γ_i = liquid phase activity coefficient, dimensionless

At high pressures, the vapor phase may also depart ideal vapor mixture behavior and the inclusion of the Poynting and other vapor correction factors may be necessary (Abbott and Prausnitz, 1987). The common approach in distillation, however, is to assume ideal vapor behavior and to correct nonideal liquid behavior with inclusion of the liquid phase activity coefficient.

The standard state for reference for the liquid phase activity coefficient is commonly chosen as $\gamma_i = 1$ for pure component i. If $\gamma_i > 1$, positive deviations from ideality in the liquid solution occur, and if $\gamma_i < 1$, there are negative deviations from ideality. The liquid-phase activity coefficient is strongly dependent upon the composition of the mixture. Positive deviations are more common and occur when the molecules of different compounds are dissimilar and exhibit repulsive forces between species. Negative deviations occur when there are attractive forces between different compounds that do not occur for either compound alone.

For nonideal liquid mixtures, Equations (2.6) and (2.7) become

$$K_i = \frac{\gamma_i P_i^{\,\circ}}{P} \qquad (2.15)$$

$$\alpha_{ij} = \frac{\gamma_i P_i^{\,\circ}}{\gamma_j P_j^{\,\circ}} \qquad (2.16)$$

In nonideal systems, K_i, K_j, and α_{ij} are dependent on composition because of the composition dependence of γ_i and γ_j, though α_{ij} is not as sensitive to composition as individual values of K_i and K_j.

Azeotropes

When *azeotropes* are encountered, vapor and liquid compositions are equal, and components cannot be separated by conventional distillation. Figure 2.5 shows binary vapor/liquid composition (*x-y*), temperature-composition (*t-x*), and pressure-composition (*P-x*) diagrams for nonazeotrope, minimum azeotrope, and maximum azeotropic systems. A minimum boiling azeotrope boils at a lower temperature than either of the components in their pure states. When separating the components of this type of system by distillation, the overhead product is the azeotrope. An example of this type of system is ethanol-water. If the more volatile component is present at low concentrations, the high-boiling component is the bottom product. If the more volatile component is present at high concentrations, it is the bottom product.

A maximum-boiling azeotrope boils higher than either component in their pure states and is the bottom product of distillation. An example of this type of system is acetone-chloroform. The overhead product is the high-boiling component when the more volatile component is present at low concentrations. If the more volatile component is present at high concentration, it becomes the overhead product. Though not shown in Figure 2.5, it is also possible to have two immiscible liquid phases that may exhibit either nonazeotropic or azeotropic behavior.

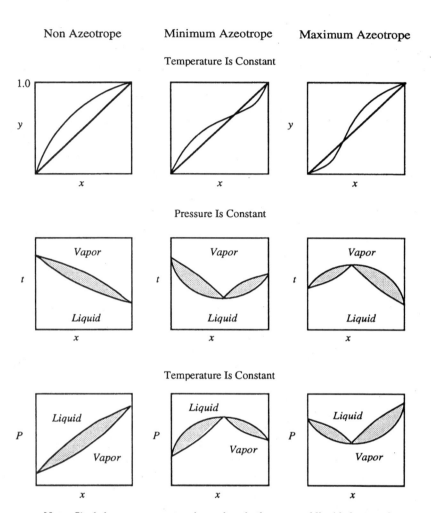

Note: Shaded areas represent regions where both vapor and liquid phases exist.

Figure 2.5 Phase equilibria for nonazeotropic and azeotropic systems.

Bubble and dew point calculations

The *bubble point* or initial boiling point of a mixture may be calculated from

$$\sum_{i=1}^{n} K_i x_i = 1.0 \qquad (2.17)$$

where n = number of components present in the mixture

The *dew point* or initial point of condensation of a mixture is calculated from

$$\sum_{i=1}^{n} \frac{y_i}{K_i} = 1.0 \qquad (2.18)$$

Procedures to determine bubble and dew points. At constant pressure, the procedures to determine bubble and dew point are:

1. Assume a temperature.
2. Determine K values.

Bubble point calculation. Calculate the sum on the left side of Equation (2.17) for determination of bubble point. If the left side of Equation (2.17) is smaller than unity, increase temperature. If greater than unity, decrease temperature. Step 2 is then repeated and the process is continued until the left side of Equation (2.17) is approximately equal to 1.

Dew point calculation. Calculate the sum of the left side of Equation (2.18) for dew point. If the left side of Equation (2.18) is smaller than unity, decrease temperature. If greater than unity, increase temperature. Step 2 is then repeated and the process is continued until the left side of Equation (2.18) is approximately equal to unity.

After bubble and dew points are determined, vapor and liquid compositions can be determined and displayed graphically as shown in Figure 2.6.

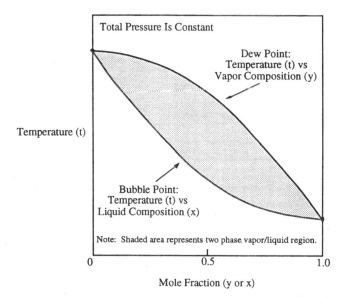

Figure 2.6 Temperature-composition diagram for vapor/liquid system.

TABLE 2.1   Key Sources of Vapor/Liquid Equilibrium Data

Source	Description	Reference(s)
Design Institute for Physical Property Data (DIPPR), American Institute of Chemical Engineers (AIChE)	Extensive data compilation project	Gess et al., 1991
Gmehling (Dechema Series)	Includes basic data plus parameters for popular models for predicting liquid phase activity coefficients. Also includes equilibria for liquid/liquid systems.	Gmehling et al., 1979 and 1994
Handbook of Applied Thermodynamics	Provides guidance on approaches to vapor liquid equilibrium (VLE) measurement and prediction.	Palmer, 1987
S. M. Walas	One of the most extensive references on phase equilibria.	Walas, 1985
Hirata et al.	Includes basic data plus parameter for models to predict liquid phase activity coefficients. Easy to use.	Hirata et al., 1975
Hala et al.	Includes references of source publications which contain basic data	Hala et al., 1967 and 1968; Wichterle et al., 1975
Chu et al.	Data for 476 systems. Vapor and liquid compositions are reported at system conditions.	Chu et al., *Vapor-Liquid Equilibrium Data,* 1950; *Distillation Equilibrium Data,* 1950
Hadden and Grayson	K values for hydrocarbon systems plus hydrogen presented in the form of nomographs	Hadden and Grayson, 1961
Natural Gas Processors Association	Comprehensive and reliable displays of hydrocarbon K values	Natural Gas Processors Suppliers Assn., 1972
Horsley	Azeotropic data	Horsley, 1973
American Petroleum Institute Research Project No. 44	Thermodynamic data on 935 hydrocarbons including vapor pressure data	American Petroleum Institute, 1953 and supplements; Chao and Seader, 1961; Gallant, 1965–1970

Sources of equilibrium data

Sources of experimental vapor/liquid equilibria are presented in Table 2.1 together with key references. Such data are generally available in the thermodynamic databases of commercially-available simulators. If necessary, equilibrium data may be determined experimentally with an apparatus like that shown in Figure 2.3.

Models for predicting equilibria

The industrial mixture is normally not a binary but rather a multicomponent system. However, most published experimental vapor/liquid equilibria are for binary systems. One approach to predict phase equilibria for multicomponent systems is to start with the data for the binary pairs and then use a model to predict multicomponent behavior. The Wilson, nonrandom two liquid (NRTL), and Universal Quasi-Chemical (UNIQUAC) models described in Table 2.2 are used for this purpose. Such models are used to predict liquid-phase activity coefficients in the multicomponent system as a function of composition and parameters based on binary pair data. Once activity coefficients are known, vapor and liquid equilibrium compositions can be predicted. The UNIQUAC Functional Group Activity Coefficients (UNIFAC) model described in Table 2.2 is used when no binary pair data are available. Most of the commercially available process simulators contain models/data for predicting vapor/liquid equilibria.

TABLE 2.2   Models for Predicting Vapor/Liquid Equilibria

Model	Description	Reference(s)
Van Laar	Fits extremely nonideal systems well such as alcohols and hydrocarbons. Cannot represent maxima or minima in liquid phase activity coefficients.	Abbott and Prausnitz, 1987
Wilson	Best for homogeneous mixtures exhibiting large positive deviations from ideal behavior such as alcohols-hydrocarbons.	Wilson, 1964
NRTL	Unlike Wilson model, it can predict interior extremes in activity coefficients and can thus predict formation of two liquid phases.	Renon and Prausnitz, 1968
UNIQUAC	Flexible and powerful model. Can predict multicomponent mixture behavior from binary data.	Abrams and Prausnitz, 1975; Anderson and Prausnitz, 1978; Maurer and Prausnitz, 1978; Prausnitz et al., 1980
UNIFAC	Used when no data are available. Functional groups are used to predict mixture behavior and liquid phase activity coefficients. Procedure is described in Fredenslund et al. (1977). Functional group parameter values are published in Gmehling et al. (1982) and Macedo et al. (1983). Special parameters are available for liquid-liquid equilibria in Magnussen et al. (1981).	Fredenslund et al., 1975; Fredenslund et al., 1977; Gmehling et al., 1982, 1994; Macedo et al., 1983; Magnussen et al., 1981; Skjold-Jorgensen et al., 1979

Equilibrium Stages

McCabe-Thiele method

The number of equilibrium stages required for separation of two components can be determined by the *McCabe-Thiele* method. The key assumption with the McCabe-Thiele method is there must be equimolar flow of the liquid and vapor phases through the column between the feed inlet and the top tray and between the feed inlet and bottom tray. For example, in Figure 2.7

$$V_n = V_{n+1} \tag{2.19}$$

and

$$L_n = L_{n-1} \tag{2.20}$$

where V_n = flow rate of vapor from stage n, mol/time
$\quad\;\; L_n$ = flow rate of liquid from stage n, mol/time

Rectifier section. In Figure 2.7 a continuous distillation column is shown with feed being introduced to the column at an intermediate point with overhead distillate and bottoms products being made. The portion of the column above the feed is called the *rectifier* or *enriching* section, the portion below the feed is called the *stripping* section. An overall material balance around the column yields

$$F = D + B \tag{2.21}$$

where F = feed rate, mol/time
$\quad\;\; D$ = overhead product rate, mol/time
$\quad\;\; B$ = bottoms product rate, mol/time

An overall material balance on the most volatile component gives

$$Fx_F = Dx_D + Bx_B \tag{2.22}$$

where x_F = concentration of most volatile component in feed, mol fraction
$\quad\;\; x_D$ = concentration of most volatile component in overhead product, mol fraction
$\quad\;\; x_B$ = concentration of most volatile component in bottoms product, mol fraction

As shown in Figure 2.7, vapor from the top tray has a composition y_1. The overhead vapor stream is condensed and the resulting liquid is its bubble point. For a total condenser, the reflux stream (L_0) and distillate (D) have the same composition ($y_1 = x_D$). Since equal molar flow is assumed, $L_1 = L_2 = L_n$ and $V_1 = V_2 = V_n = V_{n+1}$.

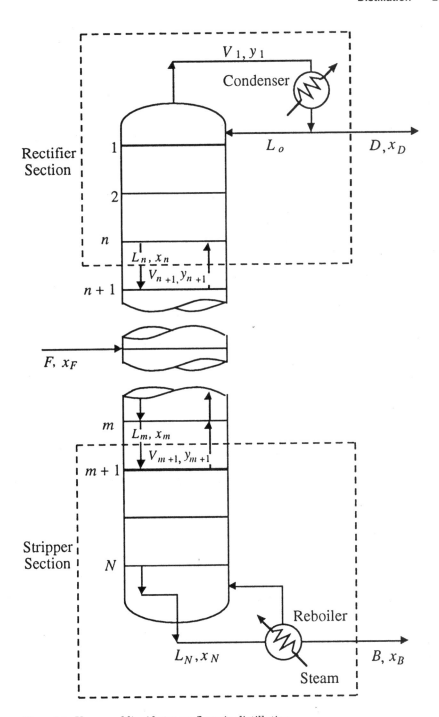

Figure 2.7 Vapor and liquid stream flows in distillation.

Making an overall material balance on the enclosed dashed-line section for the rectifier part of the column gives

$$V_{n+1} = L_n + D \tag{2.23}$$

Making a balance on the most volatile component around the rectifier section in Figure 2.7 gives

$$V_{n+1}y_{n+1} = L_n x_n + Dx_D \tag{2.24}$$

where y_{n+1} = composition of vapor from stage $n + 1$, mol fraction, dimensionless

x_n = composition of liquid from stage n, mol fraction, dimensionless

x_D = composition of overhead product, mol fraction, dimensionless

Solving for y_{n+1}, the equation relating compositions and rates for the rectifier section is

$$y_{n+1} = \frac{L_n}{V_{n+1}} x_n + \frac{Dx_D}{V_{n+1}} \tag{2.25}$$

Since $V_{n+1} = L_n + D$, then $L_n/V_{n+1} = R/(R + 1)$ and Equation (2.25) becomes

$$y_{n+1} = \frac{R}{R+1} x_n + \frac{x_D}{R+1} \tag{2.26}$$

where R is the *reflux ratio*. Reflux ratio is defined as

$$R = \frac{L_0}{D} \tag{2.27}$$

where R = reflux ratio
L_0 = amount of overhead refluxed to column, mol/time
D = amount of overhead product, mol/time

At a fixed reflux ratio, Equation (2.26) is a straight line on a graph of vapor composition versus liquid composition. It relates the compositions of two streams passing each other in the rectifier section of the column. As given in Equation (2.26), the slope of the line is $R/(R + 1)$. A plot of Equation (2.26) on a x-y graph shows that it intersects the $y = x$ line (45° diagonal) at $x_n = x_D$.

Stripping section. Making a total material balance around the stripping section in Figure 2.7 gives

$$V_{m+1} = L_m - B \tag{2.28}$$

where V_{m+1} = vapor flow rate from stage $m + 1$, mol/time
$\quad\quad L_m$ = liquid flow rate from stage m, mol/time

Making a balance on the most volatile component of the binary around the stripping section gives

$$V_{m+1}\,y_{m+1} = L_m x_m - Bx_B \tag{2.29}$$

where y_{m+1} = concentration of most volatile component in vapor
$\quad\quad\quad$ from stage $m+1$, mol fraction
$\quad\quad x_m$ = concentration of most volatile component in liquid
$\quad\quad\quad$ from stage m, mol fraction

Solving for y_{m+1}, the equation for the stripping section relating compositions and rates is

$$y_{m+1} = \frac{L_m}{V_{m+1}}\,x_m - \frac{Bx_B}{V_{m+1}} \tag{2.30}$$

Since equal molar flow is assumed, $L_m = L_n$ = constant and $V_{m+1} = V_n$ = constant. Equation (2.30) is a straight line when plotted as y versus x with a slope of L_m/V_{m+1}. It intersects the 45° diagonal line when $x_m = x_B$.

Thermal condition of feed. The thermal condition of the feed may be defined as the number of mols of saturated liquid produced on the feed tray per mol of feed. In terms of heat of vaporization, it may be expressed as

$$q = \frac{\text{heat required to vaporize 1 mol of feed}}{\text{heat of vaporization of 1 mol of liquid feed}} \tag{2.31}$$

Expressed in terms of enthalpies Equation (2.31) becomes

$$q = \frac{H_V - H_F}{H_V - H_L} \tag{2.32}$$

where H_V = enthalpy of feed at dew point, energy/mol
$\quad\quad H_L$ = enthalpy of feed at bubble point, energy/mol
$\quad\quad H_F$ = enthalpy of feed at feed conditions, energy/mol

Values of q for various conditions follow:

- $q = 1$, feed enters as liquid at its bubble point
- $q = 0$, feed enters as vapor at its dew point
- $q > 1.0$, feed enters as a subcooled liquid
- $q < 0$, feed enters as superheated vapor
- q is fractional, feed enters as part liquid and part vapor

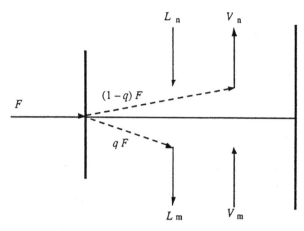

Figure 2.8 Distribution of feed between internal liquid and vapor streams.

Figure 2.8 illustrates the stream flows around the feed tray. From the definition of q, the following relationships may be developed.

$$L_m = L_n + qF \qquad (2.33)$$

$$V_n = V_m + (1-q)F \qquad (2.34)$$

The point of intersection of the rectifying and the stripping equations on an x-y graph can be determined by first rewriting Equations (2.24) and (2.29):

$$V_n y = L_n x + D x_D \qquad (2.35)$$

$$V_m y = L_m x - B x_B \qquad (2.36)$$

where the y and x values are the point of intersection of the operating lines of the rectifier and stripper sections. Subtracting Equation (2.35) from (2.36) gives

$$(V_m - V_n)y = (L_m - L_n)x - (D x_D + B x_B) \qquad (2.37)$$

The result of substituting Equations (2.24), (2.33), and (2.34) into Equation (2.37) and rearranging is

$$y = \frac{q}{q-1} x - \frac{x_F}{q-1} \qquad (2.38)$$

Equation (2.38) is called the *q-line* equation. It represents the locus of the intersection of the two operating lines. The intersection of the q-line with the 45° line is $y = x_F$, where x_F is the feed composition. The slope of the q line is $q/(q-1)$.

Stage determination by McCabe-Thiele method. One should first plot the equilibrium curve on the x-y diagram. The next steps involve

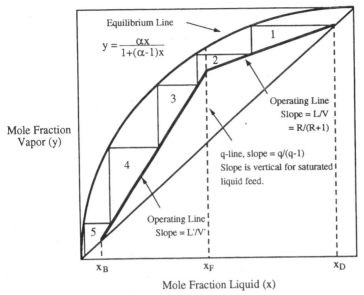

Figure 2.9 McCabe-Thiele diagram.

placement of the rectifier operating line and the q line. The stripping operating line is then located by drawing a straight line between the intersection of the q line and rectifier operating line with the bottom composition point ($y = x_B$). Beginning at the top or the bottom of the column, equilibrium stages are determined by "stepping off" the triangular steps as illustrated in Figure 2.9.

For the example shown in Figure 2.9, the shift is made from rectifier to stripper section at stage two. A total of between four and five equilibrium stages is needed with the feed located on stage two. If a partial condenser is used at the top of the column it would also be included as an equilibrium stage. Figure 2.10 illustrates the slope of the q line for various values of q.

Total reflux (minimum stages). At *total reflux* ($D = 0$), reflux ratio approaches infinity and the slope [$R/(R + 1)$] of the operating line for the rectifier section approaches unity. At total reflux, the operating lines for both the rectifier and stripper lie on the 45° diagonal and their point of intersection is not dependent on the feed composition and the location of the q line.

The number of equilibrium stages determined at total reflux gives the *minimum number of equilibrium stages* required for the separation. For the case of a total condenser, all of the overhead is condensed and returned to the column as reflux and all bottoms product is vaporized and recycled to the column. Thus, the flows of overhead

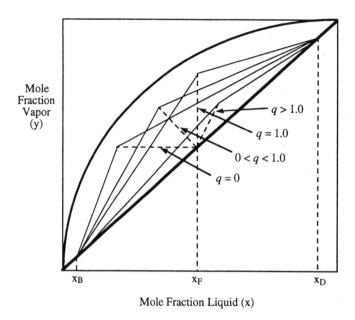

Figure 2.10   Illustration of the q line slope.

and bottoms products are zero, as is the feed to the column. Figure 2.11 illustrates the McCabe-Thiele diagram for the case of total reflux. The results show that the minimum number of equilibrium stages is about equal to three.

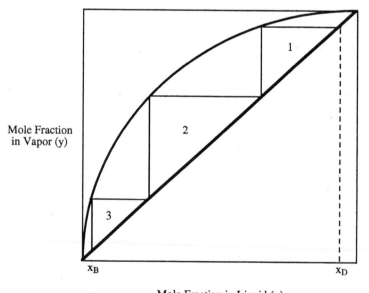

Figure 2.11   McCabe-Thiele diagram for total reflux (minimum stages).

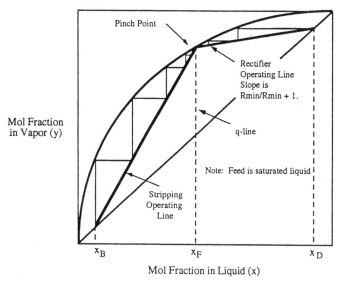

Figure 2.12   McCabe-Thiele diagram for minimum reflux (infinite stages).

Minimum reflux ratio (infinite stages).   The *minimum reflux ratio* corresponds to an *infinite number of equilibrium stages*. As R is reduced, the slope of the rectifier operating line $R/(R + 1)$ decreases, and moves closer to intersecting the equilibrium line. If reflux ratio is continued to be reduced, a "pinch point" occurs where the rectifier operating line intersects the equilibrium curve. When the pinch occurs, the number of equilibrium stages required to make the separation becomes infinite. Under these conditions, the minimum reflux ratio can then be determined from the slope of the rectifier operating line which is $R_{min}/(R_{min} + 1)$, where R_{min} is the minimum reflux ratio. The McCabe-Thiele diagram for the minimum reflux ratio case is illustrated in Figure 2.12.

Fenske-Underwood-Gilliland method

When there are three or more components, graphical procedures, such as the McCabe-Thiele method, cannot be used. It becomes necessary to use an analytical approach. The separation of a multicomponent mixture is specified in terms of two *key components* of the mixture. The *light key* will have a specified maximum limit in the bottoms product and the *heavy key* will have a specified maximum limit in the overhead product. Normally the keys are adjacent to each other in volatility. The nonkey components are called *distributed components*.

There are a number of "shortcut" methods available for determination of the number of equilibrium stages for multicomponent mixtures. The Fenske-Underwood-Gilliland (FUG) method is perhaps the most popular of these. Most of the commercially available process simulators contain optional shortcut methods as well as the more rigorous stage-to-stage algorithms, which are discussed in the section which follows. Kister (1992) presents the strengths and weaknesses of both shortcut and more rigorous stage-to-stage methods.

The FUG procedure follows:

1. The separation is specified based on the concentrations of the two key components in the overhead and bottoms products.

2. The minimum number of stages is then determined by the Fenske equation.

3. Minimum reflux ratio is determined based on the Underwood method.

4. Based on a selected reflux ratio, the Gilliland correlation is used to estimate number of equilibrium stages.

5. Concentrations of the distributed components in the overhead and bottoms products are determined by a modified Fenske equation.

6. The method of Kirkbride can then be used to determine feed tray location.

Fenske equation.   For multicomponent systems, an approximate value of the minimum number of equilibrium stages (at total reflux) may be obtained from the *Fenske* equation (Fenske, 1932). In the Fenske equation, given below as Equation (2.39), relative volatility is based on the light key relative to the heavy key.

$$N_{\min} = \frac{\ln\left[\left(\frac{x_{LK}}{x_{HK}}\right)_D \left(\frac{x_{HK}}{x_{LK}}\right)_B\right]}{\ln(\alpha_{LK/HK})_{av}} \qquad (2.39)$$

where $N_{\min}$ = minimum number of equilibrium stages
  x_{LK} = mol fraction of light key
  x_{HK} = mol fraction of heavy key
  D = when used as subscript, denotes distillate product
  B = when used as subscript, denotes bottoms product
  $(\alpha_{LK/HK})$ = average value of relative volatility of light key relative to heavy key

The average value of relative volatility is calculated from the mean value of relative volatility at the top of the column $(\alpha_{LK/HK})_D$ and at the bottom of the column $(\alpha_{LK/HK})_B$. $(\alpha_{LK/HK})_D$ is based on the dew point temperature of the overhead vapor whereas $(\alpha_{LK/HK})_B$ is based on the bubble point temperature of the bottom liquid.

$$(\alpha_{LK/HK})_{av} = \sqrt{(\alpha_{LK/HK})_D (\alpha_{LK/HK})_B} \qquad (2.40)$$

Since the distributions of the nonkey components in the distillate and bottoms are not known, determinations of the dew and bubble points are partially trial and error.

Minimum reflux by Underwood method.   For multicomponent mixtures, the Underwood method may be used for estimating *minimum reflux ratio* (Underwood, 1948). Underwood proposed Equations (2.41) and (2.42) for calculation of minimum reflux ratio:

$$\sum_{i=1}^{n} \frac{\alpha_i x_{Fi}}{\alpha_i - \theta} = 1 - q \qquad (2.41)$$

where n = number of components
α_i = average relative volatility based on component i relative to heavy key
x_{Fi} = mol fraction of component i in feed
q = number of mols of saturated liquid produced on the feed tray per mol of feed
θ = unknown parameter to be determined by trial and error

The correct value of θ will be between relative volatility of the two key components. After the value of θ is determined, Equation (2.42) may be used to determine minimum reflux ratio.

$$R_{min} + 1 = \sum_{i=1}^{n} \frac{\alpha_i x_{Di}}{\alpha_i - \theta} \qquad (2.42)$$

where R_{min} = minimum reflux ratio
x_{Di} = mol fraction of component i in distillate (D)

For more rigorous approaches to determination of minimum stages and minimum reflux ratio for multicomponent systems, consult Chien (1978).

Number of equilibrium stages by Gilliland correlation.   Once minimum stages and reflux ratio are known, the number of equilibrium stages

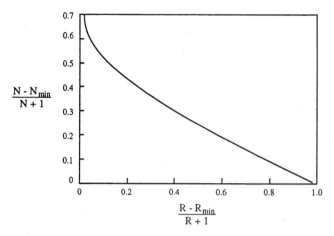

Figure 2.13 Gilliland correlation.

may be determined as a function of selected values of operating reflux ratio. Several methods have been proposed including the correlation of Gilliland (1940) to determine the number of equilibrium stages as a function of reflux ratio. The Gilliland correlation was first developed as a plot as shown in Figure 2.13, but later was transformed by Eduljee (1975) into Equation (2.43).

$$\frac{N-N_{min}}{N+1} = 0.75 \left(1 - \left[\frac{R-R_{min}}{R+1} \right]^{0.566} \right) \tag{2.43}$$

where N =   number of equilibrium stages
         R =   operating reflux ratio
         N_{min} =   minimum number of equilibrium stages
         R_{min} =   minimum reflux ratio

The optimum operating reflux ratio may be determined from capital and operating costs at various reflux ratios. Low reflux ratios require a large number of equilibrium stages which translates to high capital cost for equipment. High reflux ratios translate to high energy consumption and operating costs. The optimum operating reflux ratio usually lies between 1.1 and 1.4 times the minimum reflux ratio.

Distributed components by Fenske equation. Values for the distributed or nonkey components may be determined after the number of minimum stages has been determined. To determine concentrations of distributed components in the distillate and bottoms products, the Fenske equation may be written for any component i as follows:

$$\frac{x_{iD}}{x_{iB}} = \left(\alpha_i\right)^N_{\text{av}}{}^{\text{min}} \times \frac{(x_{\text{HK}})_D}{(x_{\text{HK}})_B} \tag{2.44}$$

where $(\alpha_i)_{\text{av}}$ = average relative volatility of component i relative to heavy key

Location of feed tray. The method of Kirkbride (1944) may be used to determine the ratio of the number of equilibrium stages above and below the feed point:

$$\log \frac{M}{P} = 0.206 \log \left\{ \frac{B}{D} \left(\frac{x_{\text{HK}}}{x_{\text{LK}}}\right)_F \left[\frac{(x_{\text{LK}})_B}{(x_{\text{HK}})_D} \right]^2 \right\} \tag{2.45}$$

where M = number of equilibrium stages above the feed tray
P = number of equilibrium stages below the feed tray

Determination of stage requirements by simulator

In industry, a simulator containing stage-to-stage calculation algorithms is almost always used to determine the number of equilibrium stages. Such simulators also include thermodynamic databases which also allow determinations of the vapor/liquid equilibria. And these simulators usually offer options, which not only allow the user to select a particular algorithm for stage calculations, but also the model/approach to determine vapor/liquid equilibria. Some simulators also allow the user to incorporate tray efficiencies, as a function of composition, so the number of actual trays can be determined as well. Users of such simulators must have a good understanding of fundamentals in order to appreciate limitations of the methods and models, and to recognize errors when they occur.

There are a number of simulators which are commercially available which will determine distillation stage requirements. Some examples of these simulators, their suppliers, and brief descriptions of their capabilities are given in the Appendix (see Table A.3, pp. 349 and 350).

Efficiency

Mass transfer—Two resistance theory

How closely the performance of an actual tray or packing in a distillation column approaches equilibrium depends on the *rate of mass transfer*. To illustrate the concept of mass transfer, consider component A diffusing from the gas phase into the liquid phase shown in

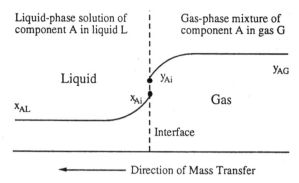

Figure 2.14 Mass transfer between gas and liquid phases.

Figure 2.14. In doing so, it must first pass through the gas phase, through the interface, and then through the liquid phase. The average or bulk concentrations of A in the gas and liquid phases are y_{AG} and x_{AL}, respectively. The concentration in the bulk gas phase (y_{AG}) decreases to y_{Ai} at the gas-liquid interface. The liquid concentration starts at x_{Ai} at the interface and declines to x_{AL} in the bulk liquid phase. According to two resistance theory (also called two-film theory) there is no resistance to mass transfer at the interface, and y_{Ai} and x_{Ai} are in equilibrium ($y_{Ai} = K_A x_{Ai}$).

For steady-state mass transfer, the rate at which component A migrates to the interface from the gas phase is equal to the rate at which it diffuses through the liquid phase. Thus there is no change in the amount of component A at the interface. The flux of component A may be expressed as

$$N_A = k_G(y_{AG} - y_{Ai}) = k_L(x_{Ai} - x_{AL}) \qquad (2.46)$$

where N_A = rate of mass transfer (or flux) of component A, mol/time-length^2

k_G = mass transfer coefficient for gas phase, mol/time-length^2

k_L = mass transfer coefficient for liquid phase, mol/time-length^2

y_{AG} = mol fraction of component A in bulk of gas phase, dimensionless

y_{Ai} = mol fraction of component A in gas phase at interface, dimensionless

x_{Ai} = mol fraction of component A in liquid phase at interface, dimensionless

x_{AL} = mol fraction of component A in bulk of liquid phase, dimensionless

The global or overall rate of mass transfer of component A through the gas and liquid phases may be represented in terms of an overall mass transfer coefficient and an overall concentration driving force.

$$N_A = K_G \left(y_{AG} - y_A^* \right) \tag{2.47}$$

where K_G = overall mass transfer coefficient, based on gas phase, mols/time-length

y_A^* = mol fraction of component A in gas phase which would be in equilibrium with x_A, given by: $y_A^* = K_A x_A$, dimensionless

The concentration driving force in Equation (2.47) may be rewritten as

$$y_{AG} - y_A^* = \left(y_{AG} - y_{Ai} \right) + \left(y_{Ai} - y_A^* \right) = \left(y_{AG} - y_{Ai} \right) + m \left(x_{Ai} - x_{AL} \right) \tag{2.48}$$

where

$$m = \left(y_{Ai} - y_A^* \right) / \left(x_{Ai} - x_{AL} \right) \tag{2.49}$$

and

m = slope of the equilibrium line, dimensionless

Combining Equation (2.49) with Equation (2.46) yields

$$\frac{1}{K_G} = \frac{1}{k_G} + \frac{m}{k_L} \tag{2.50}$$

Equation (2.50) provides the relationship between the overall mass transfer coefficient and the individual phase mass transfer coefficients based on two resistance theory. If the slope of the equilibrium line (m) is small (i.e., much less than one), then m/k_L is small and the major resistance to mass transfer is $1/k_G$. Under this condition, the resistance to mass transfer is *controlled by the gas phase*. Conversely, when the slope is large the term $1/k_G$ becomes less significant compared with m/k_L, and resistance to mass transfer is *controlled by the liquid phase*.

Efficiency of tray columns

Overall tray efficiency. Though several different efficiency parameters are used for distillation trays, the one ultimately needed for the design of sieve tray columns is the *overall tray efficiency*. Overall tray efficiency, also sometimes called *overall column efficiency*, is defined as the number of equilibrium stages divided by the actual number of trays.

$$E_o = \frac{N}{N_{act}} \times 100 \tag{2.51}$$

where E_o = overall tray efficiency, %
N = number of equilibrium stages, dimensionless
N_{act} = actual number of trays, dimensionless

In Equation (2.51), a reboiler or a partial condenser would not be included in the number of equilibrium stages (N). N includes only the stages which are contained in the column. For hydrocarbon distillations, overall tray efficiencies are 50–90%. In absorption processes, the range is 10–50%.

A number of empirical approaches have been proposed for making approximate predictions of overall tray efficiency. The correlation shown in Figure 2.15 partially takes into account property variations (O'Connell, 1946). Its use should be restricted to preliminary designs.

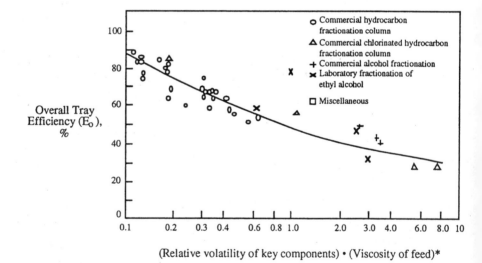

(Relative volatility of key components) • (Viscosity of feed)*

*at average column conditions

Figure 2.15 O'Connell correlation for estimating overall tray efficiency. (*Adapted from O'Connell, H. E., AIChE Trans, Vol. 42, 1946 with permission of the American Institute Of Chemical Engineers. Copyright © 1946 AIChE. All rights reserved.*)

Murphree tray efficiency. Because tray efficiency may vary from tray-to-tray within the same column, it is useful to define an efficiency for individual trays. *Murphree Tray Efficiency*, based on the vapor phase, is defined as

$$E_{mv} = \frac{Y_n - Y_{n+1}}{Y_n^* - Y_{n+1}} \times 100 \qquad (2.52)$$

where E_{mv} = Murphree tray efficiency based on vapor phase, %
Y_n = mol fraction of vapor from tray n, dimensionless
Y_{n+1} = mol fraction of vapor to tray n, dimensionless
Y_n^* = mol fraction of vapor in equilibrium with liquid from tray n, $(Y_n^* = Kx_n)$, dimensionless

Overall tray efficiency may be calculated from Murphree tray efficiency using the relationship of Lewis (1936):

$$E_o = \frac{\ln\left[1 + E_{mv}\left(\lambda - 1\right)\right]}{\ln \lambda} \qquad (2.53)$$

where

$$\lambda = \frac{m}{\left(\dfrac{L}{G}\right)} = \frac{\text{slope of equilibrium line}}{\text{slope of operating lnie}} \qquad (2.54)$$

L = liquid rate, mol/time
G = vapor rate, mol/time

Point efficiency. Because liquid concentration can vary from point to point on an individual tray, particularly on large diameter trays where a significant amount of crossflow is possible, the Murphree efficiency may have a value greater than 100%. Thus, it is useful to define *point efficiency* on a distillation tray. Point efficiency is also sometimes called *local efficiency*.

$$E_{point} = \left.\left|\frac{y_n - y_{n+1}}{y_n^* - y_{n-1}}\right|\right._{point} \times 100 \qquad (2.55)$$

where E_{point} = point efficiency on a distillation tray, %

Because liquid concentration is based on a single point, concentration changes are not involved, and point efficiency cannot have a value greater than 100%. However, because concentration gradients are complex and difficult to predict, point efficiency is used more in research than in industrial design procedures.

If the liquid and vapor are completely mixed on a tray, then

$$E_{mv} = E_{point} \tag{2.55a}$$

If the liquid moves across the tray in plug flow,

$$E_{mv} = \frac{1}{\lambda}\left[e^{\lambda E_{point}} - 1\right] \tag{2.55b}$$

Equations (2.55a) and (2.55b) represent the extremes of the mixing pattern possibilities and corrections must be made using a mixing model (Fair et al., 1983).

Efficiency of packed columns

The *height of a transfer unit (HTU)* or *height equivalent to a theoretical plate (HETP)* is used to represent the mass transfer efficiency of packed columns.

Height of a transfer unit (HTU).   Equation (2.56) is the result of completing a material balance around a differential element of a packed column (Figure 2.16).

$$V_G dy_A = K_G(y_A - y_A^*)a_i S dZ \tag{2.56}$$

where V_G = total flow of vapor phase, mol/time
dy_A = differential change in mol fraction of component, dimensionless
K_G = overall mass transfer coefficient for component A based on the gas phase, mol/time-length^2
y_A = mol fraction of component A in gas phase, dimensionless
y_A^* = mol fraction of A in equilibrium with A in bulk liquid (given by $y_A^* = K_A x_A$ where K_A is equilibrium constant), dimensionless
a_i = interfacial area between phases per unit volume of packed column, length^2/length^3
S = cross-sectional area of column, length^2
dZ = differential height of packing, length

Solving for Z results in the following equation:

$$Z = \left(\frac{V_G}{K_G a_i S}\right)\int_{y_{A,\,in}}^{y_{A,\,out}} \frac{dy_A}{y_A - y_A^*} \tag{2.57}$$

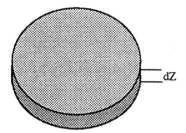

Figure 2.16 Differential height of packed column.

where Z = height of column containing packing, length

$y_{A,\text{ in}}$ = mol fraction of component A in gas phase at inlet of packed height, dimensionless

$y_{A,\text{ out}}$ = mol fraction of component A in gas phase at outlet of packed height, dimensionless

The height of a transfer unit is defined as

$$(\text{HTU})_{\text{OG}} = \frac{V_G}{K_G a_i S} \qquad (2.58)$$

where $(\text{HTU})_{\text{OG}}$ = overall height of a transfer unit, based on gas phase, length

HTU is used as a measure of efficiency in packed columns; the lower the value of HTU, the higher the efficiency.

Number of transfer units (NTU).   The number of transfer units, based on the overall gas phase $(\text{NTU})_{\text{OG}}$, is defined as

$$(\text{NTU})_{\text{OG}} = \int_{y_{A,\text{ in}}}^{y_{A,\text{ out}}} \frac{dy_A}{y_A - y_A^*} \qquad (2.59)$$

where $(\text{NTU})_{\text{OG}}$ = number of transfer units based on gas phase, dimensionless

Equation (2.59), which defines $(\text{NTU})_{\text{OG}}$, is based on the driving force between bulk vapor and the vapor composition in equilibrium with the bulk liquid. It is the ratio of the change in bulk gas composition to the average driving force and is representative of the degree of separation needed.

If y_A^* is constant or linear, as is the case when there are straight equilibrium and operating lines, Equation (2.59) can be integrated to give

$$(NTU)_{OG} = \frac{(y_A)_{out} - (y_A)_{in}}{(y_A^* - y_A)_M} \tag{2.60}$$

where

$$(y_A^* - y_A)_M = \frac{(y_A^* - y_A)_{in} - (y_A^* - y_A)_{out}}{\ln\left[(y_A^* - y_A)_{in}/(y_A^* - y_A)_{out}\right]} \tag{2.61}$$

Using Equations (2.58) and (2.59) for HTU and NTU, respectively, Equation (2.57) may be rewritten as

$$Z = (HTU)_{OG} \cdot (NTU)_{OG} \tag{2.62}$$

When gas and liquid rates are constant, one may define HTUs based on the individual vapor and liquid phases

$$(HTU)_G = \frac{V_G}{k_G a_i} \tag{2.63}$$

$$(HTU)_L = \frac{L}{k_L a_i} \tag{2.64}$$

Combining Equations (2.63) and (2.64) with Equation (2.50) based on two resistance theory yields

$$(HTU)_{OG} = (HTU)_G + \frac{m}{L/G}(HTU)_L \tag{2.65}$$

Several methods are available for predicting values of $(HTU)_G$ and $(HTU)_L$. They include the methods of Bravo and Fair (1982), Bolles and Fair (1979, 1982), Cornell et al. (1960), and Onda et al. (1968). Once values of $(HTU)_G$ and $(HTU)_L$ are determined, Equation (2.65) is used to determine $(HTU)_{OG}$.

Height equivalent to a theoretical plate (HETP). Height equivalent to a theoretical plate (HETP), also called *height equivalent to a theoretical stage (HETS)*, was introduced to allow comparisons of tray and packed columns. HETP is more widely used than HTU, even though HTU has a more fundamental basis. For packed columns, HETP is defined as

$$HETP = \frac{Z}{N} \tag{2.66}$$

where HETP = height equivalent to a theoretical plate, length
    Z = height of mass transfer zone containing packing or trays, length
    N = number of equilibrium stages

TABLE 2.3 Efficiency Parameters—Distillation Trays and Packings*

Device	Efficiency parameter
Trays	Overall tray efficiency (E_o)
	Murphree tray efficiency (E_{mv})
	Point efficiency (E_{point})
	Height equivalent to a theoretical plate (HETP)
Packings	Height of a transfer unit (HTU)
	Height equivalent to a theoretical plate (HETP)

*Kister (1992) presents correlations and experimental data for efficiencies of packings and trays for numerous distillation systems.

For tray columns, the *mass transfer zone* (Z), which is the part of the column containing the trays or packing, is equal to the tray spacing times the number of trays. Thus, Equation (2.66) becomes

$$\text{HETP} = \frac{H_s \times N_{act}}{N} = \frac{H_s}{E_o} \qquad (2.67)$$

where H_s = tray spacing, length
E_o = overall tray efficiency, dimensionless

Lewis (1936) showed that HETP may be related to HTU by the following

$$\frac{\text{HETP}}{(\text{HTU})_{OG}} = \frac{\ln \lambda}{\lambda - 1} \qquad (2.68)$$

When the slopes of the equilibrium and operating lines are parallel ($\lambda = 1$), $(\text{HTU})_{OG}$ and HETP are equal.

A form of Equation (2.68) may also be used to relate the number of transfer units to the number of theoretical stages.

$$(\text{NTU})_{OG} = N \frac{\ln \lambda}{\lambda - 1} \qquad (2.69)$$

A summary of the efficiency parameters used for distillation trays and packings is given in Table 2.3. Because HETP is the only common efficiency parameter between trays and packings, it is the one which is used to compare tray and packed columns.

Trays

In this section we present descriptions, illustrations, and scaleup information for distillation trays. More comprehensive information on tray fundamentals and performance may be found in the book by Lockett (1986).

Sieve, *valve*, and *bubble cap* trays are examples of traditional cross-distillation trays. However, the development of newer trays has virtually exploded within the last 5 years (Jantz et al., 1995). Thus, we include descriptions of some of these newer trays, including the Nye™, Max-Frac™, some of the newer multiple-downcomer trays, as well as the Ultra-Frac™ and Trutna trays.

Traditional crossflow trays

Sieve trays. Typical sieve tray geometries including guidelines for hole dimensions, free areas, and downcomer areas are given in Table 2.4.

The physical characteristics of the sieve tray are illustrated in Figure 2.17. Figure 2.18 shows the general trend between overall tray efficiency and vapor rate. It shows that at low vapor rates, the sieve tray will *weep*, and at still lower vapor rates it will totally *dump* allowing all liquid to fall through the sieve tray holes. At the dump point no liquid will flow over the weir and into the downcomer. Weeping and dumping result in a decline in tray efficiency.

At high vapor rates, *entrainment* will occur, whereby vapor physically carries suspended liquid droplets to the tray above. This will ultimately result in *entrainment flood* accompanied by a decline in

TABLE 2.4 Geometry of Sieve Trays

Parameter	Range
Hole diameter, in.	0.2–1.0
Fractional free area	0.06–0.16
Fractional downcomer area	0.05–0.30
Pitch/hole diameter ratio	2.5–4.0
Tray spacing, in.	12–36
Weir height, in.	1–3

Liquid

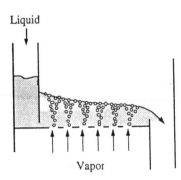

Figure 2.17 Conventional distillation sieve tray.

Vapor

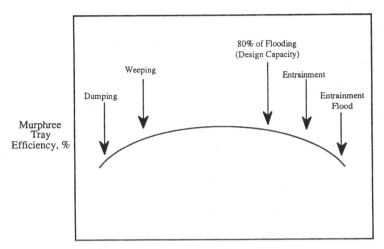

Figure 2.18 Murphree tray efficiency as a function of vapor rate (simplified).

Murphree tray efficiency. The other mode by which sieve trays flood is *downcomer flood*. Under conditions of high liquid rates, downcomers may not be able to transport all of the liquid, and the result is a backup of liquid resulting in downcomer flood. The Fractionation Research Institute (Stillwater, OK) has a videotape, showing internal flows in a 4-ft-diameter column, which illustrates these concepts. A wealth of information and guidelines for testing distillation trays has been published by the American Institute of Chemical Engineers (*Tray Distillation Columns,* 1987).

Characteristic overall tray efficiencies and some sources of efficiency data for distillation sieve trays are given in Table 2.5.

The advantages of the sieve tray are that it is well understood, is low-risk, cost is low, and it handles wide variations in flow rates. The disadvantages are its high pressure drop and low capacity relative to some of the newer trays and packings. The weeping, which is experienced at low vapor rates, may be overcome by using valve trays.

TABLE 2.5   Efficiency of Distillation Sieve Trays

System	Overall tray efficiency, %	Reference
Isobutane/n-butane	60–100	Yanagi and Sakata, 1982
n-Heptane/cyclohexane	35–80	Yanagi and Sakata, 1982
Carbon tetrachloride/benzene	60–85	Kageyama, 1969
Chlorobenzene/ethylbenzene	70–85	Kageyama, 1969
Ethylbenzene/styrene	60–80	Billet et al., 1969

Scaleup of sieve tray columns. It has been shown that a laboratory scale Oldershaw tray column produces separations close to those in large tray columns as long as the comparison is based on the same approach to the flood point (Fair et al., 1983). Internals of the Oldershaw column are illustrated in Figure 2.19.

Experimental results show that the commercial column will require no more trays than the number of trays required for the same separation in the Oldershaw column. Depending on the diameter of the column and degree of crossflow, the industrial column could indeed require fewer trays than the Oldershaw column. A comparison of point efficiency for the Oldershaw column with the overall tray efficiency for a 4-ft diameter column for the cyclohexane/n-heptane system is given in Figure 2.20.

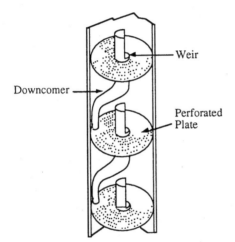

Figure 2.19 Oldershaw column.

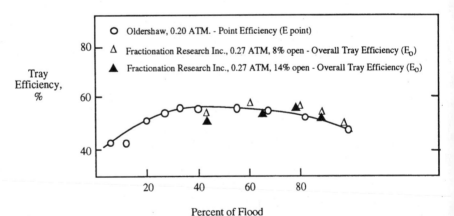

Figure 2.20 Efficiency comparison: 1-in. Oldershaw versus 4-ft-diameter column for cyclohexane/n-heptane system. (*Adapted from Fair, J. R. et al., "Scale-up of Plate Efficiency from Laboratory Oldershaw Data,"* Ind. Eng. Chem. Proc. (1983) *with permission of the American Chemical Society.*)

Scaleup procedure. The following scaleup procedure for sieve trays has been adapted from Fair et al. (1983).

In the Oldershaw column:

1. Run system to determine the flood point.
2. Establish normal operation for 60–80% of flood.
3. Run the Oldershaw at total reflux, taking compositions at the top and bottom of the column.
4. Compute the required number of theoretical stages and the Oldershaw overall column efficiency E_{ov}.
5. Assume Oldershaw $E_{ov} = E_{mv} = E_{oc}$.
6. Assume commercial $E_{ov} =$ Oldershaw E_{ov}.
7. A conservative assumption is to take commercial $E_{oc} = E_{ov}$.
8. Determine required actual column trays.

If equilibrium data are not available:

1. Run the system in the Oldershaw and find by test the combination of plates and reflux that gives the desired separation.
2. Assume that a commercial column with the same reflux ratio and number of trays will make the same separation.
3. If the commercial column is large in diameter, the number of trays may be reduced somewhat by estimating the liquid crossflow enhancement of efficiency (i.e., Murphree efficiency may be greater than point efficiency due to crossflow).

Valve trays. In valve trays, holes or slots on a tray are covered with movable mechanical valves (or caps) which can rise or fall depending on gas flow (Figure 2.21). At low gas rates, valves close to prevent weeping. At high gas rates, valves open to allow passage of vapor and to minimize pressure drop. As one would expect, the pressure drop is higher than it is with sieve trays. Tray efficiencies appear to be on the same order as sieve trays. In recent years, a number of distillation columns have been equipped with trays having minivalves (Lee, 1995). These trays are the same as conventional valve trays except the valves are much smaller allowing more valves to be incorporated on a tray, thereby improving vapor/liquid contact on each tray.

Bubble cap trays. Many older distillation columns contain bubble cap trays. Many columns used to dehydrate natural gas with triethylene glycol solvent are equipped with bubble cap trays. Because of their more complex mechanical design, bubble cap trays are much more

Sieve Tray

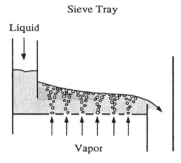

Valve Tray

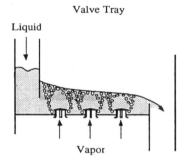

Bubble Cap Tray

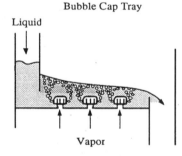

Figure 2.21 Examples of distillation crossflow trays.

expensive than sieve and valve trays. With bubble cap trays, vapor flows through the bubble caps which are submerged in the liquid on a tray. Once into the bubble caps, the vapor then changes direction and flows through the slots and into the liquid on the tray (Figure 2.21). Like valve trays, bubble cap trays are very effective for prevention of weeping at low gas rates.

New crossflow trays

Nye™ tray. Most of the newer crossflow trays are designed to increase capacity in high-pressure distillation columns (i.e., pressures greater than 50–100 psig). An example is the Nye™ Tray, which can provide 10–20% higher capacity than conventional sieve trays. Well over 200 distillation columns have been retrofitted with Nye™ trays (Lee, 1995). Particularly good performance has been reported with dethanizers, depropanizers, and debutanizers with capacity improvements ranging from 14 to 28% (Sasson and Pate, 1993). In most of these retrofits, completely new trays were installed which incorporated the Nye™ tray feature. However, in some cases, it may be possible to use a Nye™ tray insert with existing sieve trays. Each application must be evaluated to determine if completely new trays or inserts are appropriate.

Nye™ trays enable the vapor to utilize the normally inactive area beneath the downcomer as increased vapor-liquid disengaging space (see Figure 2.22). Vapor can flow through the insert area beneath the downcomer in addition to the normal sieve area. Thus, at a given rate, hole velocity of the rising vapor is lower. And with the extra disengaging space, these trays entrain less than sieve trays at the same liquid and vapor loads. Thus, these trays can handle extra liquid-vapor traffic for the same level of entrainment. Figure 2.23 provides a method to estimate the potential for increasing the capacity of a sieve tray column by retrofitting with Nye™ trays.

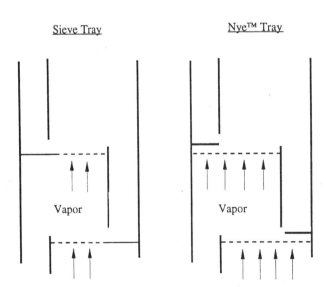

Sieve Tray   Nye™ Tray

Vapor   Vapor

Figure 2.22 Comparison of conventional sieve and Nye™ trays. (*Adapted from Glitsch Brochure, "High Capacity Tray Technology, Nye™ Trays," Bulletin 5138.*)

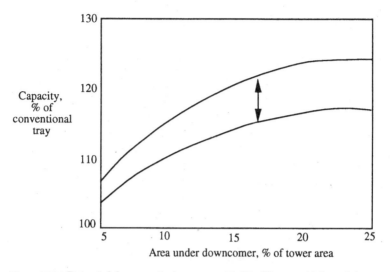

Figure 2.23   Potential for capacity increase with Nye™ trays. (*Adapted from Glitsch Brochure, "High Capacity Tray Technology, Nye™ Trays," Bulletin 5138.*)

Max-Frac™ tray.   Similar to the Nye™ tray, the Max-Frac™ utilizes a modified downcomer to increase the area under the downcomer to provide additional vapor disengaging area. In contrast to the Nye™ tray, liquid with a horizontal velocity component issues from the bottom of the downcomer. The horizontal component of the liquid velocity is achieved with the use of louvers located at the bottom of the downcomer (see Figure 2.24). Capacity increases are expected to be similar to those obtained with the Nye™ tray. Max-Frac™ trays have been installed in over 50 columns for light hydrocarbon separations and for a variety of other chemical and refinery applications (Yeoman, 1996).

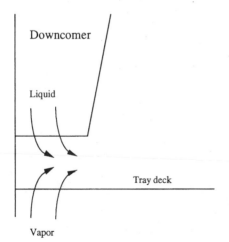

Figure 2.24   Max-Frac™ tray—flow patterns at bottom of downcomer. (*Provided by Koch Engineering Co., Inc.*)

Multiple downcomer trays.   New multiple downcomer trays are available to increase capacity of distillation columns operating at high pressures.

One of these is the Enhanced Capacity Multiple Downcomer (ECMD) tray which was commercialized in 1989. There are over 60 industrial columns equipped with ECMD trays (Summers et al., 1995). The ECMD tray utilizes slots on the tray deck whose function is to redirect some of the vertical upflowing vapor into a horizontal direction. This reduces the upflowing vapor momentum and results in a shorter froth height on the tray deck, which allows increases in vapor rate. Tray spacings of about 10 in. are used. Capacity increases over conventional trays can range from 35–50%. In this tray, plates are placed along the center of the downcomers to prevent the slots from propelling the liquid over to the other side of the tray deck. A benefit of these trays is that they assist with the disengagement of vapor from liquid as the froth enters the downcomer. These trays are used in C_3 splitters, C_2 splitters, xylene splitters, deethanizers, demethanizers, debutanizers, deisobutanizers, and depentanizers. They are in use in columns with diameters ranging from 4.5 to 19.0 ft and at operating pressures of 50–1200 psi.

Enhanced Efficiency Multiple Downcomer (EEMD) trays were commercialized in 1993. Like ECMD trays, the primary use of these trays is in the retrofit of distillation columns operating at high pressure. They have been installed in demethanizer and butadiene columns. There are two industrial columns which have been equipped with EEMD trays. These trays are configured so that the upflowing vapor always encounters liquid flow in the same direction. This parallel flow construction facilitates Lewis Case II point-to-plate efficiency enhancement. A 20% increase in the number of stages over conventional multiple downcomer trays (for the same tower shell) is possible (Shakur et al., 1996). High capacities are achieved via the elimination of receiving pans and deck slotting. Baffles are placed under the downcomers to prevent the falling liquid to enter the downcomer of the tray below.

Screen Tray™.   The Screen Tray™ is a high-efficiency tray designed to debottleneck traditional trayed columns. Screen trays produce dispersed liquid droplets in the vapor phase and operate in the froth regime over a wide range of conditions. The Screen Tray™ is expected to give about the same mass transfer efficiency as valve trays, but can provide higher capacity. When suspended solids are involved, caution should be exercised when using screen trays; they may be more susceptible to fouling than conventional trays.

P-K™ tray. P-K™ trays are suited for use in fouling environments, such as in stripping sections of atmospheric and vacuum pipe stills, and in flue gas desulfurization scrubbers where loadings of particulates are high. Tray decks are lanced and stamped to create a multiplicity of slots and vanes which impart a horizontal component to ascending vapor. High-velocity jets of vapor issuing from these slots sweep the deck, thereby reducing accumulations of solids. The slots are oriented to direct liquid away from dead areas, such as those near the support ring.

Cocurrent trays

Ultra-Frac™ tray. The concept upon which the cocurrent tray operates represents a radical departure from traditional crossflow trays. With the cocurrent tray, liquid drops are physically carried (entrained) by the vapor between the trays. Equipment is included between trays to accomplish separation of liquid drops and vapor. Capacity increases of 40–50% above conventional trays have been achieved and higher capacities are likely. At least six columns have been equipped with Ultra-Frac™ trays (Yeoman, 1996). This tray carries with it a higher pressure drop, perhaps 20–50% higher than conventional trays for many applications. A comparison between conventional trays and the Max-Frac™ and Ultra-Frac™ trays is given in Figure 2.25.

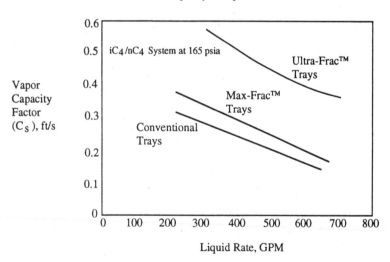

Figure 2.25 Capacity comparison of conventional, Max-Frac™, and Ultra-Frac™ trays. (*Adapted from information provided by Koch Engineering Co., Inc.*)

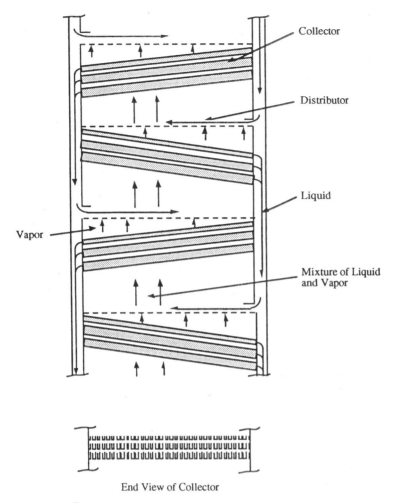

Collector

Distributor

Liquid

Vapor

Mixture of Liquid
and Vapor

End View of Collector

Figure 2.26 Trutna tray.

Trutna tray. Another important example of the cocurrent tray is the Trutna tray, named after its inventor, William R. Trutna. As of this writing, this tray is not commercial, but it is expected to be in the near future. With this device, the vapor entrains and carries liquid drops to a liquid/vapor coalescer located between trays, where liquid and vapor are separated. Though the cost of the Trutna tray is expected to be about three times the cost of the sieve tray, it can achieve capacities up to twice that for conventional trays. In retrofits of propylene/propane columns to increase capacity, Trutna trays are projected to achieve payouts of a few months (Seibert and Fair, 1995). The downside for the Trutna (and other cocurrent trays) is the increased complexity and cost due to the vapor/liquid phase separation. And cocurrent trays will lose efficiency at low turndown ratios. The Trutna tray is illustrated in Figure 2.26.

Packings

To promote mass transfer in distillation, it is desired that the liquid phase "wet" the packing and spread evenly over its surface. And the packing should have a large surface area per unit volume. Packing surface area is almost always larger than the interfacial area it actually creates between the liquid and vapor phases. An exception is that the surface area of structured gauze packing approaches that of the interfacial area it creates.

Packings should also possess desirable hydraulic characteristics. This means that the fractional void volume, or fraction of empty space, in the packed bed should be large enough to permit hydraulic passage of the liquid and gas phases, while providing a large area of contact between the phases. Guidelines for testing of packings are available from the American Institute of Chemical Engineers (*Packed Distillation Columns,* 1991). A comprehensive book on packing characteristics and performance is also available (Strigle, 1994).

Random packings

Examples of random packings are shown in Figure 2.27. Their physical characteristics are presented in Table 2.6. Note that newer random packings have thinner elements and would occupy less of the column volume than traditional packings, which are characterized by thick elements.

There is a large number of materials of construction available for random packings and all have their advantages and disadvantages. For example, stoneware may be attacked by alkali and hydrofluoric acid. To prevent breakage of stoneware packings, columns are first filled with water to reduce the velocity of fall as the packing is dumped into the column. When metals are used, concerns usually center around wettability issues and the possibility of high corrosion rates.

Plastic random packings find good applications when aqueous systems are involved. Though generally of the random type, structured plastic packings are also available. The removal of volatile organics from water by air stripping is an application where plastic packings are used. Plastic packings are lightweight, easy to install, have a low pressure drop per theoretical stage, are low cost, and corrosion is not a problem. Disadvantages include a limited upper temperature range and they are difficult to "wet." Efficiencies for some random plastic packings are given in Table 2.7.

Traditional Random Packings

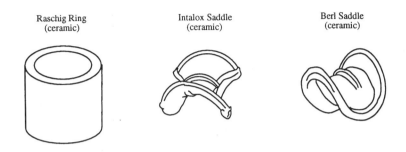

Raschig Ring (ceramic)   Intalox Saddle (ceramic)   Berl Saddle (ceramic)

Newer Random Packings

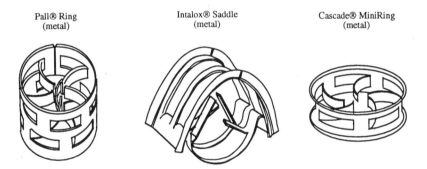

Pall® Ring (metal)   Intalox® Saddle (metal)   Cascade® MiniRing (metal)

Figure 2.27   Examples of random packings.

TABLE 2.6   Geometry of Random Packings

Packing	Surface area, sq ft/cu ft	Void fraction
1-in. ceramic Raschig rings	58	0.74
1-in. ceramic Intalox saddles	78	0.74
1-in. metal Pall rings	63	0.94
50-mm Hiflow Ring	33	0.93
50-mm Norpac (NSW)	31	0.94
25-mm Plasic Pall Rings	63	0.94

TABLE 2.7 Efficiencies of Random Packings

Packing	Chemical system	HETP, in	Reference
Raschig rings	n-Heptane/cyclohexane	24–48	Stichlmair et al., 1989
Pall rings	Methanol/water	10–12	Stichlmair et al., 1989
Pall rings	Ethylbenzene/styrene	12–24	Stichlmair et al., 1989
25-mm Norpac (plastic)	Ammonia/air/water	18–24	Billet and Mackowiak, 1982
50-mm Norpac (plastic)	Ammonia/air/water	18–34	Billet and Mackowiak, 1982
25-mm Pall rings (plastic)	Ammonia/air/water	16–20	Billet and Mackowiak, 1982

Scaleup of columns containing random packings. Of the various types of trays and packings, the scaleup of columns containing random packings is perhaps the most difficult and risky. A loss in efficiency as column diameter is increased should be expected. One should be wary of changing the size of packing from the pilot unit to the large scale plant because there will be a difference in the lateral spread of liquid. Also in pilot columns, the wetted wall contributes more to the mass transfer because of the higher proportion of the wall surface. This problem can be minimized by maintaining the ratio of column diameter to packing diameter above 10. Procedures for including a safety factor for cases where wall effects are significant have been developed (Wu and Chen, 1987). Kister (1992) provides some good guidelines for scaling small packed columns to larger ones. Guidelines for scaleup follow.

- Use same packing type and size in pilot column as in large unit.

- Use a pilot column of at least 1-ft in diameter.

- In both pilot and large columns, use a column to packing diameter ratio of at least 10 (or correct HETP for wall effects).

- Use the same number of drip points per unit cross-sectional area in the pilot and large columns.

- Make sure that liquid is distributed to the wall in the pilot and large columns.

- Use the same packing installation methods in both the pilot and large columns.

- Use a bed at least 5 ft tall (and preferably 10 ft tall) in the pilot column.

- Use acceptable sampling techniques for determining pilot column HETP. One should consult the procedures provided by the American Institute of Chemical Engineers (*Packed Distillation Columns,* 1991).

- Perform pilot tests over the entire range of vapor and liquid loads between minimum and maximum operating rates.

- Use the highest measured HETP in the pilot as the basis for the design of the large column.

- Be cautious of wetting and underwetting effects. When these occur, scaleup may be unreliable.

- Use packings of identical materials of construction and surface treatments in pilot and large scale columns.

- Ensure that plastic packings have been properly "aged" prior to the pilot test.

Structured gauze and sheet metal packings

At vacuum and low distillation pressures, structured packings have become prominent as the device that gives excellent mass transfer efficiency at high capacity. Structured packings have a regular geometric structure and, in contrast to random packings, are fabricated and fitted carefully to the dimensions of the column. When placed in the column, successive elements are oriented at 90° to each other. Each element consists of corrugated sheets positioned in a parallel arrangement.

Structured packings are available in different types (gauze, sheet) and in a variety of materials, including metals, plastics, ceramics, and carbon. The cost of structured sheet metal packing is about one-third that for gauze. However, for many applications, efficiencies for sheet metal packings are in the same range of those for gauze packings. Thus, structured sheet metal packings typically offer a more attractive performance to cost ratio than structured gauze packings. Structured gauze is generally used in high-vacuum columns where it is essential to minimize pressure drop, contact time, or column temperature. Structured sheet metal packing which is used in distillation is illustrated in Figure 2.28.

Figure 2.29 illustrates a column equipped with structured packing and the hardware required. Note that a significant part of the column is occupied by hardware. The packing occupies 60–70% of the column with hardware and space for phase disengagement occupying the other 30–40%. When structured packing is used, it is desirable to have one supplier to provide the packing, associated hardware, and be directly involved in the installation and start-up.

Applications for structured packings in distillation and absorption are illustrated in Figure 2.30 (Bomio et al., 1993). Applications in distillation are given by the dark shadings while those for absorption are given by the light ones.

Figure 2.28 Structured sheet metal packing. (*Provided by Norton Chemical Process Products Corp.*)

Advantages. For vacuum and low-pressure applications, structured packings can provide a higher mass transfer efficiency than trays. In the case of absorption, another possible benefit is a reduction in solvent to gas feed ratio. A lower solvent rate translates to smaller equipment and operating costs for downstream solvent recovery equipment.

At low column pressures, the capacity of structured packing is generally higher than that for trays or dumped packings. Increased capacity can be an attractive feature for new plants as well as for retrofits. With new columns, increased capacity means smaller column size and less weight, which is particularly important when exotic materials of construction are involved. And smaller size is important for high-pressure columns where the shell is of considerable thickness.

With regard to turndown, sieve trays are efficient at design capacity, but at low vapor rates "weeping" may produce unsatisfactory results. Random packings are less sensitive to changes in gas rates, but can be prone to gas-liquid bypass at low loads. Structured packing performance remains unaffected at gas rates down to about 10% of the design load.

With structured packings, a limiting factor turns out to be liquid distribution above the packing. Liquid should be introduced at the top

Overhead Vapor

Reflux from
Condenser

Structured Packing

Liquid Collectors

Feed

Liquid Distributor

Inlet Pipe

Column Sump

Circulation Pipe to
Reboiler

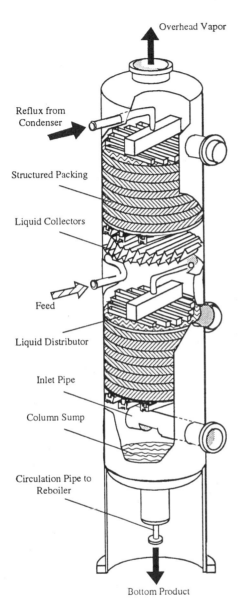

Bottom Product

Figure 2.29 Distillation column
equipped with structured packing.
(*Adapted from drawing provided
by Sulzer Chemtech Limited,
Winterthur, Switzerland.*)

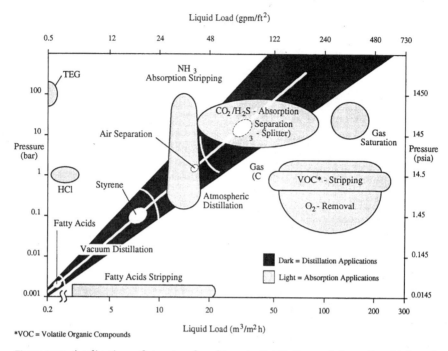

Figure 2.30 Applications of structured packings in distillation and absorption. [*Adapted from Bomio, P., B. Saner, and K. Breu (all of Sulzer Chemtech, Ltd.) "Experience with Structured Packings in High Pressure Gas Absorption," Paper presented at AIChE Spring National Meeting, Houston, TX, March 28–April 1, 1993.*]

of the packing by means of a suitable liquid distributor. Liquid distributors have to be designed for the required turn-down ratio of the liquid flow and should be tested experimentally. Most large suppliers of packings have their own in-house test systems to check the performance of their distributor designs. Liquid turn down, given by the design of the liquid distributor at the top of the tower, is usually about 30% of design capacity. The liquid distributor at the top of the column should deliver 5–10 drip points/sq ft of cross-sectional area of the column. Because of the way gas contacts liquid in corrugated sheet packings where gas is brought into tangential contact with liquid rather than forced to flow through it, entrainment of liquid is usually negligible.

Disadvantages. Disadvantages of structured packings are high costs relative to trays, criticality of initial liquid and vapor distributions, and the associated hardware required. For a column several feet in diameter, the installed cost of structured sheet metal packing plus associated hardware is about three to four times that for conventional trays (installed cost basis). Additionally, liquid redistribution is necessary every 15–20 ft of packing. Thus, a good deal of properly designed hardware is necessary for structured packings to perform effectively.

In addition, unexplained poor mass transfer efficiency has been encountered with structured packings at high column pressures (Bomio et al., 1993). Under these conditions, the densities of the liquid and vapor approach each other, and it is thought that backmixing in the vapor phase occurs, resulting in significantly lower mass transfer efficiencies. The use of trays should be considered when high column pressures (greater than 50–100 psig) are involved. Geometries of several structured packings are given in Table 2.8, while efficiencies of some packings are given in Table 2.9.

TABLE 2.8   Geometry of Structured Packings

Packing	Surface area, sq ft/cu ft	Void fraction
Flexipac 1 (Koch)	170	0.91
Flexipac 2 (Koch)	76	0.93
Gempak 4A (Glitsch)	160	0.91
Gempak 2A (Glitsch)	80	0.93
Sulzer BX (Sulzer)	152	0.90
Intalox 2T (Norton)	65	0.96
Flexeramic™ 48 (Koch)	48	

TABLE 2.9   Efficiencies of Structured Packings

Packing	Chemical system	HETP, in.
Flexipac 2 (Koch)	n-Heptane/cyclohexane	12–16*
Gempak 2AT (Glitsch)	n-Heptane/cyclohexane	16–20*
Gempak 2A (Glitsch)	n-Heptane/cyclohexane	12–16*
Sulzer BX (Koch)	n-Heptane/cyclohexane	8–12*
Flexeramic™ 48 (ceramic)	Ammonia/air/water	10–15†

*From Martin et al., 1988.
†From Lucero et al., 1990.

Scaleup for columns containing structured packings. Because of the regular structure, the scaleup of columns containing structured packing is less risky than for columns containing random packings. However, less data are available because structured packings are newer than random packings. One should expect to get some decline in efficiency as column diameter increases because of the increased opportunity for maldistribution of the liquid phase. Procedures and guidelines for the scaleup of small diameter pilot plant columns containing structured packing to large commercial units are available (Hufton et al., 1988; Kister, 1992).

Optiflow packing

Optiflow packing more closely approaches the configuration of a true monolith than traditional structured packings. Eighteen industrial columns in Europe and three in the United States have been equipped with this packing (Jancic, 1996).

Optiflow packing consists of individual diamond-shaped surfaces joined together at the corners to form pyramids. The result is symmetry and regularity of structure in all directions. Figure 2.31 shows

Figure 2.31 Optiflow packing. (*Provided by Sulzer Chemtech, Ltd.*)

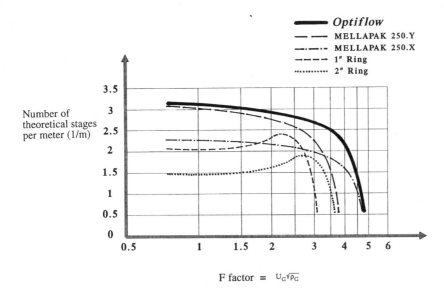

Figure 2.32 Comparison of mass transfer devices for vacuum distillation [100 millibar (mbar)]. (*Provided by Sulzer Chemtech, Ltd.*)

ordinary soda straws inserted into the structure of Optiflow packing to illustrate the four different directions of flow.

Comparisons of the mass transfer efficiency and hydraulic pressure drop of Optiflow compared with rings and traditional structured sheet metal packing are given in Figure 2.32 (Suess et al., 1994).

Results from performance tests provide the following comparison of Optiflow with structured sheet metal packings:

- Optiflow gives a 25% higher capacity than Sulzer Mellapak® 250.Y and a slightly higher number of theoretical stages per meter, and

- 50% more theoretical stages per meter are obtained with Optiflow in comparison with a structured packing of similar geometry (e.g., Sulzer Mellapak® 250.X).

A comparison of pressure drop of Optiflow with 1-in. rings and a traditional structured sheet metal packing is given in Figure 2.33.

A comparison of performance and costs for structured sheet metal packings with Optiflow for the separation of fatty acids from their esters is given in Figure 2.34 and in Table 2.10.

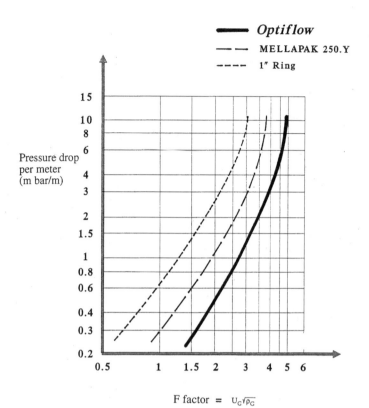

F factor $=$ $U_G\sqrt{\rho_G}$

Figure 2.33 Pressure drop comparisons: Optiflow versus other devices—1000 mm column diameter/100 mbar top pressure. (*Provided by Sulzer Chemtech, Ltd.*)

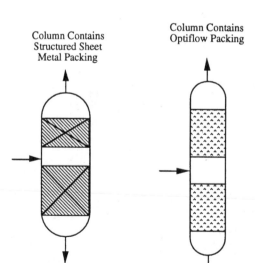

Column Contains Structured Sheet Metal Packing

Column Contains Optiflow Packing

Figure 2.34 Comparison of structured sheet metal packing with Optiflow packing—new column design for fatty acid/ester separation.

TABLE 2.10   Performance Comparison of Structured Sheet Metal Packing
with Optiflow Packing—New Column Design for Fatty Acid/Ester Separation

| | Structured packings | | Optiflow | |
	Rectifying section	Stripping section	Rectifying section	Stripping section
	Sulzer packing BX®	Mellapak® 250.Y	Optiflow	Optiflow
Section height (m)	2	4	4	4
No. of stages	9	9	9	9
Head pressure (mbar)	10			
Bottom pressure (mbar)	15.5			
Pressure drop (mbar)	5.5			
Diameter (m)	1.8		1.5	
Volume (packing)(cu m)	18		14	
Total investments ($)	280,000		224,000	

Trays Versus Packings

For any application one must determine whether a tray or packing is
the most appropriate. The following factors are an indication of when
trays or packings are favored.

Factors favoring trays

- High liquid rate (occurs when high column pressures are involved)
- Large diameter (packing prone to maldistribution)
- Complex columns with multiple feed/takeoffs
- Feed composition variation
- Scaleup less risky
- Columns equipped with trays weigh less than those equipped with
  packings

Factors favoring packings

- Vacuum conditions
- Low-pressure drop required
- In smaller diameter columns (where trays are more difficult to
  install, diameters are 2–3 ft or less)
- Corrosive systems (more materials of construction available)
- Foaming
- Low liquid hold-up

Performance comparison: Trays versus packings

A quantitative analysis of efficiency and capacity of trays versus packings, both random and structured packing, has been completed by Kister et al. (1994). Highlights of the results are presented here. Efficiency and capacity differences between trays and packings are due to the four factors below. The comparative analysis was based on the first factor, that is, differences between the efficiency and capacity of an optimally designed tray and an optimally designed packed tower.

1. *Differences between efficiency and capacity of an optimally designed tray and an optimally-designed packed tower.* In an optimal tray design, geometry is chosen to maximize efficiency and capacity. In the optimal packing design, distributors, supports, and bed heights are chosen to maximize efficiency and capacity.

2. *Deviations from optimal design of trays and packings and associated hardware.* A packing may perform much better than trays due to a poor tray design. Trays may perform better than packing due to a poor liquid distributor design.

3. *Unique system characteristics and special design features.* These characteristics include fouling, foaming, and corrosion.

4. *Capacity and separation gains due to the lower pressure drop of a packing.* The pressure drop in a packing may be 20–40% that of trays. Lower average pressure increases relative volatility, improves separation, and reduces the number of separation stages required. Reflux can be reduced making room for more feed and an increased capacity.

In making performance comparisons, data measured by independent research organizations were used. All of the data for trays, random packings, and high-pressure distillations (greater than 90 psia) for structured packings were provided by Fractionation Research, Inc. (FRI) in its 4-ft-diameter tower. The low pressure (less than 90 psia) structured packing data were provided by the Separations Research Program (SRP), University of Texas at Austin, taken in their 17-in.-diameter column.

The test systems used by FRI and SRP were cyclohexane/n-heptane at 5 and 24 psia, and isobutane/n-butane at 165 psia. A smaller amount of data were also available for the isobutane/n-butane system at 300 and 400 psia. Further details on system data are available (Kister et al., 1993).

Efficiency comparisons. Comparisons of trays versus packings are expressed as a function of two parameters. The first is the height equivalent to a theoretical plate (HETP). The second is the flow parameter (FP) defined by:

$$FP = \frac{L}{V} \sqrt{\frac{\rho_G}{\rho_L}} \qquad (2.70)$$

where FP = flow parameter
  L = liquid flow rate, lb/hr ft^2 of cross-sectional area
  V = vapor flow rate, lb/hr ft^2 of cross-sectional area
  ρ_G = gas density, lb/ft^3
  ρ_L = liquid density, lb/ft^3

Low flow parameters, less than about 0.1, are typical of vacuum distillations. High flow parameters, greater than 0.3, are typical of high pressure or high liquid rate operation. Table 2.11 shows the ranges of values of the flow parameter for each type of distillation.

Figure 2.35 is a plot of HETP against FP for optimized trays (at 24-in. tray spacing), random packing (Nutter Ring™), and structured packing (Norton Intalox® 2T). In this figure, adjustments are made for the vertical column height consumed by distributors, redistributors, and the end tray.

Figure 2.35 shows minima in the plots of HETP versus FP. At FPs up to 0.2, HETP declines with a rise in FP. In this region, the HETP of optimized trays is about the same as Nutter Rings™, and both are about 50% higher than that of the Intalox® 2T structured packing.

At flow parameters beyond 0.3, HETP rapidly rises with FP. The rate of rise is the least with random packings and the most with structured packings. At 400 psia and an FP of about 0.5, the HETP of Intalox® 2T structured packing is slightly higher than optimized trays and about 20% higher than that of Nutter Rings™.

TABLE 2.11   Values of Flow Parameter Versus Distillation Pressure Ranges

Values of flow parameter (FP)	Pressure distillation
0.02–0.1	Vacuum
0.1–0.3	Low to medium pressure
0.3–0.5	High pressure

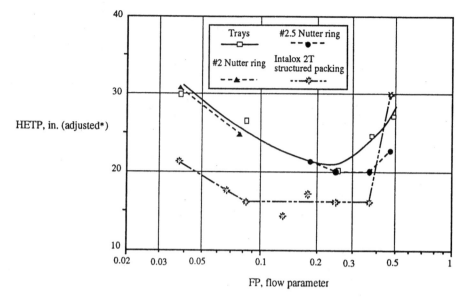

* Adjusted for vertical height consumed by distributor, redistributor and end tray

Figure 2.35 Trays versus packings—comparison of efficiency. (*Adapted from Kister, H. Z. et al.,* Chemical Engineering Progress, *February 1994 with permission of the American Institute of Chemical Engineers. Copyright © 1994 AIChE. All rights reserved.*)

Capacity. In comparing capacity of trays versus random and structured packings, results are expressed as a function of the flood capacity factor (also called the flood C-factor) and the flow parameter (FP). The capacity factor is a measure of the vapor load at flood and is defined as:

$$C_s = U_G \sqrt{\frac{\rho_G}{\rho_L - \rho_G}} \qquad (2.71)$$

where C_s = capacity factor, ft/s, based on column superficial area
U_G = gas velocity, ft/s, based on column superficial area
ρ_G = gas density, lb/ft^3
ρ_L = liquid density, lb/ft^3

Figure 2.36 illustrates that at low FPs, of about 0.02, the capacity of Intalox® 2T is greater by 30–40% than that of either trays or Nutter Rings™. However, as FP increases, the capacity of the structured packing declines much faster than the capacity of the optimized tray or Nutter Rings™. When the FP reaches 0.1, the capacity of Intalox® 2T approaches that of the optimized tray.

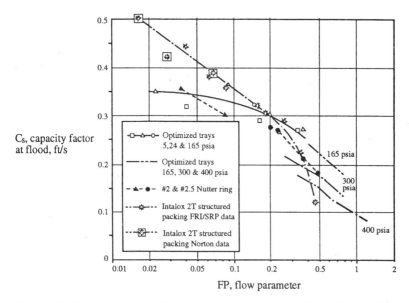

Figure 2.36 Trays versus packings—comparison of capacity. (*Adapted from Kister, H. Z. et al.*, Chemical Engineering Progress, *February 1994 with permission of the American Institute of Chemical Engineers. Copyright © 1994 AIChE. All rights reserved.*)

At FPs beyond 0.2–0.3, results show capacity of Intalox® 2T structured packing declines rapidly with a rise in FP. The capacity of the optimized trays declines more slowly, and that of Nutter Rings™ most slowly. At an FP of 0.5 and pressure of 400 psia, Intalox® 2T has the lowest capacity, while optimized trays at 24-in. tray spacing are about 20% higher, and Nutter Rings™ are 20% higher.

Summary. The results are based on data obtained for optimized designs and under ideal test conditions. To translate results to actual columns, one must consider liquid and vapor maldistribution, which is far more detrimental to the efficiency of packings than trays and also account for poor design of internals, more detrimental to the capacity of trays than packings. Comparing trays (24-in. tray spacing) with state-of-the-art random and structured packings, all optimally designed, it was found that:

At FPs of 0.02–0.1 (corresponding to vacuum distillation):

- Structured packing efficiency is about 50% higher than either trays or random packings.

- As FP increases from 0.02 to 0.1, the capacity advantage of structured packing declines from 30–40% to zero.

- Trays and random packing have about the same efficiency and capacity.

At FPs of 0.1–0.3:

- Efficiency and capacity of trays, random packing, and structured packing decline with an increase in flow parameter.

- The rate of decline in the capacity and efficiency is the most significant with structured packing, and least significant in random packings.

At an FP of 0.5 and 400 psia:

- Random packing appears to have the highest capacity and efficiency, and the structured packing the least.

Relative costs: Valve trays versus structured sheet metal packing

One must also consider relative costs in addition to the comparisons presented above. Estimates of relative costs of valve trays versus structured sheet metal packing are given in Table 2.12 (Yeoman, 1994). These relative costs were estimated based on a tower shell 5 ft in diameter and 40 ft tall.

The tray case in Table 2.12 is based on 30 valve trays, while the packing case is based on two packing beds with a total packed height of 38.5 ft. The packing case includes a liquid distributor and redistributor. All equipment is 304 stainless steel. Results in Table 2.12 show that the total installed cost of the structured packing is close to three times higher than for valve trays for the case considered. This difference is primarily due to the higher equipment cost of structured packing relative to valve trays.

TABLE 2.12   Relative Costs of Valve Trays Versus Structured Sheet Metal Packing

Device	Equipment (relative cost)	Installation (relative cost)	Total (relative cost)
Valve trays	1	1	1
Structured sheet metal packing	4.2	1.4	2.8

Membrane Phase Contactors

Though not applicable to distillation, there is a new type of contacting device for use in absorption and stripping processes. Called the *membrane phase contactor*, this device has been commercialized for absorption and stripping applications (Reed et al., 1995; Sengupta et al., 1995). In this contactor, the liquid phase invades the membrane material and in doing so, forms a film across each of the membrane pores. It is through these films in the pores that mass transfer takes place. Because of the large surface area of the films, this type of contactor can provide a large number of separation stages in a relatively short length of module. Specifically, it can provide values of the height equivalent to a theoretical plate (HETP) of 10–16 in. when used to strip oxygen from water (Sengupta, 1996). Values of HETP for a conventional packed column, for the same application, can be expected to be several times higher.

The features of the membrane phase contactor are illustrated in Figure 2.37. Hollow fiber membranes are housed in a module that is equipped with baffles. One phase is transported through the bore of the fibers while the other is transported on the shell side. The baffles divert and direct the flow on the shell side to promote contact of the shell side fluid with all of the hollow fibers. In membrane phase contactors, polyolefins, such as polypropylene, are used as the materials for the membrane fibers.

Membrane phase contactors are used commercially for the stripping of oxygen from water to produce high-purity water for use in the microelectronics industry. A vacuum and a small amount of nitrogen are used to strip oxygen from water. The flowsheet for a commercial unit is given in Figure 2.38.

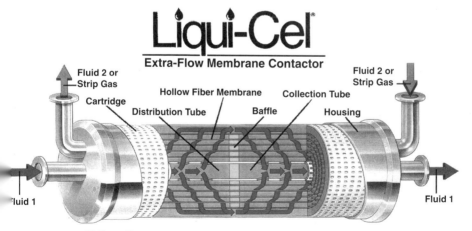

Figure 2.37   Hollow fiber membrane phase contactor. (*Provided by Hoechst Celanese Corporation.*)

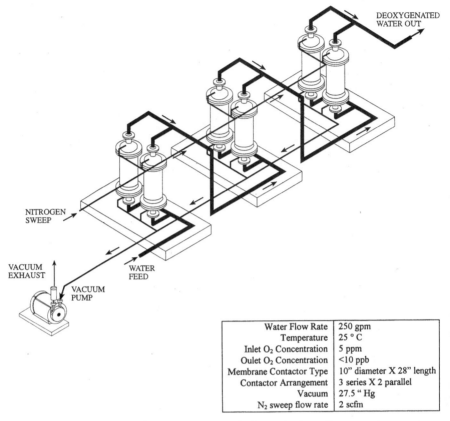

Water Flow Rate	250 gpm
Temperature	25 ° C
Inlet O_2 Concentration	5 ppm
Oulet O_2 Concentration	<10 ppb
Membrane Contactor Type	10" diameter X 28" length
Contactor Arrangement	3 series X 2 parallel
Vacuum	27.5 " Hg
N_2 sweep flow rate	2 scfm

Figure 2.38   Membrane phase contactor unit for stripping of oxygen from water. (*Provided by Hoechst Celanese Corporation.*)

In the beverage industry, membrane phase contactors have also been commercialized to add carbon dioxide to aqueous solutions (carbonation). Worldwide, over 1000 contactors are in operation in various types of absorption and stripping applications, with most being used in stripping (Pittman, 1996). This is a surprisingly large development for a technology which is relatively new to the commercial world.

Design Procedures

The procedures to design a new distillation column, based on the traditional equilibrium stage approach, are given in Table 2.13. A newer method, called the *rate-based design method,* offers advantages over the traditional equilibrium stage approach. This newer method is expected to gradually replace the traditional approach in the years to come. This method is described in the section which follows. Detailed design procedures, associated with the steps given in Table 2.13, are provided in the Appendix (pages 342 through 367).

TABLE 2.13   Procedures for Design of a New Distillation Column

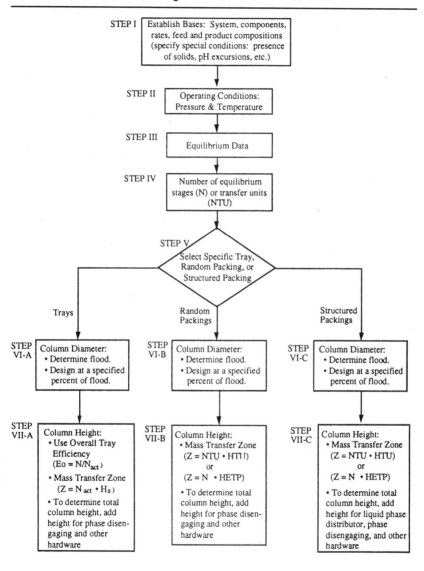

Rate-Based Design Method

An important new design method has been developed for distillation and other staged processes (Krishnamurthy and Taylor, 1985). A feature of this method, called the *rate-based design method,* is that component material and energy balances for each phase, together with mass and energy transfer rate equations, and equilibrium equations for the phase interface, are solved to determine the actual separation directly.

In the rate-based design method, calculations are done on an incremental basis as one proceeds through the column. The uncertainties of computations using average tray efficiencies, based on individual components, are entirely avoided. This method is expected to play an increasingly important role in designing distillation columns by replacing the two-step method of first determining the number of equilibrium stages and then using an "average" efficiency factor (of a tray or packing) to determine the required column height.

A review of this method, as well as a comparison with the conventional equilibrium stage approach, has been presented by Seader (1989). Experimental validation of the rate-based design method has been presented by Krishnamurthy and Taylor (1985) and Pelkonen and Gorak (1995).

Alternative Distillation Processes

Batch distillation

Most distillations are run continuously, but there are many cases where *batch distillation* is preferred, particularly in the food and pharmaceutical industries. Batch distillation is used when small amounts of product are made in a pilot plant to provide samples for product sampling or testing. Products produced in low volumes, such as some pharmaceuticals and specialty chemicals, are often purified by batch distillation. In batch distillation, the unit is campaigned where several different products are made with the vessel cleaned after each run. A special section providing references for batch distillation is included at the end of this chapter. And a very recent book has been published which presents the fundamentals of batch distillation (Diwekar, 1995).

In batch distillation, a liquid mixture is charged to a vessel where it is heated to the boiling point. After boiling begins, vapor is removed and subsequently condensed to obtain a distillate product. Lower-boiling components are concentrated in the vapor while heavier components are concentrated in the liquid. Batch distillation may be operated with no reflux (as shown in Figure 2.39). It may also be operated with a constant amount, or a variable amount, of reflux. When a vari-

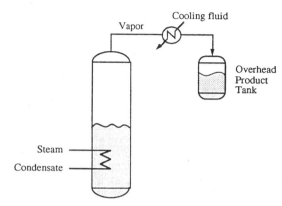

Figure 2.39 Batch distillation.

able reflux is used, it is generally increased as boiling proceeds in order to achieve a constant overhead product composition.

Advantages. The advantages of batch distillation are that several products can be made from a single unit, it is simple to control, and can effectively handle sludges and solids.

Disadvantages. The disadvantages of batch distillation are that for a given product rate, the equipment is larger. It requires more operator attention, uses more energy, and, because it is a dynamic process, is harder to control and model.

Rayleigh equation. In single-stage batch distillation calculations, it is assumed that the vapor and liquid are in equilibrium. The composition of the liquid changes continuously as boiling proceeds. The composition of the vapor also changes, and it is assumed to be in equilibrium with the liquid mixture remaining in the batch vessel.

If L is the number of mols of liquid in the vessel at any given time, and dL is the differential amount vaporized, a material balance on component i yields

$$L\,x_i = (L-dL)(x_i-dx_i) + (y_i + dy_i)dL \qquad (2.72)$$

Neglecting products of differentials, we have

$$(x_i - y_i)\,dL = L\,dx_i \qquad (2.73)$$

The result of integrating over the change in quantity of liquid from the initial condition to the final condition, and from the initial to the final concentration of liquid, yields the important *Rayleigh equation* for batch distillation (Rayleigh, 1902):

$$\int_{L_1}^{L_2} \frac{dL}{L} = \int_{x_{i1}}^{x_{i2}} \frac{dx_i}{y_i - x_i} \tag{2.74}$$

and

$$\ln \frac{L_2}{L_1} = \int_{x_{i1}}^{x_{i2}} \frac{dx_i}{y_i - x_i} \tag{2.75}$$

where L_1 = initial liquid, mols

L_2 = final residual liquid, mols

x_{i2} = mol fraction of component i in liquid at final conditions, dimensionless

x_{i1} = initial mol fraction of component i in liquid, dimensionless

y_i = mol fraction of component i in vapor, dimensionless

x_i = mol fraction of component i in liquid, dimensionless

dL = differential quantity of liquid distilled, mols

dx_i = differential change in mol fraction of liquid, dimensionless

dy_i = differential change in mol fraction of vapor, dimensionless

When equilibrium data are available, the right-hand side of Equation (2.75) can be integrated graphically for a binary system by plotting $1/(y_i - x_i)$ between the limits. This process is illustrated in Figure 2.40 for an initial concentration of $x_1 = 0.5$ and a final concentration of $x_2 = 0.2$.

By expressing vapor composition (y_i) in terms of liquid composition (x_i) and relative volatility (α_{ij}), Equation (2.75) may be expressed as

$$\ln \frac{L_2}{L_1} = \int_{x_{i1}}^{x_{i2}} \frac{dx_i}{[\alpha_{ij}/(1/x_i + \alpha_{ij} - 1)] - x_i} \tag{2.76}$$

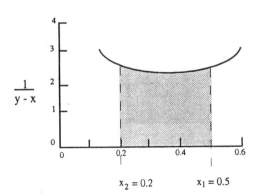

$x_2 = 0.2$ $x_1 = 0.5$

Figure 2.40 Graphical integration of Rayleigh equation.

When relative volatility is constant, Equation (2.76) can be integrated to give

$$\ln \frac{L_2}{L_1} = \frac{1}{\alpha_{ij}-1} \ln \frac{x_{i2}(1-x_{i1})}{x_{i1}(1-x_{i2})} + \ln \frac{1-x_{i1}}{1-x_{i2}} \qquad (2.77)$$

Extractive distillation

In *extractive distillation*, a solvent is added to the distillation tower to increase relative volatility of key components in the feed mixture. Alteration of volatilities is desired in systems having low relative volatilities, or those which exhibit azeotropes. The effective extractive solvent will selectively interact with one (or more) of the components, thereby increasing relative volatilities.

An extractive solvent is usually selected that boils at a much higher temperature than the components of the feed. The extractive solvent should be chosen so that no new azeotropes are formed. Because the high-boiling solvent can be recovered by distillation, and there are no azeotropes, extractive distillation is less complex and more widely used than azeotropic distillation. Azeotropic distillation is described in the section which follows.

The extractive distillation process, illustrated in Figure 2.41, separates isobutane from 1-butene using furfural as solvent. The relative

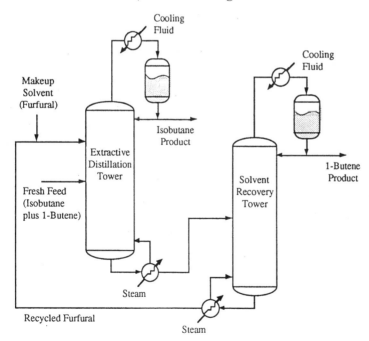

Figure 2.41 Example of extractive distillation process.

volatility of isobutane to 1-butene, in the presence of furfural, is 2.0, as opposed to a relative volatility of 1.2 in the absence of furfural. The furfural molecule selectively interacts with the double bond in the 1-butene molecule.

In order to maintain a high concentration of solvent throughout the column, solvent is generally introduced on a few trays below the top of the column, with the exact locations being determined by the necessity to reduce solvent concentration in the overhead vapor.

Because reflux at the top of the column dilutes solvent, the advantage of higher reflux must be balanced against the disadvantage of lower solvent concentration and reduction in relative volatility. Solvent solubility must be known to make sure that solvent is miscible.

The choice of solvent determines which components become the overhead product. For example, assume in Figure 2.41 that fresh feed to the extractive-distillation column is a mixture of 80 mol% ethanol in water. If ethylene glycol is the solvent, ethanol is the more volatile component. Therefore, ethanol is removed as the overhead product in the extractive-distillation column and water is separated from ethylene glycol in the solvent-recovery column. If a high-boiling hydrocarbon (such as octane) is used as solvent, water is the most volatile component so that it becomes the distillate in the extractive-distillation column.

Extractive and azeotropic distillation are described in detail by Seader and Kurtyka (1984). Algorithms for rigorous stage-to-stage energy and material balances may be used if the equilibria are known. Case studies are provided by Black and Ditsler (1972), Gerster et al. (1955), and Kumar et al. (1972). Because heavy solvents are involved, tray efficiencies are characteristically lower than in conventional distillation. Approaches to selection of solvent are presented by Berg (1993) and Berg and Yang (1994).

The extractive solvent should be:

- Effective at enhancing relative volatility
- Commercially available
- Inexpensive
- Safe for operators to handle
- Thermally stable
- Nonreactive
- Miscible
- Easily separated from other components

Azeotropic distillation

Like extractive distillation, azeotropic distillation is used to separate mixtures not separable by conventional distillation. If a solvent is added to form an azeotrope, with one or more of the feed components, the azeotrope will exit the column as the overhead or bottom product leaving behind component(s) which may be recovered in the pure state. In industrial applications, the azeotrope usually exits the column as the overhead product.

An example of azeotropic distillation is the recovery of anhydrous ethanol from aqueous solutions. Ethanol cannot be recovered and purified from aqueous solutions by conventional distillation at ordinary conditions because of a minimum-boiling azeotrope. However, organic solvents may be used to form desirable azeotropes, allowing the separation to be made. In the crude column, shown in Figure 2.42, the overhead is a 96% ethanol and 4% water azeotrope while the bottoms product is water.

The azeotrope, produced as overhead product from the crude column, is fed to the azeotropic distillation column where cyclohexane is used as solvent, to form an azeotrope with water. In the azeotropic column, ethanol product is taken as bottom product while a minimum-boiling azeotrope of water and cyclohexane is taken overhead. The overhead

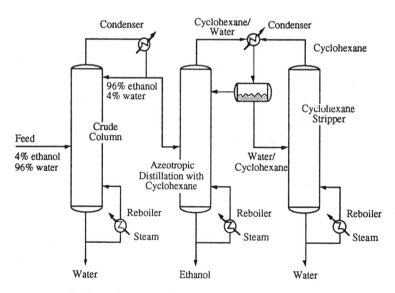

Figure 2.42 Traditional azeotropic distillation process to separate ethanol/water.

azeotropic mixture is condensed where water-rich and cyclohexane-rich liquid phases are formed. The cyclohexane-rich phase is refluxed to the azeotropic column, while the water-rich phase, which contains a small concentration of cyclohexane, is fed to the cyclohexane stripper for recovery cyclohexane. Because of concerns about worker exposure, benzene has been replaced in many plants by cyclohexane, and other less toxic solvents, capable of forming suitable azeotropes.

Selection of solvent. The azeotropic solvent should be:

- One which forms a desirable azeotrope
- Commercially available
- Inexpensive
- One which boils in the same range as the hydrocarbons
- Safe for operators to handle
- Soluble
- Separable from the azeotrope
- Thermally stable
- Nonreactive

Processes Similar to Distillation

Absorption and *stripping* are similar processes to distillation. In distillation, the mass transfer is between liquid and its vapor. In absorption and stripping the mass transfer is between a gas and a liquid phase.

Absorption

In absorption, solute(s) is transferred from the gas phase to the liquid phase. In stripping, the reverse is true; the transfer is from the liquid to the gas phase. The types of equipment and design concepts used in distillation are similar to those used for absorption and stripping. Fundamentals and design procedures for absorption processes have been published by Diab and Maddox (1982).

Absorption and stripping have important applications in product purification and in environmental applications. In the environmental area, absorption is used to remove contaminants from gases, such as absorption of ammonia from air with water (Figure 2.43). Stripping is used to remove volatile organic compounds (VOCs), such as benzene, from wastewaters.

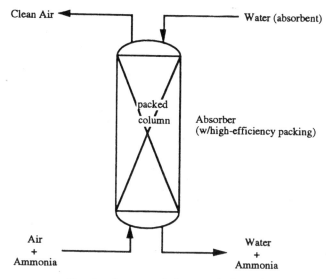

Figure 2.43   Absorber for removal of ammonia from air.

Structured packings in absorption. The use of structured packings in absorption is a relatively new commercial development, particularly in the United States. In contrast to high-pressure distillation, structured packings can be used without concern about a decline in efficiency at high column pressures.

For a given efficiency and capacity, gas pressure drop is small with structured packings. Depending on operating conditions and physical properties, pressure drop is about 10% that of trays and 20–50% of rings. In units treating very large volumes of gas, a reduction in pressure drop can represent important savings in compression power and can debottleneck a compressor train. Several comparisons of trays versus structured packings for absorption applications follow, which are based on the work of Bomio et al. (1993).

Absorption of hydrogen sulfide from natural gas. Figure 2.44 illustrates differences in size of a new column equipped with trays versus one equipped with structured sheet metal packing. The application is the absorption of hydrogen sulfide from natural gas with monodiethanolamine (MDEA) as the solvent. This is a particularly important industrial application in the gas industry where various types of solvents, particularly amines, are used to remove hydrogen sulfide and carbon dioxide from natural gas. Table 2.14 provides a comparison of tray and packed column cases for this application. Results show a smaller column height and diameter is required when structured packing is used for this application.

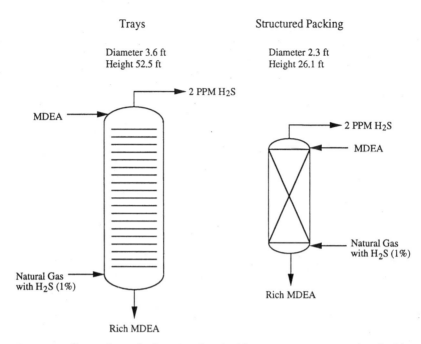

Figure 2.44 Comparison of column equipped with trays versus one equipped with structured sheet metal packing for absorption of hydrogen sulfide from natural gas.

TABLE 2.14 New Design—Comparison of Column Equipped with Trays Versus One Equipped with Structured Sheet Metal Packing—Absorption of Hydrogen Sulfide from Natural Gas

	Trays	Structured packing
Diameter, ft	3.6	2.3
Height, ft	52.5	36.1
Pressure, psia	812	812
Temperature, °F	85–95	85–95
Liquid/gas mole ratio	0.15	0.15

Gas dehydration. The traditional process for dehydrating natural gas is to absorb the water using triethylene glycol (TEG) as the solvent. Traditionally, many of these dehydration columns were equipped with bubble cap trays. More recently, some of these columns have been retrofitted with structured sheet metal packings. The differences in performance between TEG drying columns, equipped with bubble cap trays, and structured sheet metal packings, are given in Table 2.15.

TABLE 2.15   Comparison Between Bubble Cap Trays and Structured Sheet
Metal Packing—Dehydration of Natural Gas

	New design		Retrofit	
	Bubble cap trays	Structured packing	Bubble cap trays	Structured packing
Capacity	100	100	100	185
Column diameter, ft	8.2	5.9	8.2	8.2
Turndown capability (gas)	~1:5	>1:5	~1:5	>1:5
Column weight, tons	80	50		
Pressure, bar	100	100	100	100

The first use of structured packings in TEG columns was in Europe
in 1982 and they were used in North America for the first time in 1985.
Performance data and operating characteristics for TEG columns have
been published (Collins and Breu, 1992; Kean et al., 1991).

Absorption of ethylene oxide.   As shown in Figure 2.45, ethylene oxide
is recovered from reactor effluent by absorption with water. For a new

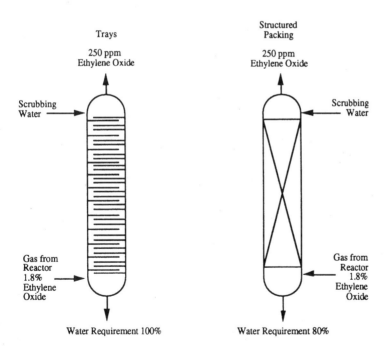

Figure 2.45   Conversion of trays to structured packing in an ethylene
oxide absorber.

TABLE 2.16   New Design—Trays Versus Structured
Sheet Metal Packing—Absorption of Ethylene
Oxide with Water

	Trays	Structured packing
Column diameter, ft	21.6	19.7
Column height, ft	125	112
Pressure drop, in. H_2O	44	6

TABLE 2.17   Retrofit—Conversion of Trays to
Structured Sheet Metal Packing—Ethylene Oxide
Absorber

Parameter	Trays	Structured packings
Mass transfer zone	35 trays	47.2 ft
Diameter, ft	15.7	15.7
Pressure, psia	275	275
Water requirement, %	100	80

ethylene oxide recovery column, the selection of structured packing rather than trays can result in a lower overall investment cost (column plus packing plus blower), since column dimensions are smaller and the pressure drop is less.

Table 2.16 presents a comparison of sieve trays versus structured packing for an ethylene oxide absorber (based on a new plant design).

In the ethylene oxide application, an existing sieve tray column can be retrofitted with structured packing to reduce energy consumption. Due to the higher efficiency of the structured packing, the amount of scrubbing water can be reduced by about 20%. Reduced water rate translates to a savings of steam for the adjacent stripper required for water recovery. Another benefit is a reduction in pressure drop. An increased capacity reserve of about 50% is also reported to be available (Figure 2.45 and Table 2.17).

Stripping

Stripping is accomplished by countercurrent contact of a liquid phase flowing downward with a stripping gas flowing upward. During contact, solute(s) is transferred from the liquid to the stripping gas. The

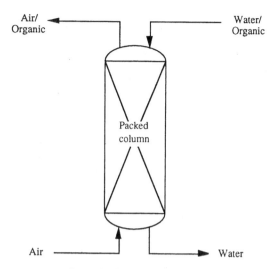

Figure 2.46   Air stripping process.

air stripping process is illustrated in Figure 2.46. With this process, organic components, such as benzene, can be removed from water using air, steam, or natural gas as the stripping gas.

In the case of air stripping, problems can be created if effluent gas is dispersed directly to the atmosphere. The gaseous effluent may have to be treated by a second process, such as carbon adsorption, to remove contaminants. Review papers on air stripping have been published by Byers and Morton (1985) and Kavanaugh and Trussel (1980). The air stripping process is illustrated in Figure 2.46.

Steam stripping.   Steam stripping is similar to air stripping except it operates at a higher temperature because of the addition of live steam. An excellent review of steam stripping, together with an extensive tabulation of Henry's constants, has been published by Hwang et al. (1992).

The higher temperature of steam stripping results in an increase in Henry's constants, from hundreds to tens of thousands. The result is that high-boiling organics can effectively be stripped from water.

As shown in Figure 2.47, the effluent containing steam and an organic, such as benzene, can be condensed, whereupon the water-rich and benzene-rich phases separate. Bulk benzene and water phases can then be recovered for disposal or recycle. For a steam stripper, operating costs are a strong function of the unit cost of steam.

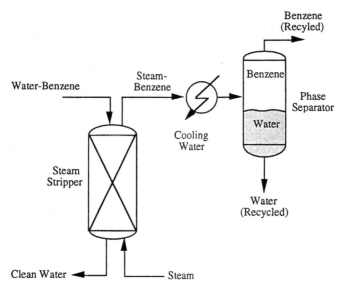

Figure 2.47   Steam stripping for removal of benzene from water.

Natural gas as a stripping fluid.   Natural gas can also be used as a stripping fluid. This process can be favored when the organic is a hazardous material which is to be thermally destroyed. In this process, stripper effluent (containing the organic contaminant) would be fed directly to a boiler where the natural gas containing the contaminant(s) can be burned for fuel value. Because natural gas is typically available at pressures of 100 psig or higher, no blower would be required for natural gas strippers.

Figure 2.48 illustrates the natural gas stripping process for removal of benzene from water. It illustrates the subsequent combustion of the natural gas/benzene effluent to destroy the benzene and to realize fuel value as well. Capital costs of natural gas and air strippers are expected to be similar (Humphrey, 1989).

Enhanced Distillation Configurations

Reactive distillation

Reactive distillation, also called *catalytic distillation*, is finding important new applications for the production of methyl tertiary-butyl ether (MTBE) and methyl acetate (DeGarmo et al., 1992). A key advantage of reactive distillation is a lower capital investment; one reactive distillation column takes the place of a reactor plus a distillation column (Figure 2.49).

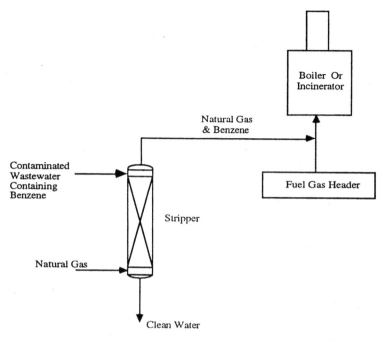

Figure 2.48 Use of natural gas as a stripping fluid for removal of benzene from water.

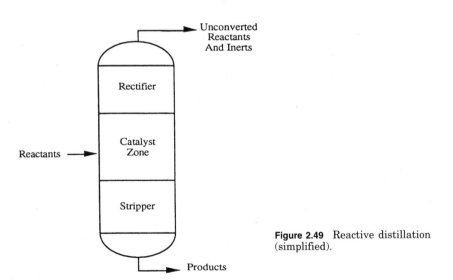

Figure 2.49 Reactive distillation (simplified).

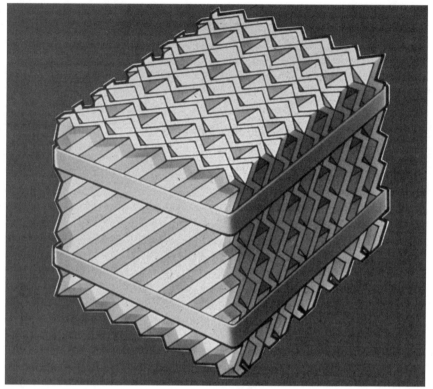

Figure 2.50 Catalyst contacting device for use in reactive distillation. (*Provided by Koch Engineering Co., Inc.*)

Factors which favor reactive distillation are:

■ Reaction and distillation temperatures overlap

■ A reversible reaction is involved—by removing product as it is formed, the reaction is driven to completion in the desired direction

■ Reaction is exothermic—heat of reaction may be used to provide the heat for distillation

A relatively new contacting device for use in reactive distillation column is illustrated in Figure 2.50. Having a monolithic configuration, it provides for good contact between the reactants and catalyst while allowing good hydraulic passage of the vapor and liquid phases.

Perhaps the most important single application of catalytic distillation is the production of methyl tertiary butyl ether (MTBE), an octane enhancer which is made by the reaction of methanol and isobutylene. Commercial plants are in operation in North America, Europe, the Middle East, and Asia (Strauss, 1990). Another important commercial application is the production of methyl acetate, which is made from the reaction of methanol and acetic acid. There are a number of pilot plants in operation testing applications for reactive distillation, including the production of cumene from benzene and propylene (Shoemaker and Jones, 1987).

Heat integration

In this process, the heat in the overhead stream of one distillation column is used to reboil an adjacent column(s) in the same system (Ho and Keller, 1987). Heat integration is generally low risk technology and the economics can be very attractive. Commercial installations have slowed in recent years in the United States due to low energy costs. Heat integration can be considered if appropriate temperature driving forces exist and integration of columns do not cause operational or control problems.

The Buckau-Walther process, used to purify and dehydrate ethanol by azeotropic distillation, is based on the heat integration concept. In this process, heat in the overhead product of one column is used to reboil adjacent columns. This process, illustrated in Figure 2.51, requires only 17,500 Btu to purify a gallon of ethanol, compared to about 28,000 Btu/gal for the conventional azeotropic process. Buckau-Walther plants are in operation in Germany, and other parts of Europe, and, to a lesser extent, in the United States.

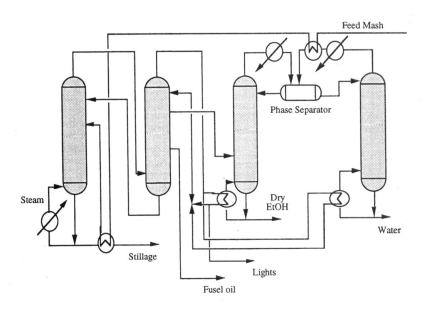

Figure 2.51 Buckau-Walther process for dehydration of ethanol.

Short-path distillation

Short-path distillation is used when the product(s) of interest thermally decompose at normal distillation temperatures. Short-path distillation is operated at high vacuum and corresponding lower temperatures. The short-path distillation system includes a double-walled heated evaporator made of either glass or metal, a roller wiper system, and a condenser which is placed within the column. The evaporator and condenser surfaces are placed directly opposite each other so the amount of time molecules are subjected to the higher temperature are minimized. Applications of short-path distillation include the purification of amino acid esters, epoxy resins, plasticizers, lube oils, high-molecular-weight alcohols and esters, and other high-value products (Billet, 1992; Clayton and Erdweg, 1987).

High-gravity distillation

Packed towers used for absorption, stripping, and distillation are vertical cylindrical shells filled with packings or trays to enhance mass transfer. If gravity could be increased, the liquid would be pulled through the packing more vigorously resulting in: (1) higher gas and liquid flow rates, (2) an increase in the interfacial area between the liquid and vapor phases, (3) thinner liquid films, and (4) a reduced tendency to flood. Such results would allow the physical size of equipment to be smaller, thus reducing space requirements.

While the Earth's gravitational field cannot be increased, *high gravity distillation* can be achieved with the use of a centrifugal field obtained by spinning a rotator containing solid particles (Figure 2.52). In this

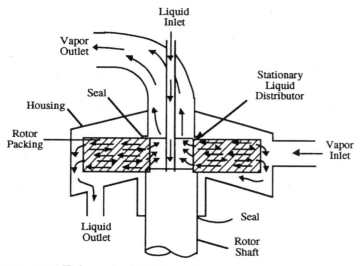

Figure 2.52 High-gravity distillation unit.

process, liquid is sprayed into the center of the rotor where it flows radially outward because of the centrifugal force. Gas is introduced at the outer surface of the rotor and is forced countercurrent to the liquid by pressure driving forces. Using this configuration, laboratory results indicate that values of the height equivalent to a theoretical plate (HETP) of less than 1-in. are possible (Martin and Martelli, 1992).

Spinning cone distillation

Developed primarily for the food industry, the *spinning cone distillation* column contains alternate rotating and stationary cones whose surfaces are wetted with liquid that flows across the surfaces of the cones. There are several commercial units in operation in the United States. The flow of liquid is forced by an applied centrifugal force. Spinning cone distillation has advantages of low liquid holdup and short residence time, as well as low pressure drop (Figure 2.53). It

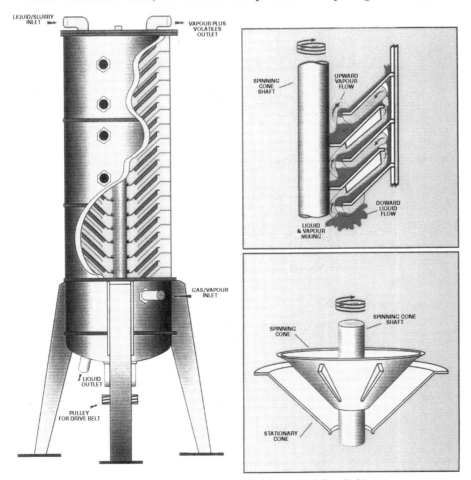

Figure 2.53   Spinning cone column. (*Provided by Flavourtech Pty. Ltd.*)

has also been reported that this process is highly tolerant of solids (Casimir et al., 1991; Sykes and Prince, 1992).

Mechanical vapor recompression

In some cases, energy efficiency of distillation columns can be increased by the use of mechanical vapor recompression (MVR). MVR is used in applications involving close boiling components, such as in the separation of propylene-propane, and to recover water from brine. In this process, heat in the overhead vapor is used to reboil the same column aided by a compressor. The process is shown in Figure 2.54 (Meili, 1990; Meszaros and Fonyo, 1986).

Though MVR is energy-efficient, the compressor and related equipment add to process complexity and costs. Applications are generally limited to: (1) those that would otherwise consume expensive utilities such as high-pressure steam, and (2) columns having overhead and bottom temperature differences of 50–60°F or less. The low-pressure drop associated with structured packings may possibly enhance applications of this process. However, MVR is a mature technology and it is likely that there will be only a limited number of new applications for this process.

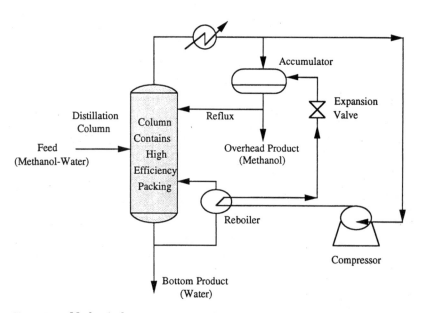

Figure 2.54 Mechanical vapor recompression.

Hybrid Systems

A second process can be used in combination with distillation to form a hybrid system, which is more efficient than either process standing alone. Some examples of hybrid systems which are used commercially are:

- Reverse osmosis/evaporation
- Reverse osmosis/distillation
- Distillation/adsorption
- Distillation/pervaporation

Reverse osmosis/evaporation

In the reverse osmosis/evaporation hybrid system, reverse osmosis is used to concentrate wastewaters from cogeneration plants prior to evaporation (see Figure 2.55). The comparison in Table 2.18 shows that a 50% energy savings is achieved with the hybrid relative to an evaporator used alone. Results in Table 2.18 are based on replacing a conventional evaporator, equipped with MVR, by a reverse osmosis/evaporation hybrid system. In the hybrid, reverse osmosis removes half of the water before evaporation, thereby reducing energy consumption and the size and cost of the evaporator. Further details may be obtained from Pankratz and Johanson (1992).

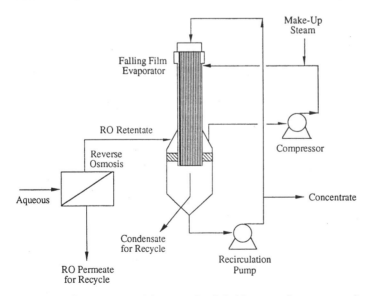

Figure 2.55   Reverse osmosis/evaporation hybrid system for recovery of water from cogeneration wastewaters. (*Adapted from Pankratz and Johanson, 1992.*)

TABLE 2.18   Energy Comparison—Hybrid System Versus
Conventional Evaporator

	Energy comparison	
	Hybrid system	Conventional evaporator
Reverse osmosis, kwh/hr	19	N/A
Evaporator, kwh/hr	145	354
Total energy, kwh/hr	164	354

Reverse osmosis/distillation

Because of low relative volatilities, the separation of carboxylic acids and water require complex distillation/extraction systems. Acetic acid, which occurs in dilute concentrations in aqueous streams, is often discarded for this reason. Reverse osmosis can be used to first concentrate acetic acid/water mixtures, thereby reducing feed volume and the the size of downstream equipment (Figure 2.56). Using reverse osmosis, concentrations of acetic acid up to 5% are possible before osmotic pressures become excessive. This approach can be used to achieve water quality, and to recover acetic acid which can be sold for $550–$700/ton. This system is being pilot tested for acetic acid/water mixtures and appears close to commercialization (Kramer et al., 1996; Wytcherley et al., 1994).

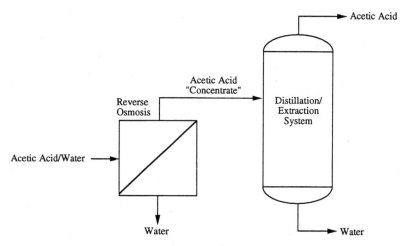

Figure 2.56   Reverse osmosis/distillation hybrid system for acetic acid/water separation.

Distillation/adsorption

Background. The traditional azeotropic distillation system to recover and purify ethanol consists of three columns. In the first column, the ethanol-water azeotrope is produced as an overhead product (96% ethanol, 4% water). In the second (azeotropic column), a solvent is added to create an azeotrope with water. Water is removed as part of an azeotrope at the top of the column and ethanol product is produced at the bottom. The third column (stripper column) is used to recover the azeotropic solvent (Figure 2.42).

During the 1980s, adsorption made significant inroads in the replacement of the traditional azeotropic and solvent recovery columns in new plants. The adsorption process, based on using 3A molecular sieves to dehydrate the azeotrope, is regenerated with a pressure swing adsorption (PSA) cycle. Other adsorbents, such as corn grits, are also used. In a new plant, the use of PSA results in a reduction in capital and energy costs. An additional benefit is that there is no handling of solvent. Also, better operability is possible because start-ups and shut downs are faster and easier. PSA can also offer advantages of increasing capacity by reducing reflux in distillation columns which operate at high reflux ratios. For the fuel ethanol industry, PSA has emerged as the process of choice to dehydrate ethanol.

Process description. In the distillation/adsorption hybrid system, sub-azeotropic ethanol/water vapor feed is preheated, raising its temperature to about 163°C. Steam serves as the heating medium in the superheater, with the steam condensate being used elsewhere for energy recovery.

The PSA unit can be a two- or multibed process. In the two-bed system, one bed is being fed, producing dehydrated ethanol, while the second one, operating under a vacuum, is being regenerated for the next adsorption cycle. Cycle times run about 5–12 min (Shroff, 1996).

The anhydrous vapor product from a PSA unit is condensed and cooled. A portion of the anhydrous vapor from the adsorbing bed is recycled to the bed undergoing regeneration, removing water from the previous cycle. The resulting ethanol/water purge mixture is condensed as recycle liquid. This recycle is sent back to the distillation process for ethanol recovery. A portion of this recycle stream serves as seal fluid for the vacuum pump.

Economic case study (for retrofit)—energy savings only. Retrofitting existing azeotropic distillation systems with PSA units may or may not be cost-effective if energy savings is the only benefit. To evaluate this type of retrofit, an economic case study was completed based on replacing the azeotropic and solvent recovery columns with a PSA unit. Return on Investment (ROI) was used as the key indicator to determine economic attractiveness. ROI is defined as:

$$\text{ROI, \%} = \frac{\text{Annual Earnings From Project, \$}}{\text{Total Capital Investment, \$}} \times 100 \qquad (2.78)$$

For ROI calculations, Annual Earnings From Project is equal to revenue minus all costs. Total Capital Investment is the total installed cost of new equipment. No costs of shutdown, tear out, rework, or removal were included. In some actual cases, the cost of shutdown, particularly the cost due to lost production, will be significant and must be included.

The case study is based on the hybrid system shown in Figure 2.57. In this system, distillation is used to concentrate the ethanol-water feed mixture up to the azeotrope concentration of 96% ethanol and 4% water. The mixture is then dehydrated with PSA beds containing 3A molecular sieves (Knoblauch, 1978; Rodrigues et al., 1989).

The total installed cost for the three-bed PSA unit to produce 7500 lb/h of fuel grade ethanol is about $1,000,000 (Howe-Baker, 1990). The cost of the PSA unit will vary depending on size and equipment options, such as inclusion of heat recovery equipment. No heat recovery equipment was included in the case here. It was assumed that the 3A molecular sieves would be replaced every 3 years.

The retrofit results in a reduction in steam consumption from 7.7 to 0.5 million Btu/h. However, there is an increase in electric energy

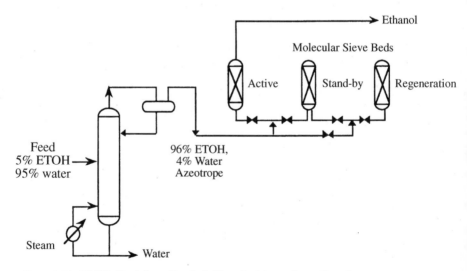

Figure 2.57 Distillation/adsorption hybrid system to produce ethanol.

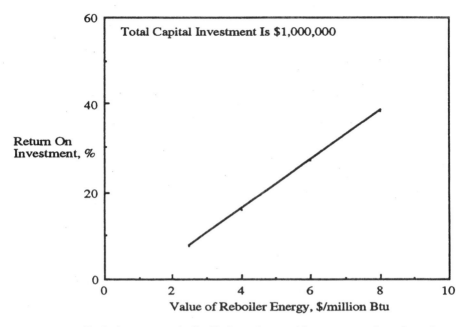

Figure 2.58 Replacing azeotropic distillation column with pressure swing adsorption unit for ethanol-water separation: energy savings only—effect of value of reboiler energy on Return on Investment.

required in the amount of 788,000 kW-h/yr (to drive the vacuum pumps for the PSA unit). Another advantage, though not quantified in this case study, is that the use of the azeotropic solvent, such as cyclohexane, is eliminated.

Results of ROI versus the value of reboiler energy saved are illustrated in Figure 2.58. It is shown that ROI increases from 10% at a reboiler energy of $3 per million Btu, to 25% at $6 per million Btu. Thus, for the retrofit case, the hybrid system gives reasonably good ROIs (greater than 20%) at values of reboiler energy of $5–$6 per million Btu and higher.

Distillation/pervaporation

With a number of commercial units in operation, pervaporation is the more mature of the membrane processes used to dehydrate azeotropes (Miyake and Matsuo, 1994). However, vapor permeation is also used commercially (Jansen et al.; Ninomiya et al., 1991). Potential benefits of using distillation/membrane hybrid systems for the dehydration of azeotropic mixtures have been discussed in some detail (Goldblatt and Gooding, 1985; Howell, 1990). The flowsheet for purification of ethanol by a distillation/pervaporation hybrid system is illustrated in Figure 2.59.

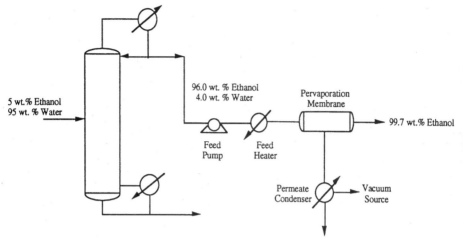

Figure 2.59 Distillation/pervaporation hybrid system for ethanol-water separation.

Economic case study—energy savings only. Similar to the case study for the retrofit with PSA, an economic case study for replacing the azeotropic and solvent recovery columns with a pervaporation membrane has been completed. Like PSA, pervaporation offers advantages of energy savings as well as elimination of solvent. It can also offer advantages of increasing capacity of columns operating at reflux ratios much higher than the minimum. For ethanol-water, commercial polyvinyl alcohol membranes have a water flux range of about 0.1–10 kg/m^2-h (Ho and Sirkar, 1992; Rapin, 1988). The average membrane flux was assumed to be 0.5 kg/m^2-h. The case study presented here is based on dehydration of a 96 wt% ethanol feed to produce 7500 lb/h of 99.7% ethanol.

The installed cost of the plate-and-frame membrane module was estimated to be $3000/m^2 of membrane surface area, including pumps, heat exchangers, piping, and instrumentation. A membrane life of 2 years is assumed with a membrane replacement cost of $800/m^2.

Energy consumption was reduced from 7.7 to 0.9 million Btu/h. New electrical energy in the amount of 1.2 million kwh/yr is required to drive the vacuum pumps for the pervaporation unit (Sander and Soukup, 1988). Based on a polyvinyl alcohol membrane and a plate-and-frame module, the total installed cost of the pervaporation unit was estimated to be $1,250,000.

Based on the energy savings, associated with shutting down the existing azeotropic and stripper columns, the distillation/membrane hybrid system gives values of ROI of 20–25% at values of reboiler energy of $7–$8 per million Btu (Figure 2.60). Other possible applications include dehydration of isopropyl alcohol, ethylene glycol, ethyl acetate, and acetone streams. The results presented here could be made more attractive by use of lower-cost membrane modules and the

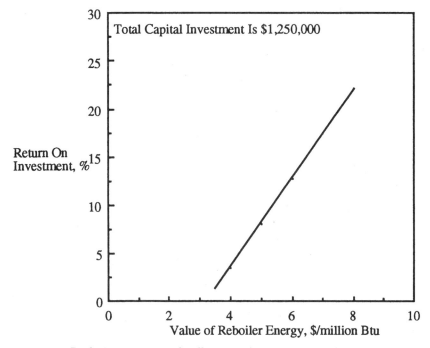

Figure 2.60 Replacing azeotropic distillation with pervaporation for ethanol-water separation: energy savings only—effect of value of reboiler energy on Return on Investment.

development of more permeable pervaporation membranes. Such improvements are expected to be commercially available in the short term (Ellinghorst et al., 1996).

Distillation Economics

Case study—Capital versus energy costs

A large-scale distillation system is presented in Figure 2.61. This flow-sheet and Table 2.19 are presented to illustrate the relative importance of capital versus energy costs in large distillation systems.

The separation of ethylbenzene and styrene is a high-energy-usage distillation (Frank et al., 1969; Ho and Keller, 1987). The end product is polymer-grade styrene produced at 500 million lb/yr. The column is large (70 trays) and separation by distillation is difficult; the average relative volatility is only 1.38. Steam usage is 1.3 lb of steam per lb of styrene.

The capital and energy charges are summarized in Table 2.19. Capital charges consist principally of the ROI income and depreciation. Typically, this annual cost is 20–30% of the capital investment.

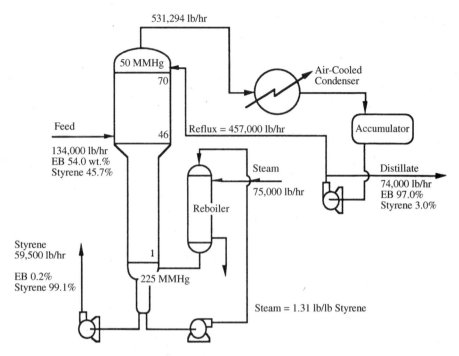

Figure 2.61 Ethylbenzene-styrene splitter. Basis: 500 million lb styrene produced per year. (*Adapted from Frank, J. C. et al., "Styrene-Ethylbenzene Separation with Sieve Trays,*" Chemical Engineering Progress, *65(2), 79–86, 1969 with permission of the American Institute of Chemical Engineers. Copyright © 1969 AIChE. All rights reserved.*)

TABLE 2.19 Capital and Energy Cost Analyses of Ethylbenzene-Styrene Distillation Column

Basis: Column and associated equipment shown in Figure 2.61, 1993 economics, 8400 h/yr operation

System installed cost = $22,000,000

Investment accuracy = ± 25%

Capital charges

Depreciation (10% of installed cost)	$2,200,000
Before tax return on investment (20% of installed cost)	4,400,000
Total	$6,600,000

Energy charges

Steam usage = 655,000,000 lb/yr	
Steam cost at $4.00/1000 lb	$2,620,000
Ratio of capital / energy charges	2.5

Results show that, despite the fact that distillation is not energy-efficient, and the use of steam is substantial, the ratio of the capital to energy charge is 2.5. Thus, based on current capital and energy charges in the United States, capital costs are more important than energy costs in large distillation systems.

Variation of capital costs of distillation columns with capacity

The "six-tenths power rule" is often used to estimate capital costs of distillation columns as a function of capacity. It has the form

$$\frac{C_1}{C_2} = \left(\frac{P_1}{P_2}\right)^{\text{Exponent}} \tag{2.79}$$

where C_1 = installed cost at capacity P_1, \$
      C_2 = installed cost at capacity P_2, \$
      P_1 = capacity for case 1, mass/time
      P_2 = capacity for case 2, mass/time
      Exponent = capacity ratio exponent, dimensionless

The exponent in Equation (2.79) for distillation systems of moderate to large volumes (2–8 ft in diameter) is about 0.6. However, as shown in Figure 2.62, this exponent declines to below 0.6 for column diame-

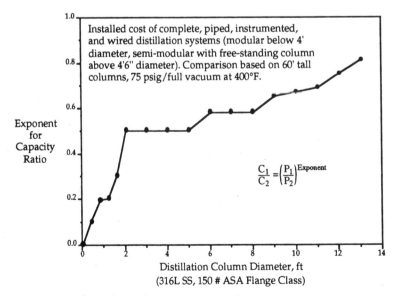

Figure 2.62   Capacity ratio exponent as a function of column diameter. (*Adapted from Licht, S., APV Crepaco, Inc., Letter Communication, June 24, 1994.*)

ters under 2 ft and increases to above 0.6 for diameters over 8 ft. The capital cost of many other processes relative to distillation, such as membrane-based processes, tend to increase linearly with capacity.

Summary

With approximately 40,000 distillation columns in operation, distillation is used to make 95% of all the separations in the process industries. Distillation consumes the energy equivalent of 1.2 million barrels of crude oil per day and dominates the separations picture. If costs are to be reduced, production increased, or quality of products improved, advances in distillation technology must be addressed. There are four approaches to improving performance and design of distillation processes:

- Improve distillation process itself through the use of high-efficiency trays, packings, and improved design tools
- Replace conventional distillation with enhanced distillation configurations
- Augment distillation with advanced processes to form efficient hybrid systems
- Replace distillation with alternate processes, such as adsorption or membranes

The approaches above are described in Table 2.20.

TABLE 2.20   Approaches to Improve Distillation Performance

Approach	Technologies	Comments
Improve distillation process through the use of high-efficiency trays, packings, and new design tools	High-efficiency packings and trays Rate-based design method	This approach should be pursued first. Technologies are commercially available. Projects involving capacity increases often have quick payouts.
Replace conventional distillation with enhanced distillation configurations	Examples of enhanced configurations include reactive distillation, heat integration, and short-path distillation	Used mostly for new plants
Augment distillation with second processes to form hybrid systems	Augment distillation with adsorption, membrane, crystallization, or other processes to form hybrid systems	Marginal payouts are to be expected if energy savings is the only benefit
Replace distillation with alternate processes	Alternate processes include extraction, adsorption, membranes, and crystallization	Replacing existing distillation columns with alternate processes will be very difficult to justify economically.

Improve distillation process through the use of high-efficiency trays, packings, and new design tools. The best projects will be those where an increase in capacity is needed and it can be achieved by retrofitting an existing column with a high-efficiency tray or packing. For most applications, such trays and packings can be installed in a relatively short period of time. Retrofitting and debottlenecking options to increase capacity of distillation columns have been presented by Fair and Seibert (1996).

Replace distillation with enhanced distillation configurations. Enhanced distillation configurations offer opportunities in specialized applications. In general, such configurations are limited to new plants. A summary of enhanced distillation configurations follows.

- Reactive distillation offers the prospect of reducing capital by combining reaction and distillation in one column. The best applications involve reactions which are reversible and whose reaction and distillation temperatures overlap.

- Heat integration concepts should be pursued on energy-intensive distillations where it is possible to integrate column operation without causing overall system problems.

- Short-path distillation may be cost-effective for high-valued products which decompose at normal distillation conditions.

- High-gravity distillation gives high efficiency and capacity with small equipment and may find applications where space is limited, such as within buildings or on offshore platforms.

- Developed for the food industry, spinning cone distillation has a short residence time and is reported to be highly tolerant of solids.

- Mechanical vapor recompression may be used to save energy.

Hybrid systems. A second process may be used in combination with distillation to form a hybrid system. Reverse osmosis may be used to concentrate wastewater feeds to evaporators as well as organic acid/water feeds to distillation columns. Pressure swing adsorption units, using 3A molecular sieves, may be used to dehydrate azeotropes. Pervaporation membranes and vapor/permeation may also be used for this purpose, both are used commercially in Japan.

Replace distillation with alternate processes. Completely replacing distillation with competing processes such as membrane, adsorption, crystallization, or other processes has the potential for significant energy savings. Over the long term, it is expected that energy prices will increase and distillation will begin to lose favor. However, replacement of distillation by competing processes will not be economically attrac-

TABLE 2.21   Technologies to Improve Distillation Processes

Key technologies	Potential benefits	Applications	Keys to effective use
High-efficiency trays	Increased capacity	High-pressure distillations (greater than 50 psig) (propylene/propane)	Proper selection and design of tray
High-efficiency packings	Increased capacity Lower pressure drop Improved product quality	Vacuum distillations and absorption (ethylbenzene/styrene)	Proper distribution and redistribution of liquid and vapor phases
Rate-based design method	More accurate column designs	Distillation and absorption applications	Reliable heat and mass transfer correlations are needed
Reactive distillation	Reduced capital cost	Reactions which are reversible (production of MTBE and methyl acetate)	Reaction and distillation temperatures must overlap
Heat integration	Reduced energy consumption	Distillation separations with high energy consumptions/costs	Need sound process control strategy for integrated columns
Short-path distillation	Minimizes thermal decomposition of high-molecular-weight products	Vitamins, high-molecular-weight alcohol and esters, epoxy resins	Use on small- to moderate-scale applications
High-gravity distillation	Reduced space requirements	Can be substituted for conventional distillation, stripping, and absorption applications	Minimize bearing and other possible mechanical failures
Spinning cone distillation	Low residence time Handles solids	Food products	Minimize possible mechanical failures due to rotating equipment
Mechanical vapor recompression	Energy savings	Distillation separations where energy costs are high. Difference between top and bottom temperature is less than 50°F	Minimize potential mechanical problems of compressor
Hybrid systems	Reduced capital and energy consumption	Azeotropes/other separations with low relative volatility	Small to moderate volume applications
Replace distillation with alternate processes (membranes, adsorption)	Reduced energy consumption	Distillations having significant energy consumption and costs	Make sure specification-grade products can really be achieved

tive for many years to come. The biggest obstacle is that, for many applications, more cost-effective replacement processes do not exist. Wholesale replacement of existing distillation columns requires that standing equipment be scrapped, an action generally not attractive to plant managers. Thus, replacement of distillation will probably have to wait until new plants are built and then will occur only if future developments lead to more cost-effective alternate processes.

A summary of technologies, which may be used to improve performance of distillation processes, is given in Table 2.21.

References

Abbott, M. M. and J. M. Prausnitz, "Phase Equilibria," in R. W. Rousseau, ed., *Handbook of Separation Process Technology,* Wiley, 1987.

Abrams, D. S. and J. M. Prausnitz, *AIChE Journal,* **21,** 116, 1975.

American Petroleum Institute, *Selected Values of Physical and Thermodynamic Properties of Hydrocarbons and Related Compounds,* Project 44, Carnegie Press, Pittsburgh, PA: 1953 and supplements.

Anderson, T. F. and J. M. Prausnitz, *Ind. Eng. Chem. Process Des. Dev.,* **17,** 552, 1978.

Berg, L. and Z. F. Yang, "The Recovery of Chloro Compounds by Extractive Distillation," presented at the AIChE Spring National Meeting, Atlanta, GA, April 1994.

Berg, L., "Industrial Applications of Extractive Distillation," presented at the AIChE Spring National Meeting, Houston, TX, March 1993.

Billet, I. R. et al., *I. Chem. E. Symposium Series No. 32,* **5,** 111, 1969.

Billet, R., *Sep. Technol.,* **2,** October 1992, pp. 183–191.

Billet, R. and J. Mackowiak, *Verfahrenstechnik,* **16**(2), 67, 1982.

Black, C. and D. E. Ditsler, "Dehydration of Aqueous Ethanol Mixtures by Extractive Distillation," in *Extractive and Azeotropic Distillation,* American Chemical Society Advances in Chemistry, No. 115, Washington, DC, 1972.

Bolles, W. L. and J. R. Fair, *Int. Chem. Eng. Symp. Ser. No. 56,* 3.3/35, 1979.

Bolles, W. L. and J. R. Fair, *Chem. Eng.,* **89**(14), 109, 1982.

Bomio, P. et al., "Experience with Structured Packings in High Pressure Gas Absorption," presented at the AIChE Spring National Meeting, Houston, TX, March 28–April 1, 1993.

Bravo, J. L. and J. R. Fair, *Ind. Eng. Chem. Proc. Des. Dev.,* **21,** 162, 1982.

Byers, W. D. and C. M. Morton, *Environ. Progress,* **4**(2), May 1985.

Casimir, D. J. et al., "The Australian Spinning Cone Column: A Novel Mass Transfer Apparatus for the Separation of Volatile Materials," presented at the 1st Conference of Food Engineering, American Institute of Chemical Engineers, Chicago, 1991.

Chao, K. C. and J. D. Seader, *AIChE Journal,* **7,** 598, 1961.

Chien, H. H., *AIChE J.,* **24,** 606, 1970.

Chu, J. C. et al., *Distillation Equilibrium Data,* Reinhold, New York, 1950.

Chu, J. C. et al., *Vapor-Liquid Equilibrium Data,* Edwards, Ann Arbor, MI, 1950.

Clayton, M. and K. Erdweg, *The Chemical Engineer Supplement,* September, 1987, pp. 20–22.

Collins, P. and K. Breu, "Structured Packings for TEG Contactors—Design Methods and Operating Characteristic," presented at the 9th Continental Meeting of the European Chapter of the Gas Processors Association, May 14, 1992.

Cornell, D. et al., *Chem. Eng. Prog.,* **56**(7), 68, 1960.

DeGarmo, J. L. et al., *Chemical Engineering Progress,* March 1992.

Diab, S. and R. N. Maddox, *Chemical Engineering,* December 27, 1982, pp. 38–56.

Diwekar, U. M., *Batch Distillation: Simulation, Optimal Design and Control,* Taylor & Francis, 1995.

Eduljee, H. E., *Hydrocarbon Processing,* **54**(9), 120, 1975.

Ellinghorst, G. et al., "Dehydration of Organics to Very Low Water Content by Pervaporation," presented at the 8th Annual Meeting of the North American Membrane Society, Ottawa, Ontario, Canada, May 1996.

Fair, J. R. and A. F. Seibert, "Understand Distillation-Column Debottlenecking Options," *Chemical Engineering Progress,* June 1996.

Fair, J. R. et al., *Ind. Eng. Chem. Proc. Des. Dev.,* **22,** 53–58, 1983.

Fenske, M. R., *Ind. Eng. Chem.,* **24,** 482, 1932.

Frank, J. C. et al., *Chem. Eng. Prog.,* **65**(2), 79, 1969.

Fredenslund, A. et al., *AIChE Journal,* **21,** 1086, 1975.

Fredenslund, A. et al., *Vapor-Liquid Equilibria Using UNIFAC,* Elsevier, Amsterdam, 1977.

Gallant, R. W., *Physical Properties of Hydrocarbons, Hydrocarbon Process,* 44–49, July 1965–January 1970.

Gerster, J. A. et al., *AIChE J.,* **1,** 536, 1955.

Gess, M. A. et al., "Thermodynamic Analysis of Vapor-Liquid Equilibria: Recommended Models and a Standard Data Base," in *Design Inst. for Phys. Prop. Data,* American Institute of Chemical Engineers, New York, 1991.

Gilliland, E. R., *Ind. Eng. Chem.,* **32,** 1220, 1940.

Gmehling, J. et al., *Vapor-Liquid Equilibrium Collection* (continuing series), DECHEMA, Frankfurt, 1979.

Gmehling, J. et al., *Ind. Eng. Chem. Process Des. Dev.,* **21,** 118, 1982.

Gmehling, J. et al., *Azeotropic Data,* DM 598, ISBN 3-527-28671-3, Germany, 1994.

Goldblatt, M. E. and C. H. Gooding, "An Engineering Analysis of Membrane-Aided Distillation," *AIChE Symposium Series,* **82**(248):51, 1985.

Hadden, S. T. and H. G. Grayson, *Petrol. Refiner.,* **40**(9), 207, 1961.

Hala, E. et al., *Vapor-Liquid Equilibrium,* 2nd ed., Pergamon, Oxford, 1967.

Hala, E. et al., *Vapor-Liquid Equilibrium at Normal Pressures,* Pergamon, Oxford, 1968.

Hirata, M. et al., *Computer Aided Data Book of Vapor-Liquid Equilibria,* Elsevier, Amsterdam, 1975.

Ho, F. and G. E. Keller II, "Process Integration," in Y. A. Liu et al., eds., *Recent Developments in Chemical Process and Plant Design,* Wiley, 1987, pp. 101–126.

Ho, W. W. S. and K. K. Sirkar, eds., *Membrane Handbook,* New York: Van Nostrand, 1992.

Horsley, L. H., *Azeotropic Data-III,* American Chemical Society Advances in Chemistry Series 116, American Chemical Society, Washington, DC, 1973.

Howe-Baker Engineers, Tyler, TX, letter communication, 1990.

Howell, J. A., Watt Committee Report, *The Membrane Alternative: Energy Implications for Industry,* Elsevier, 1990.

Hufton, J. R. et al., "Scale-Up of Laboratory Data for Distillation Columns Containing Corrugated Metal-Type Structured Packing," *Ind. Eng. Chem. Res.,* **27,** 2096–2100, 1988.

Humphrey, J. L., *Utilization of Natural Gas in Large Scale Separation Processes,* Final Report, Gas Research Institute, ID No. 89/0005, February 1989.

Humphrey, J. L. et al., "Separation Technologies—Advances and Priorities," U.S. Dept. of Energy Final Report, DOE/ID/12920-1, February 1991.

Hwang, Y. et al., *Ind. Eng. Chem. Res.,* **31**(7), 1992.

Jancic, S. J., Sulzer Chemtech Limited, personal communication, May 5, 1996.

Jansen, A. E. et al., in Vansant, E. F. and R. DeWolfs, eds., *Gas Separation Technology,* Elsevier, Amsterdam, pp. 413–427.

Jantz, R. D. et al., "Tray Development Wakes Up," presented at the '95 Chem Show Management & Technical Conference: "Practical Solutions for Process Industry Practitioners," New York, December 1995.

Kageyama, O., *I. Chem. E. Symposium Series No. 32,* **2,** 72, 1969.

Kavanaugh, M. C. and R. R. Trussel, *Journal American Water Works Association,* December 1980.

Kean, J. et al., *Hydrocarbon Processing,* April 1991.

Kirkbride, C. G., *Petrol. Refiner.,* **23**(32), 1944.

Kister, H. Z. et al., "Capacity and Efficiency: How Trays and Packings Compare," presented at the AIChE Spring Meeting, Houston, TX, March–April, 1993.

Kister, H. Z. et al., *Chem. Eng. Prog.*, February 1994, pp. 23–32.
Kister, H. Z., *Distillation Design,* New York: McGraw-Hill, 1992.
Knoblauch, K., *Chemical Engineering,* **85**(23), 87, November 6, 1978.
Kramer, R. et al., "Recovery of Carboxylic Acids from Waste Streams Using a Hybrid System," presented at the AIChE Spring Meeting, New Orleans, February 1996.
Krishnamurthy, R. and R. Taylor, "A Nonequilibrium Stage Model of Multicomponent Separation Processes," *AIChE Journal,* **31**(3), March 1985.
Kumar, R. et al., "Process Design Considerations for Extractive Distillation: Separation of Propylene-Propane," in *Extractive and Azeotropic Distillation,* 2, American Chemical Society Advances in Chemistry No. 115, Washington, DC, 1972.
Lee, A. Glitsch, Inc., personal communication, January 30, 1995.
Lewis, W. K., *Ind. Eng. Chem.,* **28**, 399, 1936.
Licht, S., APV Crepaco, Inc., letter communication, June 24, 1994.
Lockett, M. J., *Distillation Tray Fundamentals,* Cambridge University Press, Cambridge, UK, 1986.
Lucero, B. Y. et al., "Ceramic Structured Packing Ancient Material-Modern Technology," presented at the AIChE Spring National Meeting, Orlando, FL, March 1990.
Macedo, E. A. et al., *Ind. Eng. Chem. Process Des. Dev.,* **22**, 676, 1983.
Magnussen, T. et al., *Ind. Eng. Chem. Process Des. Dev.,* **20**, 331, 1981.
Martin, C. L. and M. Martelli, "Preliminary Distillation Mass Transfer and Pressure Drop Results Using a Pilot Plant Scale High Gravity Contacting Unit," presented at the American Institute of Chemical Engineers Spring National Meeting, New Orleans, LA, 1992.
Martin, C. L. et al., "Performance of Structured Packings in Distillation Service," presented at the AIChE Spring National Meeting, New Orleans, LA, March 7, 1988.
Maurer, G. and J. M. Prausnitz, *Fluid Phase Equilibria,* **2**, 91, 1978.
Meili, A., *Chemical Engineering Progress,* June 1990, pp. 60–65.
Meszaros, I. and Z. Fonyo, *Heat Recovery Systems,* **6**(6), 469–476, 1986.
Miyake, N. and Y. Matsuo, "Pervaporation Technology and Its Applications," presented at the Japanese Chemical Engineering Society, Yamaguchi Prefecture, Japan, December 1994.
Natural Gas Processors Suppliers Assn., *Engineering Data Book,* 9th ed., Tulsa, OK, 1972.
Ninomiya, K. et al., "Dehydration of Water-Alcohol Mixtures by Vapor Permeation Through Polyimide Membranes," presented at the American Institute of Chemical Engineers National Meeting, Pittsburgh, PA, August 1991.
O'Connell, H. E., *Trans. AIChE,* **42**, 741, 1946.
Onda, K. et al., *J. Chem. Eng. Japan,* **1**, 56, 1968.
Packed Distillation Columns, 2nd ed., Equip. Test. Proc. Com., American Institute of Chemical Engineers, New York, 1991.
Palmer, D. A., *Handbook of Applied Thermodynamics,* CRC Press, Boca Raton, FL, 1987.
Pankratz, T. and K. Johanson, "A Hybrid Zero Liquid Discharge Treatment System," Aqua-Chem, Inc., Milwaukee, WI, presented at the Cogen Turbo Power Conference, 1992.
Pelkonen, S. and A. Gorak, "Experimental Validation of the Rate-Based Approach to Multicomponent Distillation," presented at the American Institute of Chemical Engineers Annual Meeting, Miami Beach, FL, November 1995.
Pittman, A., Hoechst Celanese Corp., personal communication, May 6, 1996.
Prausnitz, J. M. et al., *Computer Calculations for Multicomponent Vapor-Liquid and Liquid-Liquid Equilibria,* Prentice-Hall, Englewood Cliffs, NJ, 1980.
Rapin, J. L., "The Betheniville Pervaporation Unit—The First Large-Scale Production Plant for the Dehydration of Ethanol," in *Proceedings of Third International Conference on Pervaporation Processes in the Chemical Industry,* Englewood, NJ, 1988, pp. 364–378.
Rayleigh, *Phil. Mag.,* **4**(6), 521, 1902.
Reed, B. W. et al., "Membrane Contactors," in R. D. Noble and S. A. Stern, eds., *Membrane Separations Technology-Principles and Applications,* Chapter 10, Elsevier, 1995.
Reid, R. C. et al., *The Properties of Gases and Liquids,* 3rd ed., McGraw-Hill, New York, 1977.
Renon, H. and J. M. Prausnitz, *AIChE Journal,* **14**, 135, 1968.

Rodrigues, A. E. et al., eds., *Adsorption: Science and Technology*, NATO, ASI Series E, Kluwer, The Netherlands, 158, 285, 1989.

Sander, U. and P. Soukup, *J. Mem. Sci.*, 36:463, 1988.

Sasson, R. and R. Pate, *Oil & Gas Journal*, August 2, 1993.

Seader, J. D., *Chemical Engineering Progress*, October 1989, pp. 41–49.

Seader, J. D. and Z. M. Kurtyka, Distillation, in R. H. Perry, late ed., D. W. Green, ed., *Perry's Chemical Engineers' Handbook*, 6th ed., McGraw-Hill, 1984.

Seibert, A. F. and J. R. Fair, "Performance and Economic Evaluation of the Cocurrent Tray," presented at the AIChE Spring National Meeting, Houston, TX, March 1995.

Sengupta, A., Hoechst Celanese Corp., letter communication, May 31, 1996.

Sengupta, A. et al., "Oxygen Removal from Water Using Process-Scale Extra-Flow Membrane Contactors & Systems," presented at the NAMS '95 meeting, Portland, OR, May 1995.

Shakur, M. S. M. et al., "Demethanizer Revamp Using Enhanced Efficiency MD Trays," paper presented at the AIChE Spring National Meeting, New Orleans, LA, February 28, 1996.

Shoemaker, J. D. and E. M. Jones, Jr., *Hydrocarbon Processing*, June 1987.

Shroff, B., Raphael Katzen Associates International, Inc., letter communication, Cincinnati, OH, February 8, 1996.

Skjold-Jorgensen, S. et al., *Ind. Eng. Chem. Process Des. Dev.*, **18,** 714, 1979.

Stichlmair, J. et al. *Gas Sep. Purif.*, **3**(10), 19, 1989.

Strauss, Michael J., *The Oil Daily*, July 12, 1990.

Strigle, R. F., *Packed Tower Design and Applications*, 2nd ed., Gulf, Houston, TX, 1994.

Suess, P. et al., *A New Mass Transfer Structure for Rectification and Absorption*, Sulzer Report, ID No. 16.12.1994, Winterthur, Switzerland, 1994.

Summers, D. R. et al., "Enhanced Capacity Multiple Downcomer (ECMD) Trays Debottleneck C_3 Splitter," presented at the AIChE Spring National Meeting, Houston, TX, March 1995.

Sykes, S. J. and R. G. H. Prince, "The Design of Spinning Cone Distillation Columns," *I. Chem. E. Symposium Series No. 128,* 1992.

Tray Distillation Columns, 2nd ed., Equip. Test. Proc. Com., American Institute of Chemical Engineers, New York, 1987.

Underwood, A. J. V., *Chem. Eng. Prog.*, **44,** 603, 1948.

Walas, S. M., *Phase Equilibria in Chemical Engineering*, Butterworth, Boston, 1985.

Wichterle, I. et al., *Vapor-Liquid Equilibrium Data Bibliography*, Elsevier, Amsterdam, 1975.

Wilson, G. M., *Journal of the American Chemical Society*, **86,** 127, 1964.

Wu, K. Y. and G. K. Chen, "Large-Scale Pilot Columns and Packed Column Scale-Up," *I. Chem. E. Symposium Series No. 104*, B225, 1987.

Wytcherley, R. W. et al., In Malloy, D., ed., *The 1995 Membrane Technology Reviews*, Business Communications, Norwalk, CT, November 1994, pp. 1–8.

Yanagi, T. and M. Sakata, *Ind. Eng. Chem. Process Des. Dev.*, **21,** 712, 1982.

Yeoman, N., Koch Engineering Co., letter communication, May 9, 1996.

Yeoman, N., Koch Engineering Co., letter communication, June 9, 1994.

Batch Distillation—Suggested References

Al-Tuwaim, M. S. and W. L. Luyben, "Multicomponent Batch Distillation," *Ind. Eng. Chem. Res.*, **30**(3), 507–516, 1991.

Boston, J. F. et al., "An Advanced System for the Simulation of Batch Distillation Operations," in *Foundations of Computer-Aided Chemical Process Design*, **2**, Engineering Foundation, New York, 1981, pp. 203–238.

Diwekar, U. M., "How Simple Can It Be? A Look at the Models for Batch Distillation," *Comput. Chem. Eng.*, **18**(Suppl.), S391–S395, 1993.

Diwekar, U. M., *Batch Distillation: Simulation, Optimal Design and Control*, Taylor & Francis, 1995.

Diwekar, U. M. and K. P. Madhavan, "Batch-Distillation: A Comprehensive Package for Simulation, Design, Optimization, and Optimal Control of Multicomponent, Multifraction Batch Distillation Columns," *Comput. Chem. Eng.*, **15**(12), 833–842, 1991.

Diwekar, U. M. and K. P. Madhavan, "Multicomponent Batch Distillation Column Design," *Ind. Eng. Chem. Res.*, **30**(4), 713–721, 1991.

Galindez, H. and A. Fredenslund, "Simulation of Multicomponent Batch Distillation Processes," *Comput. Chem. Eng.*, **12**(4), pp. 281–288, 1988.

Henley, E. J. and J. D. Seader, *Equilibrium Stage Separation Operations in Chemical Engineering*, Wiley, New York, 1981.

Koppel, P., "Fast Way to Solve Problems for Batch Distillations," *Chem. Eng.*, October 16, 1972, pp. 109–112.

Kumana, J. D., "Run Batch Distillation Processes with Spreadsheet Software," *Chem. Eng. Prog.*, **86**(12), 53–57, 1990.

Lang, P. et al., "Batch Extractive Distillation Under Constant Reflux Ratio," *Comput. Chem. Eng.*, **18**(11–12), 1057–1069, 1994.

Sundaram, S. and L. B. Evans, "Shortcut Procedure for Simulating Batch Distillation Operations," *Ind. Eng. Chem. Res.*, **32**(3), 511–518, 1993.

Treybal, R. E., "A Simple Method for Batch Distillation," *Chem. Eng.*, October 5, 1970, pp. 95–98.

Yatim, H. et al., "Dynamic Simulation of a Batch Extractive Distillation Process," *Comput. Chem. Eng.*, **17**(Suppl.), S57–S62, 1992.

Chapter

3

Extraction

Introduction

Because extraction requires the use of a solvent, it is a more complex process than distillation. In distillation the simplest system is a binary, in extraction it is a ternary. Extraction is used when the separation cannot be made by distillation. Whereas there are approximately 40,000 distillation columns in the United States, there are only about 1000 extractors. Despite process complexity, extraction offers advantages over distillation in a number of applications.

Extraction lags behind distillation in available design methods. Much of the development of design methods in extraction, particularly for mechanically aided extractors, has been done by equipment suppliers and has not been published.

Detailed information about extraction is presented in *Handbook of Solvent Extraction* (Lo et al., 1983); *Handbook of Separation Techniques for Chemical Engineers* (Lo, 1988); *Liquid-Liquid Extraction Equipment* (Godfrey and Slater, 1994); and *Kirk-Othmer Encyclopedia of Chemical Technology* (Lo and Baird, 1994). The *Journal of Solvent Extraction and Ion Exchange* also contains articles on extraction. The *Proceedings of the International Solvent Extraction Conference '96* (Shallcross et al., 1996) contains some of the most recent technical papers on extraction.

This chapter describes both liquid/liquid and supercritical extraction, discusses stage calculations and equipment, and presents new extraction processes. Lists of equipment suppliers for extraction are included in Appendix B.

Liquid/Liquid Extraction

In liquid/liquid extraction, component(s) of a liquid mixture are separated based on their different solubilities in a solvent. This process is operated at near atmospheric pressure and ambient temperature. There are two requirements for liquid/liquid extraction to be feasible:

- Component(s) to be removed from the feed must preferentially distribute in the solvent.
- The feed and solvent phases must be substantially immiscible.

The simplest extraction system involves the *solute*, or the material to be extracted; the *solvent*; and the *carrier*, or nonsolute portion of the feed mixture. In extraction, one must differentiate between the *light phase* and the *heavy phase*; the *dispersed phase* and the *continuous phase*; and the *feed, raffinate*, and *extract phases*.

To maximize the dispersion of one phase in the other, and to minimize backmixing, extractors are equipped with trays, packings, or mechanical internals. When only one or two equilibrium stages are needed, a spray tower, which contains no internals, can be used.

Figure 3.1 illustrates a sieve tray extractor for the case of a light solvent phase being dispersed. The feed phase contains the carrier plus the solute. The extract phase contains solvent and solute, and to the degree it is soluble in the solvent, a small amount of carrier. The

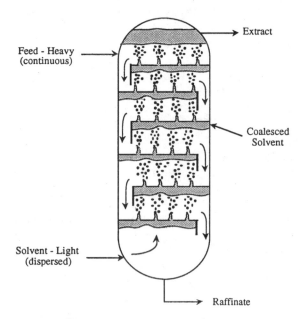

Figure 3.1 Sieve tray extractor.

raffinate contains the carrier and a small amount of solute left over after extraction. The raffinate may also contain a small amount of solvent if there is some degree of solubility of the solvent in the carrier.

In extraction, the location of the principal interface depends upon which phase is dispersed. When the light phase is dispersed, the interface is located at the top of the extractor. When the heavy phase is dispersed, the interface is located at the bottom. The solvent can be the heavy or light phase, and it can be the dispersed or continuous phase. Usually the phase which is fed at the lowest rate (normally solvent) is the dispersed phase.

An extraction system always includes at least one distillation column (or other separation process) to recover solvent from the extract phase. If the solvent exhibits some degree of miscibility in the feed, then a second separation process (normally distillation) is required to recover solvent from raffinate. Figure 3.2 illustrates an extraction system where two distillation columns are needed, one to recover solvent from extract, and the other to recover solvent from raffinate.

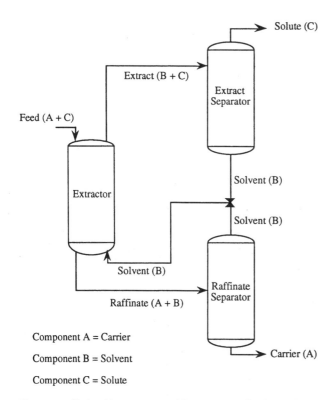

Figure 3.2 Extraction system with recovery of solvent from extract and raffinate phases.

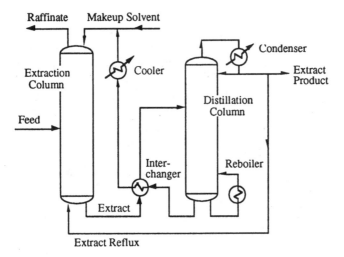

Figure 3.3 Industrial extraction system.

An industrial extraction system is shown in Figure 3.3. The feed enters the extraction column toward the middle of the column. The column is provided with reflux in the form of extract product that has been separated from solvent. The role of reflux in extraction is to relieve equilibrium limitations on the purity of the extracted material. This example is based on a solvent which is the heavy phase.

The purpose of the distillation column, shown in Figure 3.3, is to separate the solute and solvent, with the solvent being recycled. This example is based on a *low-boiling solvent* which is recovered as the top product of the distillation column. A *high-boiling solvent* would be recovered as the bottom product of the distillation column. If the solvent and solute have close boiling points, the distillation column may require many trays and a high reflux ratio, resulting in a costly process. If there is a significant difference in the boiling points, it may be possible to replace the distillation column with one or two flash steps with no reflux. In this case, extraction economics would improve considerably.

Advantages

Liquid/liquid extraction is used to separate azeotropes and components with overlapping boiling points. It can offer energy savings and can be operated at low to moderate temperatures for recovery of thermally sensitive products in the food and pharmaceutical industries.

Disadvantages

The necessity of a solvent increases the complexity of the process. The extraction system consists of an extractor plus at least one distillation column (or other separation process) for recovery of solvent. If solvent has some degree of miscibility in the carrier, then a second distillation column must be included to recover solvent from raffinate. Solvent storage tanks and a distribution system are also required.

Factors favoring extraction

The extraction process gains in favor when:

- Azeotropes or low relative volatilities are involved (and distillation cannot be used)
- Low to moderate processing temperatures are needed
- Solvent recovery is easy
- Energy savings can be realized

Applications

Example applications of extraction are presented in Table 3.1.

A family of applications involves the use of solvent extraction in the processing of nuclear fuels (Narratil and Shulz, 1980; Shallcross et al., 1996; Watson, 1995). Biotechnology and environmental applications also employ extraction processes (Shallcross et al., 1996; pp. 1381–1524; Shallcross et al., 1996; pp. 1525–1560).

The most common method of treating wastewaters to reduce the level of organic contaminants is steam stripping, particularly when the contaminant's boiling point is lower than the boiling point of water. However, there are difficult cases where liquid/liquid extraction is a viable alternative. Table 3.2 lists some organics which are best handled by liquid/liquid extraction (Cusack, 1996).

TABLE 3.1 Example Applications of Extraction

Carrier phase	Solute(s) recovered from feed phase	Solvent
Water	Phenol	Methyl isobutyl ketone (Cusack, 1996)
Heavy oil	Naphthalenes	Furfural
Water	Methyl ethyl ketone	Toluene
Liquefied natural gas	Hydrogen sulfide	Monoethanolamine
Gasoline reformate	Benzene/toluene/xylene	Diethylene glycol
Coffee beans	Caffeine	Supercritical carbon dioxide
Vacuum residue	Liquid fuels	Supercritical butane (ROSE process)

TABLE 3.2 Organics Which Are Best Removed from Wastewaters by Liquid/Liquid Extraction

| Organic compound | Boiling point, °C | Azeotrope | | | Typical reduction levels by liquid/liquid extraction (ppm) |
		Solubility, %	Boiling point, °C	Water concentration, %	
Formaldehyde	−19.0	—	—	—	<800
Formic acid	100.8	—	107.1	22.5	<1200
Acetic acid	118.0	—	—	—	<800
Pyridine	115.5	57	92.6	43.0	<10
Aniline	181.4	3.6	99.0	80.8	<10
Phenol	181.4	8.2	99.5	90.8	<10
Nitrobenzene	210.9	0.04 (approx.)	98.6	88.0	<10
2, 4-Dinitrotoluene	300.0	0.03	99–100 (est.)	90 + (est.)	<10

New developments

A summary of new developments in extraction follows:

- Use of structured packings to increase extractor capacity
- New mass transfer and hydraulic models to predict performance in tray and packed columns
- Membrane phase contactors offer advantages of lower solvent to feed ratios
- New software, based on group contribution methods, is available to aid solvent selection

Supercritical Fluid Extraction

In *supercritical fluid extraction* (SFE), a supercritical fluid such as carbon dioxide is used as the solvent. From a conceptual point of view, SFE is the same as liquid/liquid extraction. Several review papers have been written on the process (Irani and Funk, 1977; Johnston, 1984; Paulaitis et al., 1982; Williams, 1981). Recent papers on SFE are included in the *Proceedings of the International Solvent Extraction Conference '96* (Shallcross et al., 1996, pp 967–1024).

SFE has advantages due to the nontoxic nature of carbon dioxide. Supercritical solvents offer enhanced transport properties because solutes diffuse more rapidly through a supercritical solvent than a liquid solvent. Supercritical fluids also exhibit other desirable solvent properties, with equilibrium ratios and separation factors that are quite high. Since the solvent is in the supercritical state, it can be recovered as a gas merely by reducing the pressure and by changing the temperature (Figures 3.4 and 3.5).

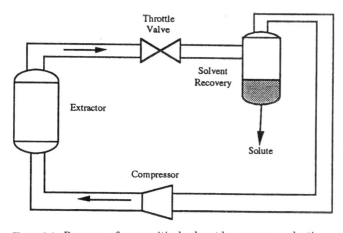

Figure 3.4   Recovery of supercritical solvent by pressure reduction.

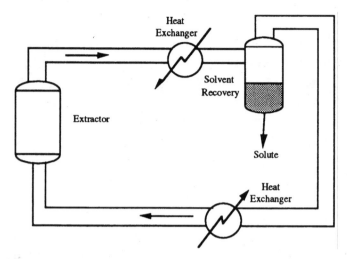

Figure 3.5 Recovery of supercritical solvent by temperature change.

The primary disadvantage of SFE is that the extractor must be operated at the high pressures required to maintain the solvent in the supercritical state. In contrast to liquid/liquid extraction, which is operated at near-atmospheric pressure, SFE must be operated at pressures of 1000–5000 psia. The result is higher capital and operating costs. Higher costs are the primary reasons why SFE has not been adopted for many separations in the process industries.

Applications. Decaffeination of coffee and tea in the food industry, using carbon dioxide as solvent, make up the bulk of commercial applications for SFE. A partial listing of the worldwide commercial applications is given in Table 3.3 (Gallagher-Wetmore, 1993).

The Residuum Oil Supercritical Extraction (ROSE) process uses supercritical pentane to recover hydrocarbon liquids from heavy oils. This process was developed by Kerr-McGee Refining Corp. in the 1950s (Gearhart and Nelson, 1983; Hood, 1994). With this process, hydrocarbon liquids can be recovered from several types of feedstocks, including atmospheric and vacuum residue.

Supercritical carbon dioxide is also used in tertiary oil recovery. In this process carbon dioxide is injected into the oil reservoir where it dissolves in the crude oil and reduces its viscosity. The lower-viscosity oil flows to the producing well more easily. Carbon dioxide is an excellent solvent for heavy hydrocarbons above 4000 psi pressure.

TABLE 3.3   Supercritical Fluid Extraction—Partial Listing of World Usage

Material	Location	Capacity (million lb/yr)
Coffee	Bremen, Germany	60
Coffee	Houston, TX	50
Coffee	Veragro, Italy	35
Tea	Munchmeunster, Germany	12
Hops	Munchmeunster, Germany	30
Hops	Sydney, Australia	Large (but unknown)
Tobacco	Hopewell, VA	Unknown
Essential oils	Reigate, England	Unknown

SOURCE: Adapted from Gallagher-Wetmore, 1993.

Carbon dioxide is also used as a substitute for toxic organic solvents and cleaning fluids. One application involves replacing chlorofluorocarbons (CFCs) used for degreasing mechanical parts with supercritical carbon dioxide. In a second application, conventional organic solvents in spray painting are replaced by carbon dioxide (Unicarb process). When the object is painted, carbon dioxide flashes, thereby reducing worker exposure and atmospheric contamination.

Perhaps the biggest single advantage of SFE is the nontoxic nature of supercritical solvents, such as carbon dioxide. This feature will likely guide SFE into important new applications where there are concerns about the environment, worker exposure, or products for human consumption. Because of the exponential increase in the cost of high-pressure equipment with an increase in capacity, SFE will be more cost-effective for small to moderate volume applications.

Promising applications for SFE include separation of specialty products which are difficult to separate by other methods. For example, SFE may find applications in purification of vitamins and other high-boiling specialty products where molecular distillation is used. In these cases, SFE has the advantage because multiple stages can be used, whereas in molecular distillation, staging is difficult and expensive. A supercritical extraction test system is shown in Figure 3.6.

Figure 3.6 Supercritical extraction test system. (*Provided by the Separations Research Program, The University of Texas at Austin.*)

Liquid/Liquid Equilibria

Experimental determination of liquid/liquid equilibria

Liquid/liquid equilibria are determined by vigorously mixing two liquid phases, allowing them to coalesce, and then sampling to determine their compositions (Figure 3.7). As in distillation, reliable equilibrium data are essential for design of extraction columns.

Determinations of liquid/liquid equilibria are easier than determining vapor/liquid equilibria. In spite of this, the amount of liquid/liquid equilibria that are available pales in comparison with the amount of available vapor/liquid data. The publications of D'Ans and Lax (1967),

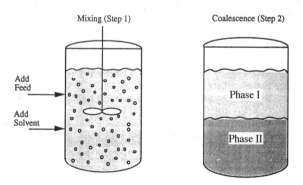

Figure 3.7 Steps to determine liquid/liquid equilibria.

Francis (1963, 1972), Lewis (1953), Prausnitz et al. (1980), Robbins (1984), Schafer and Lax (1964), Sorenson and Arlt (1979, 1980, 1987), Van Brunt (1996), and Wisniak and Tamir (1980, 1987) provide liquid/liquid equilibrium data or furnish guidance for locating such data.

Thermodynamic relationships

For a simple ternary system, the equilibrium distribution of solute and carrier between the phases is critical to the process. A *distribution coefficient* or *equilibrium ratio*, analogous to the equilibrium ratio or K value in distillation, is defined as:

$$K_C = \frac{X_{C,E}}{X_{C,R}} \qquad (3.1a)$$

where K_C = distribution coefficient for component C
$X_{C,E}$ = mol fraction of component C in extract phase, dimensionless
$X_{C,R}$ = mol fraction of component C in raffinate phase, dimensionless

and

$$K_A = \frac{X_{A,E}}{X_{A,R}} \qquad (3.1b)$$

where K_A = distribution coefficient for component A
$X_{A,E}$ = mol fraction of component A in extract phase, dimensionless
$X_{A,R}$ = mol fraction of component A in raffinate phase, dimensionless

The *separation factor*, or *selectivity*, equivalent to relative volatility in distillation, is defined as the ratio of K values:

$$\beta_{CA} = K_C/K_A = \frac{(X_C/X_A)_E}{(X_C/X_A)_R} \qquad (3.2)$$

where β_{CA} = separation factor

When liquid/liquid equilibria are determined in the presence of vapor, the vapor is in equilibrium with both of the liquid phases. Thus, Raoult's law corrected for nonideality of the liquid phases leads to:

$$\beta_{CA} = \left(\frac{\gamma_C}{\gamma_A}\right)_R \cdot \left(\frac{\gamma_A}{\gamma_C}\right)_E \qquad (3.3)$$

where γ is the liquid phase activity coefficient for the component with the solvent. Thus, liquid phase activity coefficients from vapor/liquid equilibria can be used to determine liquid/liquid equilibria. Equation (3.3) also shows that nonideality must be present for liquid/liquid extraction to be effective. In addition, there is a significant composition influence on the value of the separation factor (β_{CA}).

As illustrated in Figure 3.8, the simplest ternary system is the Type I system for one immiscible-liquid pair. For such a system, the carrier and the solvent are essentially immiscible, while the carrier-solute and the solvent-solute pairs are miscible. The *plait point* is the intersection of the raffinate-phase and extract-phase boundary curves. It is analogous to the azeotrope in distillation. No separation can be made at the plait point. The tie lines shown in the two-phase region in Figure 3.8 connect the extract and raffinate equilibrium compositions and thus can be used to determine K values and separation factors.

The Type II system is also illustrated in Figure 3.8. In this system, there are immiscibilities between two pairs: solvent-solute and solvent-carrier. The tie lines are indicated. Note, in this type of system there is no plait point. It is also possible, though it is not shown in Figure 3.8, to have a Type III system, which is characterized by immiscibilities between all three pairs.

Type I System

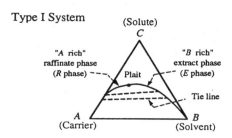

Type II System

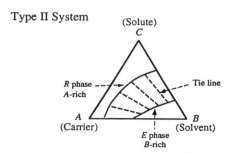

Figure 3.8 Liquid/liquid equilibria for Type I and Type II systems. (*Adapted from Humphrey, J. L. et al., "The Essentials of Extraction,"* Chemical Engineering, *September 17, 1984, pp. 76–95. Used with permission.*)

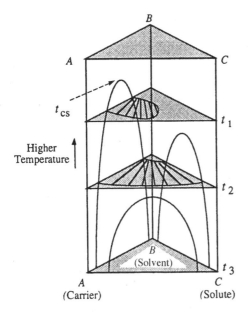

Figure 3.9 Effect of temperature on liquid/liquid equilibria. (*Adapted from Humphrey, J. L. et al., "The Essentials of Extraction," Chemical Engineering, September 17, 1984, pp. 76–95. Used with permission.*)

Liquid/liquid equilibria are a function of temperature. Figure 3.9 shows that a Type III system at temperature t_3 can become a Type II system at t_2, a Type I system at t_1, and a completely miscible system at the critical solution temperature, t_{CS}.

Equilibrium Stage Calculations

For multiple-stage extractions, both crosscurrent and countercurrent configurations are employed, though the countercurrent configuration is more common (Figure 3.10).

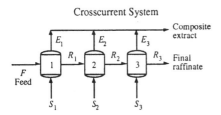

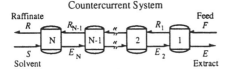

Figure 3.10 Stage arrangements in multiple stage extraction.

Minimum solvent rate, infinite stages

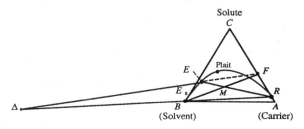

Higher-than-minimum solvent rate, 4 equilibrium stages

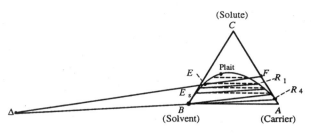

Figure 3.11 Triangular diagram for determining equilibrium stages in liquid/liquid extraction. (*Adapted from Humphrey, J. L. et al., "The Essentials of Extraction," Chemical Engineering, September 17, 1984, pp. 76–95. Used with permission.*)

Figure 3.11 shows the triangular diagram method for determining equilibrium stages. It is the counterpart to the McCabe-Thiele method in distillation. For a given feed (F) and solvent (B), values of the extract (E) and raffinate (R) can be obtained graphically, as shown. When extract and feed are in equilibrium (the line connecting them coincides with a tie line), the resulting solvent rate is the minimum rate. The *minimum solvent rate* corresponds to the *minimum reflux ratio* in distillation.

If solvent-to-feed ratio is increased (segment MF increases in proportion to segment BM), stages can be stepped off, using the difference point as the pivot. At the minimum solvent rate, an infinite number of stages are required to make the separation.

Since feed and extract pass each other on the bottom stage, equilibrium may limit the purity of the extract. To increase the purity, extract reflux is used. Returning a portion of extract is equivalent to refluxing in distillation. Reflux overcomes the limits of equilibrium and permits a purer solute to be obtained.

Rigorous models have been developed for handling multicomponent systems, and computer programs are available, via either direct purchase or process-simulation services (Seader et al., 1977). Spreadsheet approaches are also convenient to complete stage calculations (Leonard and Regalbuto, 1994). One difficulty is in getting reliable multicomponent liquid/liquid equilibrium data. To overcome this obstacle, it is acceptable to use a pseudoternary system, with pseudocomponents representing the properties of the design extract and raffinate streams.

Efficiency

Mass transfer fundamentals

In extraction, the solute is transferred from the raffinate to the extract phase. Figure 3.12 shows the mass transfer process with resistances in the raffinate and extract phases. A resistance at the interface could also be considered, but based on two-resistance theory, this resistance is assumed to be negligible.

The equations for solute flux, N_R, are:

$$N_R = k_R(x_R - x_{Ri}) = k_E(x_{Ei} - x_E) \tag{3.4}$$

$$N_R = K_R(x_R - x_R^*) = K_E(x_E^* - x_E) \tag{3.5}$$

where k_R and k_E are individual-phase transfer coefficients, and K_R and K_E are overall transfer coefficients that incorporate resistances of both phases. The x's in Equations (3.7) and (3.8) are concentrations expressed as mol fractions. x_R^* is the concentration of solute in equilibrium with x_E and v_E^* is the concentration in equilibrium with x_R. On the basis of the extract phase, Equations (3.4) and (3.5) may be combined as follows:

$$1/K_E = 1/k_E + m_E/k_R \tag{3.6}$$

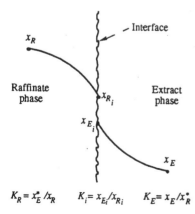

$K_R = x_E^* / x_R$ $K_i = x_{Ei} / x_{Ri}$ $K_E = x_E / x_R^*$ **Figure 3.12** Mass transfer between two liquid phases.

Concentrations refer to the transferring solute

where m_E is the slope of the equilibrium curve plotted with the extract mol fraction on the ordinate scale. If the equilibrium curve is actually a straight line through the origin, then $x_E^* = m_E x_R = K_E x_R$. Over a limited concentration range, and for a curved equilibrium relationship, it is often possible to assume $K_R \approx K_i \approx K_E$ and $m_R \approx m_i \approx m_E \approx m$. If the extract is the dispersed phase, Equation (3.6) becomes:

$$1/K_d = 1/k_d + m/k_c \tag{3.7}$$

For low concentrations of solute, and a straight equilibrium line, the number of overall transfer units in a packed extractor (raffinate-phase basis) is:

$$\text{NTU}_{OR} = \int_{x_{R2}}^{x_{R1}} \frac{dx_R}{x_R - x_R^*} + 1/2 \ln \frac{(1-x_{R1})}{(1-x_{R2})} \tag{3.8}$$

and

$$\text{HTU}_{OR} = \frac{R_m}{K_R A (1-x_R)_m} \tag{3.9}$$

Heights of individual-phase transfer units are:

$$\text{HTU}_R = \frac{R_m}{k_R A (1-x_R)_m} \tag{3.10}$$

$$\text{HTU}_E = \frac{E_m}{k_E A (1-x_E)_m} \tag{3.11}$$

where $(1-x_E)_m$ is mean value of mol fraction of nonsolute (carrier) in the extract phase

$(1-x_R)_m$ is mean value of mol fraction of nonsolute (carrier) in the raffinate phase

On a combination basis,

$$\text{HTU}_{OR} = \text{HTU}_R + \frac{R_m}{E_m m_E} \cdot \text{HTU}_E \frac{(1-x_E)_m}{(1-x_R)_m} \tag{3.12}$$

Tray efficiency in extraction columns may be expressed as:

$$E_{mR} = \frac{x_{R,n} - x_{R,n-1}}{x_{R,n}^* - x_{R,n-1}} \tag{3.13}$$

Or alternately, on the basis of the dispersed phase, tray efficiency may be expressed as:

$$E_{md} = \frac{x_{d,n} - x_{d,n-1}}{x_{d,n}^* - x_{d,n-1}} \qquad (3.14)$$

Solvent Selection

In extraction, selection of the solvent is the key decision. The following criteria should be used for solvent selection.

- *Distribution coefficient*—A high value of distribution coefficient indicates high solvent affinity for solute, which permits lower solvent/feed ratios.

- *Separation factor*—A high value of the separation factor (β_{CA}) reduces the number of equilibrium stages required. A procedure for screening potential solvents to determine activity coefficients at infinite dilution (solute concentration approaches zero) for the solute/solvent and carrier/solvent pairs, has been suggested by Gerster et al. (1960). The equation proposed by Gerster to screen solvents and their separation factors follows:

$$\beta_{CA}^\circ = \gamma_{AB}^\circ / \gamma_{CB}^\circ \qquad (3.15)$$

Computer software is also available to predict distribution coefficient and separation factors to aid in solvent selection (Naser, 1994; Naser & Fournier, 1991).

- *Density and viscosity.* A large difference in density between extract and raffinate phases permits high capacities in extraction devices using gravity for phase separation. High viscosities lead to difficulties in pumping, dispersion, and reduces rate of mass transfer.

- *Recoverability of solvent.* A clean and efficient separation of solute and solvent is desirable for solvent recovery. A solvent which boils much higher than the solute leads to better results, though solvents boiling lower than the solute are used commercially.

- *Solubility of solvent.* Mutual solubilities of carrier and solvent should be low. If there is significant solubility of solvent in the carrier, an additional distillation column may be required to recover solvent.

- *Interfacial tension.* A low interfacial tension aids dispersion of the phases and improves contacting efficiency. Systems with low interfacial tension are known as "easy" systems. On the other hand, they are slow to coalesce, and may require longer contact times for phase separation. To better understand coalescing mechanisms is an important challenge in extraction research.

- *Availability and cost.* One should make sure that the solvent of interest is commercially available. And solvent cost may represent a large initial expense for charging the system, as well as a heavy continuing expense for replacing solvent losses.

- *Toxicity and flammability.* Toxicity and flammability of the solvent are important occupational health and safety considerations.

- *Stability.* Considerations of the stability of the solvent are significant, especially if the solvent is prone to decomposition or polymerization, or it tends to react with any of the components in the feed.

- *Corrosivity.* Corrosivity of solvent can lead to problems with materials of construction. Corrosion rates should be determined if corrosion is suspected to be a problem.

- *Compatibility.* Some solvents will contaminate food or pharmaceutical products, and thus would not be suitable in such applications.

Extraction Equipment

Extraction involves intimate contacting of two liquid phases to achieve the maximum approach to equilibrium. After contacting the phases, provisions must be made to separate them.

Extraction equipment may be classified into two groups. In simple extractors, one phase is dispersed in the other with countercurrent flow achieved by flow of two phases under the influence of gravity. In mechanical extractors, dispersion is enhanced by mechanical agitation and shear. Examples of both groups of equipment are given in Table 3.4.

TABLE 3.4   Examples of Liquid/Liquid Extractors

Extractor	Comment
Simple Extractors	
Spray, sieve, and packed columns	Can be used when only a few equilibrium stages are required
Mechanical Extractors	
Mixer-settlers	Reliable scaleup procedures are available; handles liquids with high viscosity; system becomes complex and costly when more than three equilibrium stages are needed
Reciprocating-plate column (Karr)	Reliable scaleup procedure available; handles liquids containing suspended solids
Centrifugal extractor	Short contact time for unstable materials; limited space required; handles emulsified materials; handles systems of low density difference; expensive

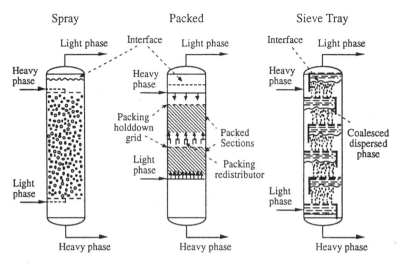

Figure 3.13 Simple extractors.

Simple Extractors

As indicated in Table 3.4, examples of simple extractors are the spray, sieve tray, and packed columns. Recent performance data taken in an 18-in. diameter column have been published for both sieve and packed column extractors (Seibert and Fair, 1995). These extractors are illustrated in Figure 3.13.

Spray extractor

The *spray extractor* is an empty vertical vessel which includes a distributor for the phase to be dispersed. The diameter of the column is obtained by using 50% of the flooding velocity (Minard and Johnson, 1952). Column height is obtained by using the method of Pratt (1952).

The spray extractor is inexpensive but suffers low efficiency because, for larger-diameter towers, there is considerable backmixing in the continuous phase. This backmixing lowers the available concentration driving force for mass transfer. Also, there is a lack of breakage and reformation of drops in a staged fashion. Spray extractors generally do not produce the equivalent of more than one or two equilibrium stages. Because of their simple construction, spray columns are used for basic operations, such as washing and neutralization (Letan and Kehat, 1967). A good comparison of performance of spray versus packed extractors is available (Seibert et al., 1990).

Sieve tray extractor

The *sieve tray extractor* resembles sieve tray distillation, but little information has been published on the performance of large-scale units. Engineering firms have experience with proprietary sieve tray extractors, but generally do not publish their design procedures. However, some data from industrial installations have been reported. For example, it has been reported that a 80-ft-high sieve tray extraction column, used for the extraction of aromatics, has the equivalent of ten theoretical stages (Reman, 1966).

As shown in Figure 3.13, weirs are not needed with extraction sieve trays. Dispersed phase drops are generated at each tray perforation. The continuous phase moves in a crossflow fashion in the "compartments" between trays. Interfacial contact between the rising drops of dispersed phase, and the crossflowing continuous phase, occurs in these compartments. Downcomers are used if the light phase is dispersed, and upcomers are used if the heavy phase is dispersed. In Figure 3.13, downcomers are used because the light phase is dispersed. Characteristic geometries of extraction sieve trays are presented in Table 3.5.

Research has confirmed that sieve tray efficiency is a strong function of the Weber number (Blass et al., 1986; Eldridge et al., 1986; Rocha et al., 1986) A minimum Weber number of about 4 is required to achieve maximum tray efficiency. The Weber number is defined as:

$$\text{We} = \frac{\rho_d U_0^2 d_0}{\sigma} \tag{3.16}$$

where We = Weber number, dimensionless
ρ_d = density of the dispersed phase, mass/length^3
U_0 = hole velocity, length/time
d_0 = hole diameter, length
σ = interfacial tension, force/length

TABLE 3.5   Geometry of Extraction Sieve Trays

Parameter	Range
Hole diameter, cm	0.32–0.64
Fractional free area	0.03–0.08
Fractional downcomer area	0.03–0.10
Pitch/hole diameter ratio	3–4
Tray spacing, cm	15–45

Overall tray efficiency of extraction sieve trays can be approximated by the correlation of Treybal (1963).

$$E_0 = \frac{3.12 \, Z^{0.5} \left(U_d / U_c \right)^{0.42}}{\sigma} \quad (3.17)$$

where E_0 = overall tray efficiency, dimensionless
Z = tray spacing, feet
U_d, U_c = superficial velocity of the dispersed and continuous phases, respectively, ft/s
σ = interfacial tension, dynes/cm

More comprehensive models for prediction of the efficiency of extraction sieve trays have been presented (Blass et al., 1986; Rocha et al., 1986; Seibert and Fair, 1993; and for supercritical extraction, Lahiere, 1986). Examples of experimental values of the height equivalent for a theoretical plate (HETP) for extraction sieve trays, based on laboratory data, are given in Table 3.6.

Packed extractor

Like the sieve tray extractor, the *packed extractor* is more efficient than the spray column. The packing elements reduce backmixing in the continuous phase, and promote mass transfer due to jostling and breakup of dispersed phase drops, as they contact packing elements. Because the packing elements reduce the cross-sectional area for flow, the column diameter for a given rate will be greater than for a spray tower. However, this disadvantage is generally offset by the advantage of increased efficiency (Gayler and Pratt, 1957).

The packings in extraction are similar to the ones used in distillation. They include random and structured packings. Devices for phase collec-

TABLE 3.6   Efficiencies of Extraction Sieve Trays

Chemical system	Tray spacing, ft	HETP, ft	Reference
Toluene/acetone/water	0.5–1.5	3–8	Blass et al., 1986
Toluene/benzoic acid/water	0.08–0.6	5–12	Mayfield and Church, 1953; Treybal, 1963
Gasoline/methyl ethyl ketone/water	0.25	4–5	Moulton and Walkey, 1944
Toluene/diethyl amine/water	0.5	4–10	Garner et al., 1953
Methyl isobutyl ketone/adipic acid/water	0.5	2–3	Garner et al., 1956
Methyl isobutyl ketone/acetic acid/water	0.5–1.5	2–4	Blass et al., 1986
Kerosene/benzoic acid/water	0.4	4–6	Allerton et al., 1943
Diethyl ether/acetic acid/water	0.4–1.7	1.5–3	Pyle et al., 1950

tion and liquid distribution are also used. The maximum rates of phase flows are obtained from flood correlations (Nemunaitis et al., 1971). The packed extraction column is generally used when the separation requires just a few equilibrium stages. A packed extractor should be designed for the case in which the continuous phase preferentially wets the packing, with the dispersed phase moving through the column as spherical drops. A decline in drop population and interfacial area between phases would result should the dispersed phase drops wet the packing. Thus, wetting of the packing by the dispersed phase should be avoided. Table 3.7 gives typical geometries and efficiencies for random packings.

For some applications, extraction columns containing structured packings have about the same efficiency as columns containing traditional packings or sieve trays, but give higher capacities. The features of the structured packing used in extraction are illustrated in Figure 3.14. Structured packings used in extraction are similar to those used in distillation. In distillation, the surface characteristics of the packing, such as perforations or embossing, affect the efficiency, but the same effect is not observed in extraction due to the difference in the phase contacting mechanism.

Examples of efficiencies and capacities of structured packings in liquid/liquid extraction columns are given in Table 3.8. The capacity is the sum of the maximum rates of the dispersed and continuous phases. As in the design of columns in distillation, good initial distribution and redistribution of the phases is necessary to optimize performance.

Correlations for the design of the packed columns with structured packings have been presented by Seibert et al. (1990), Seibert and Humphrey (1995), and Streiff and Jancic (1984). These correlations are based on a data bank comprised of data obtained from small-diameter columns with relatively short packing heights. Such correlations do not address the effects of phase maldistribution which may be pre-

TABLE 3.7 Characteristics and Efficiencies of Extraction Columns Containing Random Packings*

Type of packing	Packing void fraction	Packing surface area, sq ft/cu ft	Chemical system	HETP, ft
Pall rings	0.93	109	n-Butanol/succinic acid/water	2–4
Pall rings	0.93	109	Toluene/acetone/water	3–5
Raschig rings	0.64	110	n-Butanol/succinic acid/water	2–4
Raschig rings	0.64	110	Toluene/acetone/water	3–5
Intalox saddles	0.72	92	n-Butanol/succinic acid/water	2.5–3
Intalox saddles	0.72	92	Toluene/acetone/water	4–7

*Data are contained in Seibert (1986) and Seibert and Fair (1988).

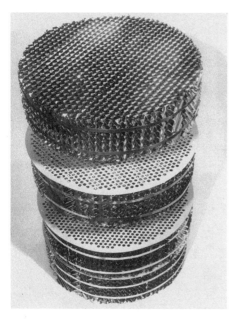

Figure 3.14 Structured packing for extraction. (*Provided by Koch Engineering Co., Inc.*)

TABLE 3.8   Efficiencies of Extraction Columns Containing Structured Packings

Packing	Chemical system	HETP (cm)	Capacity (cu m/sq m-h)	Mass transfer direction	Reference
Sulzer SMV	Toluene(d)/ acetone/water(c)	60–120	90	c → d	Seibert, 1986; Streiff and Jancic, 1984
Sulzer SMVP	Toluene(d)/ acetone/water(c)	20–30	60	c → d	Streiff and Jancic, 1984
Sulzer BX	Toluene(d)/ acetone/water(c)	50–60	30	c → d	Seibert, 1986
Intalox 2T	Toluene(d)/ acetone/water(c)	100–160	120	c → d	Seibert et al., 1990
Montz B1	Toluene(d)/ acetone/water(c)	220	170	d → c	Billet and Mackowiak, 1985
Sulzer SMV	Butanol(d)/ succinic acid/ water(c)	80–100	50	c → d	Seibert, 1986; Streiff and Jancic, 1984
Sulzer SMV	Carbon tetrachloride (d)/propionic acid/water(c)	60–80	18	c → d	Streiff and Jancic, 1984
Sulzer SMV	Toluene(d)/ acetone/water(c)	150–180	150	d → c	Seibert, 1986; Streiff and Jancic, 1984
Sulzer BX	Toluene(d)/ acetone/water(c)	100–130	180	d → c	Seibert, 1986
Sulzer BX	Butanol(d)/succinic acid/water(c)	50–70	35	c → d	Seibert, 1986

Abbreviations: c = continuous phase, d = dispersed phase.

TABLE 3.9   Examples of Liquid/Liquid Extractors Equipped with Structured
Packings

Applications for liquid/liquid extraction with structured packings	Column Diameter (mm)	Packing height (m)
Extraction of nitrotoluene after reaction of HNO_3 with toluene in H_2SO_4	600	12
Extraction of methylacrylate from organic solution with perchlorethylene	150	9
VI-reaction column: cyclohexanone/ammonium hydrogen-sulfate solution	1100	18
Extraction of furan from an aqueous solution with perchlorethylene	300	10
Benzylalcohol from a salt solution with toluene	500	12
Extraction of sulfolane from C_6C_7-mix with water	420	9
Removing H_2S from LPG with MDEA	2350	12
Extraction of caprolactam from ammonium sulfate solution with benzene	990	10
Removing ammonium sulfate from benzene/caprolactam solution with water	1450	10
Extraction of imidazole from an aqueous solution with an organic solvent	650	14 + 8
Extraction of acrylic acid from waste water with butanol	800	10 + 6
Washing column oil/water	700	6
Extraction of propylenechlorhydrin from dichloropropane with water	1000	22
Removing H_2S from LPG with MDEA	1200	10
Removing H_2S from LPG with DEA	870	4
Extraction of C-4/C-4 HCl from C_5/heptane with aqueous hydrochloric acid	470	7
Extraction with water	150	6.1
Removing H_2S from LPG with MDEA	1100	8.5
Extraction of flavor agent from an aqueous solution with alkane	250	5
Removing residual alkali from dichlorohydazobenzene with water	450	4.5
Extraction of methanol from LPG with water	730	11.4
Extraction of methanol from LPG with water	1470	7.8
Removing H_2S from LPG with MDEA	1700	3.2
Extraction of chloroacetic acid from methylchloroacetate with water	300	3

Abbreviations: DEA, diethanolamine; LPG, liquid petroleum gas.

SOURCE: Adapted from information provided by Koch Engineering Co., Inc.

sent in larger columns. As indicated in Table 3.9, structured packings have been installed in a number of commercial extraction columns.

Pulsed extractor

The *pulsed extractor* is based on the concept that the efficiency of a packed or tray column can be increased by applying an oscillating pulse (Van Dijck, 1935). Although pulsed extractors have been popular in the past, interest in them seems to have declined. In addition to mechanical problems, there are difficulties in propagating pulses throughout larger columns. On a pulse frequency/amplitude equivalency, the reciprocating-plate (Karr) column has been found to have similar operating characteristics with less maintenance cost and appears to have found broader acceptance in industry.

Membrane phase contactor

The *membrane phase contactor* is a relatively new commercial development for use in extraction, absorption, and stripping processes (Reed et al., 1995; Seibert et al., 1996). In membrane extraction, the mass transfer interface is a liquid film which forms across individual membrane pores. In the extraction application, the solute moves from the raffinate to the extract phase through these interfaces. For this type of contactor, raffinate and extract phases can have identical densities.

To maintain mass transfer efficiency in conventional extractors, such as tray or packed columns, a minimum rate of dispersed phase is necessary to provide interfacial area between dispersed and continuous phases. However, membrane phase contactors are not subject to this limitation and can be operated at lower solvent rates, for example, than a tray or a packed column. This feature can be important, particularly when expensive solvents, such as crown ethers are required (Bonnesen et al., 1994). And lower solvent rates translate to a reduction in the size and cost of downstream equipment for solvent recovery.

Electrically enhanced extraction

Electric fields have been used to achieve dispersion of one liquid phase in another (Scott, 1989; Scott and Byers, 1989; Scott and Sission, 1988; Scott and Wham, 1989). In *electrically enhanced extraction*, high-intensity pulsed electric fields can be used to break up a polar liquid phase, such as water, into small drops.

As illustrated in Figure 3.15, an oscillating electric field creates small drops. the result would be an increase in the interfacial area between two liquid phases when the smaller drops are dispersed in a continuous phase. And increased interfacial area translates to higher rates of mass transfer. The limitations of this technology are: (1) one phase must be highly polar, and (2) there is a lower limit on how

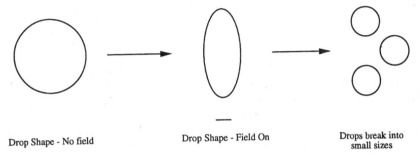

Drop Shape - No field   Drop Shape - Field On   Drops break into
                                                                      small sizes

Figure 3.15   Drop breakage achieved by electric field.

small the droplet of dispersed phase can be before it is backed out of the extractor by the continuous phase. This technology is still in the very early stages of commercial development.

Mechanical Extractors

There are many types of mechanical extractors varying in complexity from the simple mixer-settler system to the more complex centrifugal extractors. We will present the features and characteristics of several of these devices.

Mechanically aided extractors have higher efficiencies—but lower capacities—than packed or tray extractors. The exception, centrifugal extractors, provide both high efficiency and capacity, but have capital and maintenance costs that are higher than for simple extractors.

Mixer-settler systems

Mixer-settler systems involve a mixing vessel for phase dispersion, followed by a settling vessel for phase separation. Both vertical and horizontal designs are available. Design specifications for a vertical design are available (Steiner et al., 1996). The horizontal design saves headroom, while the vertical design saves floor space. Dispersion in the mixing vessel can be achieved by pump circulation, air agitation, or mechanical stirring (Figure 3.16).

While intense agitation in the dispersion vessel leads to high rates of mass transfer and close approach to equilibrium, the result can be dispersions that are difficult to separate. Because of this, designs of mixer-settler systems must strike a balance between intensity of dispersion and time of settling.

Mixer-settler systems have a number of attractions, including high efficiency, reliable scaleup, and operating flexibility. They can handle difficult-to-disperse systems, such as those having high interfacial tension and/or large phase-density difference. They can also cope with viscous liquids and solid-liquid slurries which are typical in the mining industry.

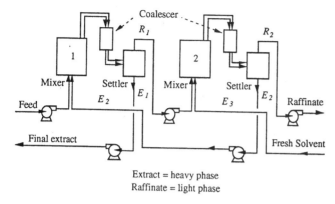

Figure 3.16   Mixer-settler system.

The disadvantages of mixer-settler systems are high capital cost per stage and high inventory of material in the vessels. Because considerable capital is required for pumps and piping, mixer-settler systems are generally limited to applications requiring just a few stages, though some of the multistage systems used in the mining industry are exceptions. Compact systems can be designed to minimize space, piping, and pumping requirements. Design procedures for mixer-settler systems are available (Godfrey and Slater, 1983).

Many of the mechanically aided extractors are proprietary. Examples are the Scheibel (also called the York-Scheibel), rotating-disc contactor (RDC), reciprocating-plate (Karr), Oldshue-Rushton, and the Kühni (Figure 3.17).

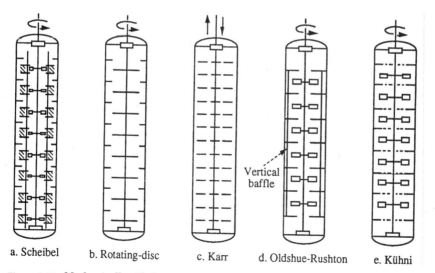

a. Scheibel   b. Rotating-disc   c. Karr   d. Oldshue-Rushton   e. Kühni

Figure 3.17   Mechanically aided extractors.

Scheibel extractor

The *Scheibel extractor* is designed to simulate a series of mixer-settler extraction units, with self-contained mesh-type coalescers between each contacting stage (Scheibel, 1995). The influence on mass transfer efficiency of speed and size of the agitators, geometry of the mixing compartments, and system physical properties has been evaluated by Scheibel and Karr (1950), Karr and Scheibel (1954), and Scheibel (1983). Although moderately expensive, the Scheibel column offers a very high efficiency. Scaleup requires small-scale experimental data. Scaleup procedures are based on providing a large-scale efficiency equal to the small-scale efficiency, with appropriate adjustment of power input. A complete example of scaleup for a Scheibel column is given in a report based on the removal of phenol from wastewater (French, 1983).

Rotating-disc contactor

The RDC was introduced in the 1950s by the Shell companies (Strand et al., 1962). It has been used in the petroleum industry for extractions involving hydrocarbon systems. Rotors on a central shaft create a dispersion and movement of the phases, while stators provide the countercurrent staging. As with the Scheibel, RDC effectiveness can be controlled to some extent by varying the speed of rotation of the disc dispersers. This contactor has been used successfully for cases where the continuous phase is actually a liquid-solid slurry.

Reciprocating-plate extractor (Karr column)

The *reciprocating-plate extractor* is a descendant of the pulsed-plate column which was popular in the 1950s. Where the pulse column has fixed plates with pulsed liquid, the Karr column involves moving an assembly of plates (Karr, 1995 and 1959; Lo et al., 1992). The Karr column relies on perforated plates having high open areas (see Figure 3.17). The plate assembly is given a reciprocating movement by an overhead drive. The scaleup of the Karr column may be accomplished with the use of Equation (3.18) (Holmes et al., 1987).

$$\frac{(\text{HETP})_2}{(\text{HETP})_1} = \left(\frac{D_2}{D_1}\right)^{0.38} \tag{3.18}$$

where D_1 = diameter of the laboratory extractor, length
D_2 = diameter of large-scale extractor, length
$(\text{HETP})_1$ = height equivalent to a theoretical plate of the laboratory extractor, length
$(\text{HETP})_2$ = height equivalent to a theoretical plate of the large-scale extractor, length

Oldshue-Rushton extractor

Designed in the 1950s, the *Oldshue-Rushton extractor* comprises several compartments separated by horizontal baffles (Oldshue, 1995; Oldshue and Rushton, 1952; Oldshue et al., 1974). Each compartment contains vertical baffles and an impeller mounted on a central shaft. Columns of up to 9 ft in diameter have been reported in service. The extractor is said to have predictability in scaleup, but only few data are available. Guidelines for scaleup and design have been developed.

Kühni extractor

The *Kühni extractor* is similar to the Scheibel column, but without the coalescing sections (Fisher, 1971). A baffled turbine impeller promotes radial discharge within a compartment, while horizontal baffles of variable hole arrangement separate the compartments. Mass-transfer studies involving drop size, holdup, and backmixing characteristics for a 6-in.-diameter column have been reported (Ingham et al., 1974). The supplier recommends pilot plant studies in a 4- to 6-in.-diameter column. The important scaleup parameters are compartment geometry, column capacity, and column efficiency. Mugli and Buhlmann (1983) have presented a scaleup procedure based on similarity of geometric and hydrodynamic parameters.

Centrifugal extractor

Unlike column contactors, which depend upon the force of gravity to bring about phase separation, the *centrifugal extractor* applies centrifugal force to increase rates of countercurrent flow and separation of the phases. The result is a more compact unit providing very short contact times. Quickness of contacting is an important feature when unstable materials are being processed. For this reason, centrifugal devices have found applications in the pharmaceutical and food industries. Another attraction is that it can handle systems that tend to emulsify, as well as those with low density difference. On a unit-throughput basis, centrifugal extractors do not need much space. Examples of these contactors are the Podbielniak (Todd and Davies, 1974), Alfa-Laval (Hanson, 1968), Quadronic (Doyle et al., 1968), and Robatel (Miachon, 1971).

The first centrifugal unit to gain widespread use in industry was the Podbielniak extractor. This extractor consists of a drum rotating around a shaft equipped with annular passages at each end for feed and raffinate. The light phase is introduced under pressure through the shaft and then routed to the periphery of the drum. The heavy phase is also fed through the shaft but is channeled to the center of

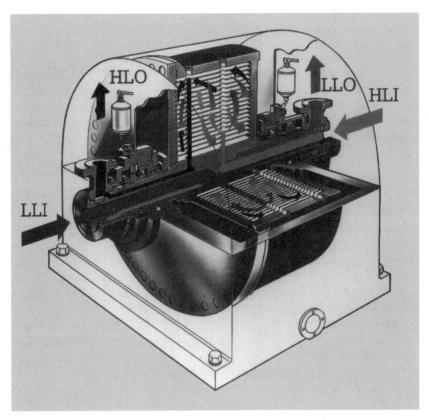

Figure 3.18 Centrifugal extractor. (*Provided by B&P Process Equipment & Systems.*)

the drum. Centrifugal force acting on the phase-density difference causes dispersion as the phases are forced through the perforations (Figure 3.18).

Advantages of this type of extractor include short contact time for unstable materials, low space requirement, easy handling of emulsified materials, and ability to cope with fluids having small density differences. The disadvantages are complexity and high capital and operating costs.

Comparisons of extractor performance

Table 3.10 gives values of HETP of several types of mechanically aided extractors.

Efficiency versus capacity for various liquid/liquid extractors is given in Figure 3.19. Spray towers may be used when only one or two

TABLE 3.10   Efficiencies of Mechanically Aided Extractors

Extractor	Chemical system	HETP, ft	Reference
Reciprocating-plate	Toluene/acetone/water	0.5–1	Holmes et al., 1987
Reciprocating-plate	Methyl isobutyl ketone/ phenol/water	2–6	Karr and Ramanujam, 1987
Rotating-disc contactor	Toluene/acetone/water	0.5–2	Holmes et al., 1987
York-Scheibel	Toluene/acetone/water	0.3–0.8	Holmes et al., 1987
Centrifugal	Toluene/acetone/water	0.6	Gebauer et al., 1983

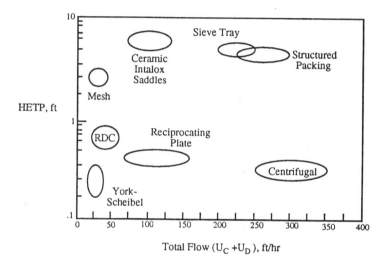

Figure 3.19 Extractors—efficiency versus capacity—toluene/acetone/water system (*Adapted from Stichlmair, 1980.*)

stages are needed. When more stages are required, packings or trays are needed. When six to eight stages or more are required, mechanically aided extractors should be considered.

Extractor selection

Selection of an extractor is somewhat complicated by the large number of extractors available, and amount and complexity of the design variables, including potential complications due to emulsions, solids, and unstable materials. To help address extractor selection, Pratt and Hanson (1983) developed the guidelines shown in Table 3.11.

TABLE 3.11 Guidelines for Extractor Selection

Type extractor	Efficiency*	Capacity†	Maintenance demands	Space requirements		Ability to handle		Residence time
				Height	Floor	Emulsions	Solids	
Mixer-settler	3	5	3	5	0	0	1	0
Spray	1	3	5	0	5	1	1	0
Packed	3	3	5	1	5	1	0	0
Sieve-tray	3	3	5	1	5	1	0	0
Pulsed-plate	3	3	3	1	5	0	1	1
York-Scheibel	3	3	3	1	5	0	0	0
Reciprocating-plate	3	3	3	1	5	0	1	1
Centrifugal	3	3	1	5	5	3	0	5

Key: 5 = most desirable, 0 = least desirable.

*For one to five theoretical stages.

†Within range of 2.5–25 cu m/h.

SOURCE: Adapted from Pratt and Hanson, 1983.

TABLE 3.12   F.O.B. Equipment Costs for Mechanical Extractors*

Equipment	Material of construction	Cost equation	Range of applicability
Rotating disc contactor	Carbon steel	$\$ = 55{,}000 \, (ZD^{1.5}/10)^{0.61}$	$ZD^{1.5} = 0.5\text{--}75$
Oldshue-Rushton, York-Scheibel	304 stainless steel	$\$ = 148{,}000 \, (ZD^{1.5}/10)^{0.66}$	$ZD^{1.5} = 0.5\text{--}55$
Pulsed-plate	316 stainless steel	$\$ = 164{,}000 \, (ZD^{1.5}/10)^{0.81}$	$ZD^{1.5} = 0.5\text{--}100$
Reciprocating-plate	316 stainless steel	$\$ = 164{,}000 \, (ZD^{1.5}/10)^{0.75}$	$ZD^{1.5} = 0.5\text{--}100$
Centrifugal	316 stainless steel	$\$ = 84{,}000 \, (Q_{raf}/2.2)^{0.25}$	$Q_{raf} = 0.03\text{--}2.2$
		$\$ = 84{,}000 \, (Q_{raf}/2.2)^{0.58}$	$Q_{raf} = 2.2\text{--}36$
Mixer-settler	Carbon steel	$\$ = 122{,}000 \, (Q_{raf}/10)^{0.22}$	$Q_{raf} = 1\text{--}10$
		$\$ = 122{,}000 \, (Q_{raf}/10)^{0.60}$	$Q_{raf} = 10\text{--}100$

Abbreviations: D = diameter of column, m; Q_{raf} = flow of aqueous feed to be treated, L/s; Z = height of column, m.

*Basis: Second Quarter 1994 Equipment Costs†/Accuracy: ± 40%

†Updated to 1994 cost basis using Marshall and Swift Equipment Cost Index (see *Chem. Eng.*, September, p. 198)—second quarter 1994 = 990.8.

Extractor Costs

Equations to determine equipment costs for mechanically aided extractors are given in Table 3.12. These equations give f.o.b. equipment costs. For example, the cost equation for the reciprocating plate extractor includes the extraction column and plates, drive mechanism, motor, and piping. The cost equations in Table 3.12 are based on the original paper by Woods (1983), but they have been updated to a 1994 cost basis. Further details may be obtained in the original paper.

A case study of costs

Scaleup procedures for a Scheibel column were studied by French (1983) in connection with reducing the phenol concentration in a wastewater stream from 0.003 to 0.000150 weight fraction. The feed wastewater flow was 820 gal/min and the solvent was diisopropyl ether. French developed the following commercial-scale design data:

Scheibel Column

Column diameter	8.0 ft
Column height	30.0 ft

Theoretical stages	2.7
Actual stages	9
Agitator speed	79 rpm
Power input per actual stage	234 ft-lb$_{force}$/S

Using the same bases as above, a packed extractor and a sieve-tray extractor were designed (Humphrey et al., 1984).

Packed Column

Column diameter	8.0 ft
Height of packing	40.0 ft
Type of packing:	1.0-in. porcelain saddles

Sieve-Tray Column

Column diameter	8.0 ft

Height of contacting section (alternatives):

35 trays, 1.0-ft spacing	34.0 ft
28 trays, 1.5-ft spacing	40.5 ft
23 trays, 2.0-ft spacing	44.0 ft

A summary of costs of these columns (based on 1994 costs) follows.

Scheibel column	$296,000
Packed column	$239,000
Sieve-tray column (28 trays)	$200,000

The results show that simple extractors compare favorably with mechanically aided extractors for the case presented. The cost data were taken from the general correlations (Hall et al., 1982; Mulet et al., 1981; Pikulik and Diaz, 1977; Woods, 1983). Firm bids by equipment suppliers would be necessary for a completely valid comparison.

Summary

Extraction trails distillation in the amount of research and development and in number of applications. However, it can sometimes offer advantages over distillation because it can be operated at moderate temperatures, thus avoiding temperature damage to thermally sensitive products. Extraction also finds applications involving systems with low relative volatilities or those exhibiting azeotropes. Approaches to improve the design and performance of extractors are discussed below and are summarized in Table 3.13.

TABLE 3.13 Extraction Summary

Technologies	Potential benefits	Key applications	Keys to effective use	Comments
Supercritical fluid extraction	Replace hazardous organics, such as CFCs, with supercritical carbon dioxide	For processing products for human consumption and environmental protection	Minimize operating costs associated with high-pressure equipment	Process(es) operate at 1000–5000 psig
Structured packings	High capacity	Retrofit of sieve-tray extraction columns	If metal packing is used, disperse organic phase rather than aqueous phase	Obtain complete retrofit from one equipment supplier
Membrane phase contactors (not commercial)	Minimum solvent rate	Cases where expensive solvents are used or recovery costs are high	Solutes must have high distribution coefficient	Exhibits good mass transfer, but has limited capacity
Computer software for solvent selection	Better solvent selected	All new designs	Accurate bases for problem	Best solvent may not be commercially available
Electrically enhanced extraction	Small drops of dispersed phase leads to improved mass transfer performance	In applications where maximum mass transfer efficiency is essential	Minimizing costs of electrical components	Still in research & development stages

The future of SFE appears to be in applications where there are concerns about toxicity or the environment. Small to moderate volumes are more realistic because of the exponential increase in the cost of high-pressure equipment with capacity. Candidate applications include mixtures involving specialty products which are difficult to separate by other methods. SFE may find new applications in purification of vitamins and other high-boiling specialty products where molecular distillation is currently used. SFE allows a multi-staged effect, whereas molecular distillation allows the use of only one or two stages.

Structured packings can offer advantages of increased capacity over sieve trays, and a significant number of industrial columns have been equipped with these packings.

Membrane phase contactors eliminate the need to disperse one phase in the other. The process may be operated at lower solvent to feed rates than conventional tray and packed columns, with the pay-off being a reduction in the size and cost of the downstream equipment required for solvent recovery.

Solvent selection is key in the design of the extraction system. Software based on group contribution methods is commercially available to aid in solvent selection.

References

Allerton, J. et al., *Trans. AIChE*, **39**(361), 1943.

Billet, R. and J. Mackowiak, *Fette Seifen Anstrichmittel*, **87**(5), 205, 1985.

Blass, E. et al., *Ger. Chem. Eng.*, **9**(222), 1986.

Bonnesen, P. V. et al., "Removal of Technetium from Alkaline Waste Media by a New Solvent Extraction Process," ACS Symposium Series, Atlanta, GA, September 1994.

Cusack, R. W., "Solve Wastewater Problems with Liquid/Liquid Extraction," *Chemical Engineering Progress*, April 1996, pp. 56–63.

D'Ans, J. and E. Lax, Eds., *Taschenbuch für Chemiker und Physiker*, 3rd ed., Vol. 1, Springer-Verlag, Berlin (1967).

Doyle, C. M. et al., *Chem. Eng. Progr.*, **64**(12), 68, 1968.

Eldridge, R. B., "Mixing Characteristics of a Crossflow Sieve Tray Extractor," The University of Texas at Austin, Ph.D. dissertation, 1986.

Fisher, A., *Verfahrenstechnik*, Vol. 5, 1971, p 360.

Francis, A. W., *Liquid-Liquid Equilibriums*, Interscience, New York, 1963.

Francis, A. W., *Handbook for Components in Solvent Extraction*, Gordon and Breach, New York, 1972.

French, W. E., *Scaleup Procedures for a Scheibel Extraction Column*, Report DOE/MC/16547-1420, U.S. Dept. of Energy, Morgantown, WV, June 1983.

Gallagher-Wetmore, P., "Solving Real Problems—An Industrial Perspective," presented at the NIST Supercritical Fluid Extraction Workshop: Technology of the Future, Boulder, CO, 1993.

Garner, F. H. et al., *Trans. Inst. Chem. Engrs.*, **31**(348), 1953.

Garner, F. H. et al., *Trans. Inst. Chem. Engrs.*, **34**(223), 1956.

Gayler, R. and H. R. C. Pratt, *Trans. Inst. Chem. Engrs.*, **35**, 273, 1957.

Gearhart, J. A. and S. R. Nelson, "Proc. 5th Annual Industrial Energy Conservation Technology Conf.," Vol. 2, 1983, p. 823.

Gebauer, K. et al., *Ger. Chem. Eng.,* **6**(381), 1983.

Gerster J. A. et al., *J. Chem. Eng. Data,* Vol. 5, 1960, p. 423.

Godfrey, J. C. and M. J. Slater, Chapter 9.1 in T. C. Lo et al., eds., *Handbook of Solvent Extraction,* Wiley, New York, 1983.

Godfrey, J. C. and M. J. Slater, *Liquid-Liquid Extraction Equipment,* Wiley, New York, 1994.

Hall, R. S. et al., *Chem. Eng.,* **89**(7), April 5, 1982, p. 77.

Hanson, C., *Chem. Eng.,* **75**(18), 98, 1968.

Holmes, T. L. et al., "Performance Characteristics of Packed and Agitated Extraction Columns," presented at the AIChE Summer National Meeting, August 1987.

Hood, R. L., *Hydrocarbon Technology International Quarterly,* March 1994.

Humphrey, J. L. et al., *Chem. Eng.,* September 17, 1984, pp. 76–95.

Ingham, J. et al., *Proc. Int. Solv. Extr. Conf.,* Vol. 2, 1974, p. 1299.

Irani, C. A. and E. W. Funk, *Recent Developments in Separation Science,* Vol. 3, CRC Press, W. Palm Beach, FL, 1977, Part A, p. 171.

Johnston, K., *Encyclopedia of Chemical Technology,* Suppl., 3rd ed., John Wiley, New York, 1984, pp. 872–893.

Karr, A. E., *AIChE J.,* **5,** 446, 1959.

Karr, A. E., "History and Development of Karr Extraction Column," presented at AIChE Annual Meeting, Miami Beach, FL, November 1995.

Karr, A. E. and S. Ramanujam, "Scaleup and Performance of 5 Ft. Diameter Reciprocating Plate Extraction Column," presented at the St. Louis AIChE Symposium, March 1987.

Karr, A. E. and E. G. Scheibel, *Chem. Eng. Progr. Symp. Ser.,* **50**(10), 73, 1954.

Lahiere, R. J., "Mass Transfer and Hydraulic Characteristics of a Supercritical Fluid Sieve Tray Extractor," Ph.D. dissertation, The University of Texas at Austin, 1986.

Leonard, R. A. and M. C. Regalbuto, "A Spreadsheet Algorithm for Stagewise Solvent Extraction," *Solvent Extraction and Ion Exchange,* **12**(5), 909–930, 1994.

Letan, R. and E. Kehat, *AIChE J.,* **13,** 443, 1967.

Lewis, J. B., *Extraction: A Critical Review,* Atomic Energy Research Establishment Report CE/R 910, H. M. Stationery Office, London, 1953.

Lo, T. C., "Commercial Liquid-Liquid Extraction Equipment," P. A. Schweitzer, ed., *Handbook of Separation Techniques for Chemical Engineers,* McGraw-Hill, 1988, pp. 1-303–369.

Lo, T. E. and M. H. I. Baird, "Extraction (Liquid-Liquid)," in J. I. Kroschwitz and M. Howe-Grant, eds., *Kirk-Othmer Encyclopedia of Chemical Technology,* 4th ed., Vol. 10, John Wiley, 1994, pp. 125–180.

Lo, T. C. et al., eds., *Handbook of Solvent Extraction,* Wiley, New York, 1983.

Lo, T. C. et al., "The Reciprocating Plate Column—Development and Applications," *Chem. Eng. Comm.,* **116,** 67–88, 1992.

Mayfield, F. D. and W. L. Church, *Ind. Eng. Chem.,* **44**(2253), 1953.

Miachon, J. P., *La Technique Moderne,* March/April, 1971.

Minard, G. W. and A. I. Johnson, *Chem. Eng. Progr.,* Vol. 2, p. 62, 1952.

Moulton, R. W. and J. E. Walkey, *Trans. AIChE,* **40**(695), 1944.

Mugli, A. and U. Buhlmann, in T. C. Lo et al., eds., *Handbook of Solvent Extraction,* Wiley, New York, 1983.

Mulet, A. et al., *Chem. Eng.* **88**(26), 77, Dec. 28, 1981.

Naser, S. F., *LLE-Designer™ 4.3 User Guide,* Novel Advanced Systems Corp., Ann Arbor, MI, 1994.

Naser, S. F. and R. L. Fournier, *Computers Chem. Eng.,* **15**(6), 331–339, 1991.

Navratil, J. D. and W. W. Schulz, eds., *Actinide Separations,* American Chemical Society, 1980.

Nemunaitis, R. R. et al., *Chem. Eng. Progr.,* **67**(11), 60, 1971.

Oldshue, J. Y., "Fluid Mixing and Scale Up in the Oldshue-Rushton Multi-Stage Extraction Column," presented at the American Institute of Chemical Engineers Annual Meeting, Miami Beach, FL, November 1995.

Oldshue, J. Y. and J. H. Rushton, *Chem. Eng. Progr.,* **48**(6), 297, 1952.

Oldshue, J. Y. et al., *Proc. Int. Solv. Extr. Conf.,* Vol. 2, 1974, p. 1651.

Paulaitis, M. E. et al., *Rev. Chem. Eng.,* Vol. I, 1982, p. 179.

Pikulik, A. and H. E. Diaz, *Chem. Eng.,* **84**(21), October 10, 1977, p. 106.

Pratt, H. R. C., *Ind. Eng. Chem., Process Des. Devel.,* **14,** 74, 1952.

Pratt, H. R. C. and C. Hanson, Chapter 16 in T. C. Lo et al., ed., *Handbook of Solvent Extraction,* Wiley, New York, 1983.

Prausnitz, J. et al., *Computer Calculations for Multicomponent Vapor-Liquid and Liquid-Liquid Equilibria,* Prentice-Hall, Englewood Cliffs, NJ, 1980.

Pyle, C. et al., *Ind. Eng. Chem.,* **42**(1042), 1950.

Reed, B. W. et al., "Membrane Contactors," in R. D. Noble and S. A. Stern, eds., *Membrane Separations Technology—Principles and Applications,* Elsevier, 1995.

Reman, G. H., *Chem. Eng. Progr.,* **62**(9), 56, 1966.

Robbins, L. A., "Liquid-Liquid Extraction," in R. H. Perry and D. Green, eds., *Perry's Chemical Engineers' Handbook,* 6th ed., McGraw-Hill, New York, 1984, pp. 15-9 to 15-14.

Rocha, J. A. et al., *Ind. Eng. Chem. Proc. Des. and Dev.,* **25**(862), 1986.

Schafer, K. and E. Lax, eds., *Landolt-Bornstein Zahlenwerte und Fupktionen aus Naturwissenschaft aus Physik, Chemie, Astronomie, Geophysik, and Technik,* 6th ed., Vol. II (2c), Springer-Verlag, Berlin, 1964.

Scheibel, E. G., Chapter 13.3, in T. C. Lo. et al., eds., *Handbook of Solvent Extraction,* Wiley, New York, 1983.

Scheibel, E. G., "Development of Scheibel Extraction Columns," presented at AIChE Annual Meeting, Miami Beach, FL, November 1995.

Scheibel, E. G. and A. E. Karr, *Ind. Eng. Chem.,* **42,** 1048, 1950.

Scott, T. C., *Separation and Purification Methods,* **18**(1), 65, 1989.

Scott, T. C. and C. H. Byers, *Chem. Eng. Comm.,* **77**(67), 1989.

Scott, T. C. and W. G. Sission, *Separation Science and Technology* **23**(12), 1541, 1988.

Scott, T. C. and R. M. Wham, *Ind. Eng. Chem. Res.,* **28**(94), 1989.

Seader, J. D. et al., *FLowtran Simulation—An Introduction,* 2nd ed., Ulrich's Bookstore, Ann Arbor, MI, 1977.

Seibert, A. F., " Hydrodynamics and Mass Transfer in Spray and Packed Liquid-Liquid Extraction Columns," Ph.D. dissertation, University of Texas at Austin, 1986.

Seibert, A. F. and J. R. Fair, " Hydrodynamics and Mass Transfer in Spray and Packed Liquid-Liquid Extraction Columns," *Industrial and Engineering Chemistry Research,* **27**(3), 470, 1988.

Seibert, A. F. and J. R. Fair, "Mass Transfer Efficiency of a Large-Scale Sieve Tray Extractor," *Industrial & Engineering Chemistry,* **32**(2213), 1993.

Seibert A. F. and J. R. Fair, "Performance of Sieve Trays and Commercial-Scale High Efficiency Packing in Liquid-Liquid Extraction," presented at AIChE Annual Meeting, Miami Beach, FL, November 1995.

Seibert, A. F. and J. L. Humphrey, *Separation Science and Technology,* **30**(7–9), 1139–1155, 1995.

Seibert, A. F. et al., "Performance of a Large-Scale Packed Liquid-Liquid Extractor," *Ind. Eng. Chem. Res.,* **29**(9), 1901, 1990.

Seibert, A. F. et al., "Hydraulics and Mass Transfer Efficiency of a Commercial-Scale Membrane Extractor," *Separation Science and Technology,* **28**(1–3), 343, 1993.

Seibert, A. F. et al., "Liqui-Cel Membrane Contactors for Liquid-Liquid Extraction," *Proceedings of ISEC '96,* University of Melbourne, 1996, pp. 1137–1142.

Shallcross, D. C. et al., eds., *Proceedings of ISEC '96,* Vols. 1 and 2, University of Melbourne, 1996.

Shallcross, D. C. et al., eds., *Proceedings of ISEC '96,* Vol. 2, University of Melbourne, 1996. pp. 967–1024.

Shallcross, D. C. et al., eds., *Proceedings of ISEC '96,* Vol. 2, University of Melbourne, 1996, pp. 1285–1380.

Shallcross, D. C. et al., eds., *Proceedings of ISEC '96,* Vol. 2, University of Melbourne, 1996, pp. 1381–1524.

Shallcross, D. C. et al., eds., *Proceedings of ISEC '96,* Vol. 2, University of Melbourne, 1996, pp. 1525–1560.

Sorensen, J. M., and W. Arlt, *Liquid-Liquid Equilibrium*: Chemistry Data Series, Vol. 5, (Part 1, Binary Systems, 1979; Part 2, Ternary Systems, 1980; Part 3, Ternary and

Quaternary Systems, 1980; Supplement 1, 1987, by Macedo, E. A. and Rasmussen, P.), DECHEMA, Frankfurt.

Steiner, L. et al., "Design Specifications of a Vertical Multistage Mixer-Settler Column," *Proceedings of ISEC '96*, Vol. 2, University of Melbourne, 1996, pp. 1173–1178.

Stichlmair, J., *Chem.-Ing.-Tech.*, **52**, 3, 1980.

Strand, C. P. et al., *AIChE J.*, **8**, 252, 1962.

Streiff, F. A. and S. J. Jancic, *Ger. Chem. Eng.*, **7**(178), 1984.

Todd, D. B. and G. R. Davies, "Proceedings of the International Solvent Extraction Conference, 1974," Society of Chemical Industry, London, 1974.

Treybal, R. E., *Liquid Extraction*, 2nd ed., McGraw-Hill, New York, 1963.

Van Brunt, V., "Extraction," in R. C. Dorf, ed., *The Engineering Handbook*, CRC Press, 1996, pp. 593–603.

Van Dijck, W. J., U. S. Patent 2,011,186 (1935).

Watson, J., "Present and Future Roles of Solvent Extraction in Treatment of Nuclear Wastes," presented at the AIChE Annual Meeting, Miami Beach, FL, November 1995.

Williams, D. F., *Chem. Eng. Sci.*, **36**, 1769, 1981.

Wisniak, J. and A. Tamir, *Liquid-Liquid Equilibrium and Extraction: A Literature Source Book*, (vol. 1, 1980; vol. 2, 1980; Supplement 1, 1987), Elsevier, New York.

Woods, D. R., Chapter 29.1 in T. C. Lo et al., eds., *Handbook of Solvent Extraction*, Wiley, 1983.

4

Adsorption

Introduction

There is a natural tendency for components of a liquid or a gas to collect—often as a monolayer but sometimes as a multilayer—at the surface of a solid material. This phenomenon is called *adsorption*. Adsorption processes consist, first, of the selective concentration (adsorption) of one or more components (called *adsorbates*) of either a gas or a liquid on the surface of a microporous solid (called an *adsorbent*). The adsorbed components are then desorbed in a second step, usually called the regeneration step. In general, these adsorbents will have high surface areas per unit weight—typically greater than 100 sq m/g and sometimes in excess of 1000 sq m/g—to facilitate a reasonable loading of adsorbates onto the adsorbent. The amount of adsorbate adsorbed per unit weight of adsorbent is called the adsorbate loading and will be given the units of weight of adsorbate per unit weight of adsorbent.

This selective adsorption followed by regeneration becomes the basis for a separation by removing more of one component from a gas or liquid than another. In the vast majority of cases, the forces which bind adsorbates to adsorbents are weaker than those of covalent forces which bind atoms together, and this fact allows the adsorption to be reversed by either raising the temperature of the adsorbent or reducing the concentration or partial pressure of the adsorbate. Sometimes both effects are used. A simple process flowsheet is shown in Figure 4.1. This figure shows an overall process consisting of two steps: first, the adsorption step, in which the adsorbate is selectively removed from the feed stream, and second, the regeneration or desorption step, in which the adsorbate is removed from the adsorbent to prepare it once again for the adsorption step.

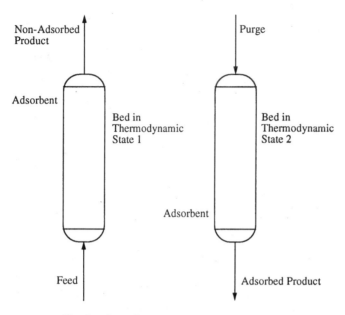

Figure 4.1 Simple adsorption process.

An adsorbent is an example of a mass separating agent (MSA)—an agent which is used to facilitate the separation. In previous chapters we have encountered mass-separating agents in the form of azeotroping and extraction solvents in distillation, as well as solvents in both absorption and extraction processes. All MSAs (except for membranes, as we shall see later) necessarily operate in an unsteady-state manner, that is, they are at one time sorbing materials and at another time being regenerated. In all previous examples of MSA-based processes, the MSA physically moves from one location in the process to another, such as the movement of an absorbing liquid between the absorber and the stripper. In adsorption processes, because the MSA is a solid, the ability to move it from one location to another is severely limited. Therefore, to create the functions of adsorbing and desorbing, we are often limited to changing the thermodynamic conditions which a fixed bed of adsorbent undergoes as a function of time to create both adsorption and desorption. Such unsteady-state operation in turn can create a great deal of process complexity, which can lead to unacceptably high capital costs in some cases.

A number of adsorbent-adsorbate forces can be used to effect selective separation. These are shown in Figure 4.2. From this diagram it can be seen that an optimal range of bond strengths exists. If the

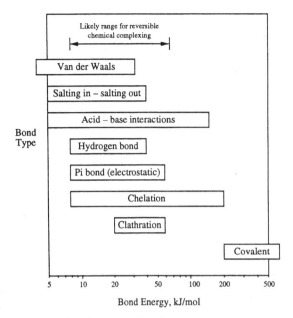

Figure 4.2   Bond energies for adsorption.

bond strength is too weak, then little material will adsorb; conversely, if the bond strength is too strong, desorption will be difficult. By creating a desorption or regeneration step in the process, the adsorbent can be reused many times, and this normally is a highly economically desirable feature. The downside of the need to regenerate the adsorbent is that the overall process is necessarily cyclic in time.

Adsorption processes have a wide range of applicability throughout a wide range of industries. In Table 4.1 are listed representative uses in the chemical, petroleum, and allied industries. In this table the separations are broken down into bulk separations and purifications. This is an important distinction and is caused by the fact that large amounts of heat are liberated when a material adsorbs and are consumed when that material desorbs. For gas adsorptions, this release is usually two to three times the heat of vaporization, and for liquid adsorptions, this release is only a little less. For purifications, this fact is of little consequence, but for bulk separations, the heat release issue becomes an important parameter in process design. This issue will be discussed later in the section on "Design Considerations".

Adsorption processes have a technological kinship with chromatographic processes, in that both use selective concentration at a solid

TABLE 4.1   Some Examples of Commercial Adsorption Separations*

Separation	Adsorbent
Gas bulk separations	
Normal paraffins/isoparaffins, aromatics	Zeolite
N_2/O_2	Zeolite
O_2/N_2	Carbon molecular sieve
CO, CH_4, CO_2, N_2, A, NH_3/H_2	Zeolite, activated carbon
Hydrocarbons/vent streams	Activated carbon
H_2O/ethanol	Zeolite
Chromatographic analytical separations	Wide range of inorganic and polymeric agents
Gas purification	
H_2O/olefin-containing cracked gas, natural gas, synthesis gas, air, etc.	Silica, alumina, zeolite
CO_2/C_2H_4, natural gas, etc.	Zeolite
Hydrocarbons, halogenated materials, solvents/vent streams	Activated carbon, silicalite, others
Sulfur compounds/natural gas, hydrogen, LPG, etc.	Zeolite
SO_2/vent streams	Zeolite
Hg/chlor-alkali cell gas effluent	Zeolite
Indoor air pollutants—VOCs	Activated carbon, silicalite
Tank-vent emissions/air or nitrogen	Activated carbon, silicalite
Odors/air	Silicalite, etc.
Liquid bulk separations	
Normal paraffins/isoparaffins, aromatics	Zeolite
p-Xylene/o-xylene, m-xylene	Zeolite
Detergent-range olefins/paraffins	Zeolite
p-Diethyl benzene/isomer mixture	Zeolite
Fructose/glucose	Zeolite
Chromatographic analytical separations	Wide range of inorganic, polymeric, and affinity agents
Liquid purifications	
H_2O/organics, oxygenated organics, halogenated organics, etc.—dehydration	Silica, alumina, zeolite, corn grits
Organics, oxygenated organics, halogenated organics, etc./H_2O—water purification	Activated carbon, silicate
Odor and taste bodies/H_2O	Activated carbon
Sulfur compounds/organics	Zeolite, others
Decolorizing petroleum fractions, syrups, vegetable oils, etc.	Activated carbon
Various fermentation products/ fermentor effluent	Activated carbon, affinity agents
Drug detoxification in the body	Activated carbon

Abbreviations: LPG, liquified petroleum gas.

*Adsorbates are listed first.

Bulk separations: adsorbate concentration is greater than about 10 wt% in the feed.

Purifications: adsorbate concentration is generally less than about 2 wt% in the feed.

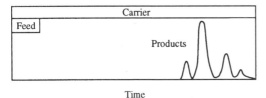

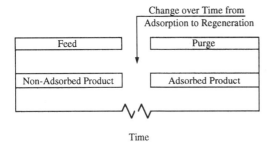

Figure 4.3 Comparison of adsorption and chromatographic processes.

surface as a means of effecting a separation. Simple flowsheets are compared in Figure 4.3. Chromatographic separations are practiced in the laboratory and in relatively small separations associated with the pharmaceutical and biotechnology industries. Truly large-scale chromatographic processes have seldom been commercialized, and in the industries we are dealing with, they are essentially absent. Therefore, in this chapter we will mention them only in passing.

In the sections to follow, we will discuss, first, the nature of adsorption and then describe the range of commercial adsorbents. Then we will discuss adsorption process types, design considerations, various adsorption processes, and finally new directions for the technology.

Adsorption and Desorption

In this section we will deal with the adsorption phenomenon in a rudimentary way. The intent is to show the implications which these phenomena have on the way in which processes are organized and designed.

The Langmuir Isotherm represents a useful way of looking at adsorption on a surface. Consider a surface onto which a certain material can adsorb. For purposes of illustration we will consider a gas in contact with a surface for now. The adsorption will have the following characteristics:

- There is only one material which can adsorb.
- Molecules adsorb at specific sites on the surface and do not move around.
- Molecules adsorb only as a monolayer on the surface.
- All sites bind molecules with the same degree of tenacity.
- There is no interaction among the adsorbed molecules.

If we define p as the partial pressure of the adsorbate in the gas phase and Q as the fraction of sites on the surface which are occupied by adsorbate, then the rate, r_{ads}, of molecules adsorbing per unit area will be equal to a constant times the partial pressure of the adsorbate times the fraction of sites unoccupied, or

$$r_{ads} = k_{ads}\, p\, (1-Q) \tag{4.1}$$

The rate of desorption is assumed to be equal to a constant, k_{des}, times the fraction of sites occupied, or

$$r_{des} = k_{des} Q \tag{4.2}$$

At equilibrium, these two rates are equal, that is,

$$k_{ads}\, p(1-Q) = k_{des} Q \tag{4.3}$$

Solving for Q:

$$Q = k_{ads}\, p / \left[k_{ads}\, p + k_{des} \right] \tag{4.4}$$

If we now define an equilibrium constant $K = k_{ads}/k_{des}$, the expression reduces to:

$$Q = Kp / (Kp + 1) \tag{4.5}$$

This expression is the Langmuir isotherm (Ruthven, 1984), and it is pictured in Figure 4.4. When the quantity K_p is considerably less than 1, then the expression reduces to

$$Q = Kp \tag{4.6}$$

This expression is the same as Henry's law, which describes a linear relationship between the concentrations of two phases in equilibrium. At higher partial pressures, the Langmuir isotherm ceases to become

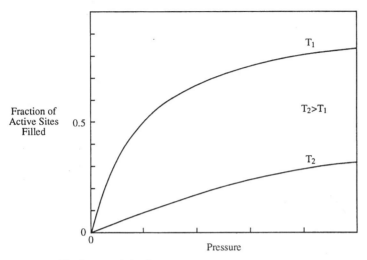

Figure 4.4   The Langmuir isotherm.

linear and eventually reaches an asymptotic value. In this region, Kp is much larger than 1, and

$$Q \rightarrow 1 \tag{4.7}$$

This value corresponds to a state in which all sites contain adsorbate.

For liquid phases in contact with an adsorbent, the Langmuir isotherm becomes

$$Q = Kc/(Kc + 1) \tag{4.8}$$

where c is the concentration of the adsorbate in the liquid phase.

The effect of temperature on a Langmuir isotherm is also shown in Figure 4.4. An inevitable thermodynamic consequence of adsorption's being exothermic is the fact that the adsorbate-adsorbent bond weakens as temperature is increased, that is, the value of K decreases. Thus, for a given partial-pressure or concentration driving force, the amount adsorbed will decline as the temperature is raised. The nature of the Langmuir isotherm does predict that the maximum loading will be the same at all temperatures, but that the approach to the asymptote will be slower at higher temperatures.

The Langmuir isotherm can also be extended to apply to multicomponent mixtures. An extension of the basic derivation gives the following:

$$Q_a = K_a p_a / (K_a p_a + K_b p_b + K_c p_c + \cdots + 1) \tag{4.9}$$

where subscript a refers to component a, b to component b, etc. This expression is also useful in determining the selectivity of the adsorbent. As in distillation, we can define a selectivity between two gas-phase and adsorbed-phase components as follows:

$$a_{ab} = (Q_a/p_a)/(Q_b/p_b) \qquad (4.10)$$

Substituting the Langmuir isotherm for the fractional surface coverages:

$$a_{ab} = K_a/K_b \qquad (4.11)$$

This expression describes a constant selectivity over the range of gas-phase compositions. A similar expression exists for liquid-phase mixtures in contact with an adsorbent. As in distillation, the higher the value of the selectivity, the easier it is to carry out the separation, and in general a_{ab} must be:

- Greater than 2 for a separation to be viable in separations involving pressure changes
- Greater than 5 in separations involving temperature changes

Unfortunately, as in distillation, ideal equilibrium is not all that common in real life. Although a number of important mixtures follow the Langmuir isotherm approximately, there are some systems which vary widely from it. Five basic types of isotherms have been identified and are shown in Figure 4.5. Isotherms II–V result from major violations of the Langmuir assumptions, including multilayer adsorption and interactions between adsorbed molecules. Many models have been formulated (see, e.g., Ruthven, 1984) to describe systems with these isotherms and provide the basis for detailed modeling of various processes. In practice, however, such detailed modeling is almost never carried out for a new-process development unless it is a small extrapolation from an old and well-understood process.

Lest one despair too quickly regarding the complexity of the isotherm picture, it needs to be noted that adsorption processes are almost always carried out at pressures (or concentrations) and tem-

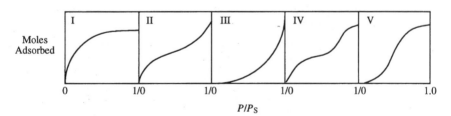

Figure 4.5 The Brunauer classification of isotherms (I–V).

peratures which favor attainment of significant changes in loading with realistic changes in these variables. These operating conditions are found toward the lower-left-hand corner of the isotherm, and in this corner, the vast majority of isotherms exhibit the Langmuir isotherm characteristics of (1) a Henry's law region, followed by (2) a concave downward general shape, and then followed by (3) a region in which the loading approaches an asymptote.

From this discussion we can begin to see how basic adsorption processes can be organized. As we saw earlier, nearly all adsorption processes involve an adsorption step followed by a desorption step. Referring to Figure 4.4, we can see that three options are open to the engineer. First, by changing the partial pressure of a component, that component can first be adsorbed and then desorbed. Second, by adsorbing at a relatively low temperature, much of the adsorbate can be removed by raising the temperature of the adsorbent. And third, another component which competitively adsorbs with the original adsorbate can be added to effect desorption. In later sections of this chapter we will see how these three types of processes are organized and carried out.

An important consideration must be added to this discussion of adsorption. The Langmuir isotherm, as presented, was derived on the basis of a unit area of adsorbent. That being the case, it is clear that the weight or molar loading of an adsorbate is highly dependent on the amount of surface area per unit weight of adsorbent. Since in most cases a loading as high as possible of adsorbate is desirable for a given partial pressure or concentration of adsorbate in the fluid phase, then a high priority must be placed on creating as high a surface area per unit weight of adsorbent as possible. Basically, any atoms of adsorbent which are not at an interface are nonfunctional, except as they are needed for structural strength of the particle or perhaps for a heat source or sink during the process's cycle. To achieve adsorbate loadings of a reasonable value, adsorbent areas, as mentioned earlier, are typically in the order of a few hundred to a few thousand square meters per gram. At the upper end of this range, this surface area amounts to more than the area of a baseball diamond per gram of adsorbent, and nearly half a square mile per kilogram! With such surface areas, practical loadings of a few wt% up to 20 or more wt% can usually be achieved.

A second consideration must be dealt with in creating a high surface area per unit weight: the surface area must be truly accessible to the adsorbate. Thus micropores can be created which are so small that various molecules cannot diffuse into them, and as a result these pores are useless for providing capacity for the larger molecules. In the case of molecular sieves, we will see that this feature can sometimes be used in a positive way to create molecular geometry–based selectivities which cannot otherwise be achieved. But with other adsorbents, the emphasis is almost always on the creation of easily accessible surface areas.

With just a few exceptions, this surface area must also be accessible in such a way that phase equilibrium can be achieved in a reasonable length of time, for example, a few seconds to a very few minutes. This means that, in addition to creating the largest possible pore diameters, commensurate with achieving high surface areas, intraparticle diffusion paths must be kept as short as possible. It is also possible to create particles with a multiplicity of small, high-surface-area pores connected by large-diameter pores to provide avenues of easy access to the small pores. In general, though, effecting rapid intraparticle diffusion is a major challenge for adsorbent manufacturers. This often leads to the use of adsorbent particles which are as small as possible—consistent with other considerations such as fluid pressure-drop considerations and maintenance of desired flow patterns through the adsorbent bed. This intraparticle diffusion problem is especially prevalent when the feed is a liquid, and in such cases even powder-size particles are sometimes used. A substantial amount of creativity for overcoming the intraparticle diffusion problem has been exerted in recent years, and we will see some fruits of this creativity in the section on "Design Considerations."

Polymer-based "adsorbents" are different from the others in that a substantial fraction of their capacity can come from the dissolution of the adsorbate into the polymer. Thus with these "adsorbents," the equilibrium achieved is both between the bulk phase of the polymer—plus the polymer surface—and the fluid phase. Despite this difference, we will refer to them as adsorbents because they are used in essentially the same way as true adsorbents in the processes we will be discussing. Intraparticle diffusion problems with polymer particles are at least as severe as with regular adsorbents since molecular-diffusion coefficients in the polymers are so low. This creates the need for short diffusion paths in the polymer, and this need is usually satisfied by using particles with as small a diameter and as much internal surface area per unit weight as possible. Internal surface areas can rival those for other adsorbents.

Adsorbents

Adsorbents most often effect separations by chemically binding one component more strongly than another. The weak forces employed in adsorption processes and the approximate values for reversibility were shown in Figure 4.2. However, two other important separating mechanisms can also be employed: exclusion of certain molecules (but not others) in the feed because they are too large to fit into the pores of the adsorbent (the molecular sieving effect), and differences in the diffusion rates of different adsorbing species in the pores of the adsorbent.

One surprising aspect of adsorbent technology is that only four types dominate in usage: activated carbon, molecular sieve zeolites (MSZs), silica gel, and activated alumina. Estimates of worldwide sales are (Humphrey, 1996; Keller, 1995):

Activated carbon	$1 billion
Molecular sieve zeolites	$100 million
Silica gel	$27 million
Activated alumina	$26 million

Among these four types, activated carbon is the workhorse for removing relatively hydrophobic species out of streams. Thus, activated carbon is often used for removing organic species from both gas and liquid (usually aqueous) streams. In these applications, the carbon may be in the form of a fine powder (typically for aqueous streams) or as hard pellets, granules, cylinders, spheres, or fibers. A small but rapidly growing market exists for so-called monolithic structures (see "Design Considerations"). Typical activated carbon properties are given in Table 4.2.

TABLE 4.2   Typical Properties of Activated Carbon Adsorbents

	Liquid-phase carbons		Vapor-phase carbons	
Physical properties	Wood base	Coal base	Granular coal	Granular coal
Mesh size (Tyler)	−100	−8 + 30	−4 + 10	−6 + 14
CCl_4 activity, %	40	50	60	60
Iodine number	700	950	1000	1000
Bulk density, kg/m^3	250	500	500	530
Ash, %	7	8	8	4

Adsorptive properties	Vapor-phase carbons, wt%
H_2O capacity at 4.6 mm Hg, 25°C	1
H_2O capacity at 250 mm Hg, 25°C	5–7
n-C_4 capacity at 4.6 mm Hg, 25°C	25

A material called carbon molecular sieve (CMS) (Knoblauch, 1978), which exhibits a very narrow pore size distribution compared to regular activated carbons, has also been commercialized. Although the volume of sales of CMSs is minuscule compared to that of regular activated carbons, nevertheless in some cases CMS facilitates separations based on different intraparticle diffusion rates, in spite of the fact that the adsorbent surface may show essentially no selectivity for one material over another. The efficient separation of air to recover nitrogen has provided a secure and somewhat growing market for CMS.

Molecular sieve zeolites (MSZs) (Breck, 1974; Breck and Anderson, 1982) are highly crystalline aluminosilicate structures. Typical geometries are pictured in Figure 4.6, and typical physical properties

Zeolite A

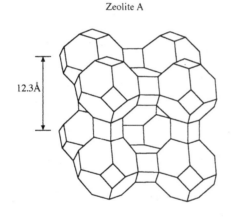

12.3Å

Zeolite B

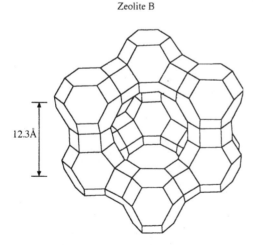

12.3Å

Figure 4.6 Schematic diagrams of framework structures of two common zeolites.

TABLE 4.3   Major Commercial Molecular Sieve Adsorbent Products* and Some Physical Properties

Zeolite type	Designation	Cation	Pore size, nm	Bulk density, kg/m³
A	3A	K	0.3	670–740
	4A	Na	0.4	660–720
	5A	Ca	0.5	670–720
X	13X	Na	0.8	610–710
Mordenite, small port	AW-300 Zeolon-300	Na + mixed cations	0.3–0.4	720–800
Chabazite	AW-300	Mixed cations	0.4–0.5	640–720

Typical Adsorption Data on Molecular Sieve Adsorbent Products (Capacity Expressed in wt% on Basis of Activated Pellet)

Designation	Type	O_2 at 100 mm Hg, $-183°C$	H_2O at 4.6 mm Hg, 25°C	CO_2 at 250 mm Hg, 25°C	n-C_4H_{10} at 250 mm Hg, 25°C
3A	LiA	NA	20	NA	NA
4A	NaA	22	23	13	NA
5A	CaA	22	21	15	10
13X	NaX	24	25	16	12
AW-300	Mordenite	7	9	6	NA
AW-500	Chabazite	17	16	12	NA

Abbreviations: NA, not adsorbed.

*All products available in pellets, beads, and granulated mesh in different size ranges. They are supplied activated with ~1.5% by weight residual water.

are given in Table 4.3. A host of varying structures can be produced by varying the silicon to aluminum atom ratio in the feed solution and the crystallization and drying conditions. In addition, other cations such as sodium, potassium, lithium, and calcium must be added to the solution to facilitate charge neutrality in the final adsorbent product. Other cations may also be substituted for these cations by ion exchange in a subsequent step. These new cations can be used in some cases to modify the adsorption characteristics of the MSZ. As can be seen from the figure, zeolites consist of highly regular channels and cages. MSZs are selective for polar, hydrophilic species, and tenacious bonds are created with water, carbon dioxide, hydrogen sulfide, and other such species, while weaker bonds are created with organic species. Among organic species, polarity can be an important consideration in separating two materials. Molecular sieving is possible when a channel size is larger than one material and smaller than another. In Figure 4.7, a range of molecular sizes can be seen as well as channel diameters of various MSZs. From this figure one could speculate that LiA (also called 3A) could be used to remove water from air streams, since water could penetrate the zeolite, while oxygen and nitrogen could not.

There is a special and relatively new type of molecular sieve which consists almost entirely of silica with virtually no aluminum or other cations present (Flanigen et al., 1978; Shultz-Sibbel et al., 1982). Physical properties are given in Table 4.4. In contrast to MSZs, these zeolites are hydrophobic, and their separation pattern more closely resembles those of activated carbons rather than MSZs (Figure 4.8). Though more costly than activated carbons, hydrophobic zeolites, because of their extremely high thermal stability, can be regenerated under much more severe conditions than can activated carbons. This feature becomes important when fouling of the adsorbent surface with polymers or other difficult removable materials is a problem. Additionally, they can be operated in the gas phase at higher relative humidities than can activated carbon.

Silica gels represent an intermediate stance between highly hydrophilic and highly hydrophobic surfaces. Most often these adsorbents are used for removing water from various gases, but hydrocar-

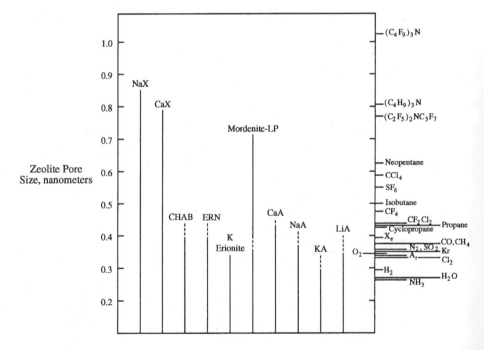

Figure 4.7 Comparison of molecular dimensions and zeolite pore sizes.

TABLE 4.4   Properties of Silicalite Adsorbent

Chemical composition: SiO_2

Adsorption pore diameter: approximately 6 Å (0.6 nm)

Crystal porosity: 33%

Adsorption characteristics at room temperature		
Adsorbate	Kinetic diameter, Å	Molecules adsorbed/unit cell
H_2O	2.65	15.1
CH_3OH	3.8	38.0
n-Butane	4.3	27.6
n-Hexane	4.3	10.9
C_6H_6	5.85	8.7
Neopentane	6.2	1.4

SOURCE: Abstracted from Flanigen et al., 1978.

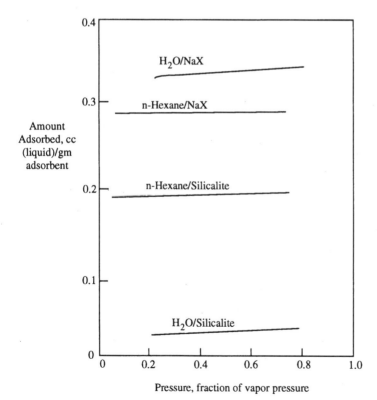

Figure 4.8   Comparison of adsorption isotherms for silicalite and sodium X. (*Adapted from Flanigan et al., 1978.*)

TABLE 4.5   Typical Properties of Adsorbent-Grade Silica Gel

	Physical properties
Surface area, m^2/g	830
Density, kg/m^3	720
Reactivation temperature, °C	130–280
Pore volume, % of total	50–55
Pore size, nm	1–40
Pore volume, cm^3/g	0.42

Adsorption properties	Percent by weight
H_2O capacity at 4.6 mm Hg, 25°C	11
H_2O capacity at 17.5 mm Hg, 25°C	35
O_2 capacity at 100 mm Hg, -183°C	22
CO_2 capacity at 250 mm Hg, 25°C	3
n-C_4 capacity at 250 mm Hg, 25°C	17

bon separations are also sometimes feasible. Typical physical properties are shown in Table 4.5.

Activated aluminas have quite a high affinity for water and are often used in drying applications for various gases. Physical properties are shown in Table 4.6. Like silica gels, the water bond with the alumina surface is not as tenacious as those of MSZs, so that regeneration of aluminas can often be accomplished at somewhat milder tem-

TABLE 4.6   Typical Properties of Adsorbent-Grade Activated Alumina

	Physical properties
Surface area, m^2/g	320
Density, kg/m^3	800
Reactivation temperature, °C	150–315
Pore volume, % of total	50
Pore size, nm	1–7.5
Pore volume, cm^3/g	0.40

Adsorption properties	Percent by weight
H_2O capacity at 4.6 mm Hg, 25°C	7
H_2O capacity at 17.5 mm Hg, 25°C	16
CO_2 capacity at 250 mm Hg, 25°C	2

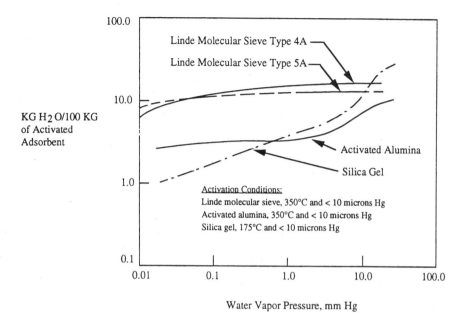

KG H₂O/100 KG of Activated Adsorbent

Water Vapor Pressure, mm Hg

Figure 4.9 Comparison of water adsorption on various adsorbents.

peratures than for MSZs. As can be seen in Figure 4.9, activated alumina can have a higher ultimate capacity for water than a typical MSZ, but the MSZ will have a higher capacity at low partial pressures of water. Thus MSZs are typically chosen if very high removals of water would be necessary, while activated alumina would be chosen if adsorbent capacity were more important than ultrahigh removal. It is also possible to mix two adsorbents or to place zones of them in series. For example, a zone of alumina to remove the bulk of the water followed by a zone of MSZ to effect very high removal might prove useful in some cases.

Relatively recently a series of "irreversible" or highly reactive adsorbents has been marketed by ICI Katalco (1993). These materials, which will be discussed in more detail later, can react tenaciously with various components in a gas or liquid stream, for example, hydrogen sulfide in a hydrocarbon, to effect virtually complete removal. Because the bond between the adsorbate and this type of adsorbent is much stronger than those of other adsorbents (Figure 4.2), on-site regeneration is not possible, and the loaded adsorbent must be returned to the manufacturer for reconstitution or regeneration. Obviously such an operation would only make sense when the adsorbate is present in minute quantities—parts per million—in the feed, because frequent replacements of the adsorbent bed, such as every few days, would become prohibitively expensive. More quantita-

TABLE 4.7   Impurities Which Can Be Removed from Gas Streams by Various
Irreversible Adsorbents

- Sulfur compounds: H_2S, COS, SO_2, organic sulfur compounds
- Halogen compounds: HF, HCl, Cl_2, organochlorides
- Organometallics: AsH_3, $As(CH_3)_3$
- Mercury and its compounds, metal carbonyls
- Nitrogen compounds: NO_x, HCN, NH_3, organonitrogen compounds
- Unsaturated hydrocarbons: olefins, di-olefins, acetylenes
- Oxygenates: O_2, H_2O, methanol, carbonyls, organic acids
- Miscellaneous: H_2, CO, CO_2

SOURCE: ICI Katalco, 1993, literature.

tive information will be given later in the chapter. Some applications
for irreversible adsorbents are given in Table 4.7.

Biosorbents (Fouhy, 1992; Togna et al., 1993) are another example
of a reactive adsorbent. Biosorbents first sorb materials such as
organic molecules and then oxidize them to carbon dioxide, water, and
other species such as hydrogen chloride if other atoms besides carbon,
hydrogen, and oxygen exist in the original molecules. In fact, the biomass in biotreatment ponds that treat municipal and industrial
waste streams can be thought of as a biosorbent. This same biomass
can also be mounted on a porous or high-surface-area substrate such
as wood or small beads and then be used as a reactive adsorbent for
treating gas streams containing organic species.

Polymer-based adsorbents have recently become a prime area of
interest. They are presently being used in a few operations removing
organic constituents of vent gas streams, such as the removal of acetone from air, as well as in a few other cases. These materials are
usually styrene-divinylbenzene copolymers which in some cases have
been derivatized to give the adsorption properties desired. In this
respect, such polymers are close in composition to typical ion
exchange resins. Macroreticular forms of these resins, which give surface areas approaching those of inorganic adsorbents, are the logical
choices for adsorption duty. Other polymers are also under investigation. Little is known quantitatively about the details of the polymer-
based adsorbents now being used commercially, however.

Affinity adsorbents are extremely selective (and extremely high-
priced) materials for use in recovering specific biomaterials or organic
molecules out of a complex mixture of organic molecules. These adsorbents consist of active centers which react reversibly with several
centers on the molecule of interest. The selectivity derives from the

fact that the active centers of the adsorbing site line up geometrically with the active sites on the molecule. Affinity adsorbents are not economically practical for use except in the recovery of very high-priced pharmaceuticals and biomaterials.

Basic Adsorption Processes

In the previous section we learned that there are three basic ways to influence the loading of an adsorbate on an adsorbent by:

- Changing pressure in a gas or concentration in a liquid
- Changing temperature
- Adding a component which competitively adsorbs with the adsorbate of interest

Certain other variables such as electric charge on the adsorbent (Eisinger and Keller, 1990) and pH can also be used in a few cases, but for all intents and purposes, pressure (or concentration), temperature, and competitive adsorption are the only three means worth considering for virtually all industrial adsorption-based separations. Ion exchange processes can also use pH as a means of removing species from streams, but these separations will not be covered here. Therefore, in this section we will discuss what the three basic adsorption processes look like and comment briefly on the other processes.

Prior to this discussion, though, we must deal with the issue of the large heat effects associated with adsorption (exothermic) and desorption (endothermic). When a molecule adsorbs from the gas phase, the heat it liberates is perhaps two to three times larger than the heat of vaporization, while a molecule adsorbing from the liquid phase liberates an amount of heat nearly as great. There are only two ways in which the heat of adsorption can be dissipated: by heating up the adsorbent (and the vessel containing it) or by heating up the fluid exiting from the vessel. When the concentration of adsorbate in the feed is low, say only a very few wt% or less, then a large fraction of the heat of adsorption passes out with the effluent stream, and the adsorbent remains close to the temperature of the incoming fluid. We shall call such a separation purification. When the concentration of adsorbate is high, say 10 wt% or more (we will call this a bulk separation), a large fraction of the heat will be trapped in the bed and cause large temperature increases with time. These increases in turn limit the amount of material which can be adsorbed and cause lowered capacity of the process. Dealing with the heat release problem is a major consideration in determining which type of regeneration cycle should be used for performing a given separation. Feeds whose adsorbate concentrations fall between the bulk separation and purification ranges constitute a gray area, and it is more difficult to make hard-and-fast conclusions as to which process is the best for these separations. Detailed calculations will probably have to be done to help in the selection, and a simple method of making an initial determination is given in the following section.

A major difference in processes based on temperature change and those based on pressure or concentration change is the time required to change the bed from an adsorbing to a desorbing or regenerating condition. In short, pressure and concentration can be changed much more rapidly than can temperature. In Figure 4.10 are given typical ranges of process cycle times for the various processes. What we can conclude, even before looking at the processes in detail, is that temperature-swing processes will be limited almost exclusively to situations in which the adsorbate concentration is quite low in the feed, that is, purifications. This is because, for higher concentrations of adsorbate in the feed, the adsorption time would become quite short compared to the regeneration time, and the bed would be unavailable for adsorption during a very large fraction of the cycle time. In addition, if bulk separations were attempted, then large exotherms would occur during the adsorption cycle, and the bed capacity would be severely limited. On the other hand, pressure-swing or concentration-swing processes can accept a short period of feed, change to a regeneration condition, and be back to the feed condition in such a way that a reasonable fraction of the cycle—say one-fourth to one-half or more—can be devoted to the feed condition. Such a situation makes pressure-swing and concentration-swing processes much more suitable for performing bulk separations.

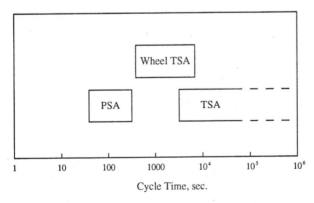

Figure 4.10 Comparison of cycle times for adsorption processes.

Temperature-swing cycles

A simple schematic drawing of a temperature-swing cycle is shown in Figure 4.11. In this process, the heat of adsorption is mostly carried out of the bed with the effluent, and in such a situation the bed approaches isothermal operation. A very important characteristic of temperature-swing-adsorption (TSA) processes is that they are used virtually exclusively for treating feeds with low concentrations of adsorbates. Typically adsorbate concentrations in the feed are less than a very few wt%. TSA processes can be designed to accept either liquid or gas feeds. We will first concentrate on cycles in which gases are fed and then make a few comments later to cover liquid feeds. During the feed period, a gas stream with an adsorbate partial pressure p_a enters the bed. The adsorbate is removed and builds to an equilibrium value of X on the adsorbent. This loading is typically expressed in terms of weight of adsorbate per unit weight of adsorbent. In an ideal case, we assume that there is no adsorbate on the adsorbent at the start of the feed period and that all adsorbent contains the equilibrium amount of adsorbate. Given these assumptions, then the length of the feed step, called the ideal feed time, can be calculated as follows:

$$\text{Ideal feed time} = WX/FxM \qquad (4.12)$$

where W = weight of adsorbent
$\quad\ \ X$ = equilibrium weight of adsorbate/unit weight of adsorbent
$\quad\ \ F$ = moles of feed/unit time
$\quad\ \ x$ = mole fraction of adsorbate in the feed
$\quad\ \ M$ = molecular weight of adsorbate

In practice, neither assumption is followed, and the practical time is usually 10–50% less than the time calculated here. One modification is to use ΔX, the "delta loading" or the difference in loading between the equilibrium loading and the residual loading at the end of the regeneration step. A second modification, which corrects for the fact that mass transfer rates are not infinitely great and that there can be deviations from plug flow, is to assume that part of the bed has essentially no adsorption taking place on it. The net effect of this assumption is to reduce the estimate of the average loading of the entire bed. The time based on these two modifications will be the true value. These points will be discussed in "Design Considerations."

Several options are available for the regeneration step. The regeneration gas could either be an inert and relatively nonadsorbing gas, or steam or a fuel gas or even occasionally the feed gas itself. In most cases the bed is heated by the regeneration gas, which—again in most cases—is fed countercurrently to the direction of the feed. Heating the bed directly through the wall with electrical or steam heating is

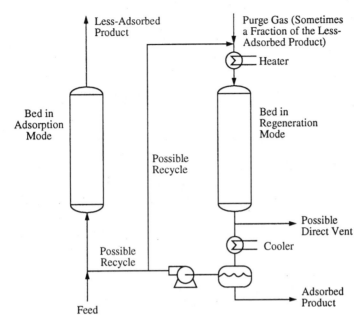

Figure 4.11   Combined temperature-swing and inert-purge cycle.

generally not practical, especially if the bed diameter is greater than a very few feet, because the rate of heating which can be accomplished by this method is very slow. The adsorbate will be removed by virtue of the higher temperature (see Figure 4.4 for the effect of temperature on the adsorption isotherm) and the reduced partial pressure of adsorbate caused by the diluting effect of the regeneration gas. At the end of the regeneration step, typically a cool gas is fed for a brief period to return the bed to the feed temperature.

As shown in Figure 4.11, if it is desirable to recover the adsorbate from the regeneration gas, then this can be done by cooling this stream to condense the adsorbate. The use of steam as the regeneration gas can facilitate condensation of the desorbed product by reducing the gas volume to near zero. The downside of using steam, of course, is that the adsorbate may have to be separated from the water phase. This becomes especially costly if the desorbed product is miscible with water.

For a process involving a liquid feed, the bed is typically dry or nearly so at the start of the feed step. Liquid is then introduced, typically in an upflow. At the end of the feed step, the bed is typically drained, and then steam or a hot gas is fed to regenerate the bed.

An obvious complication of the temperature-swing process is that there is a need for more than one bed if the feed stream is truly continuous. This arises from the fact that a bed requires a finite time to regenerate, and that during this time feed must still be processed.

In those cases in which a bed can be on-line for a few months or more before having to be regenerated, it may be preferable to have no regeneration facility at all on site. Instead, the adsorbent supply company can often supply a service which consists of dumping the bed, recharging it with fresh adsorbent, and carrying off the loaded adsorbent to a central regeneration facility. Such an arrangement can substantially reduce the investment required for the process. Extra operating costs are incurred, however, by virtue of the fact that the adsorbent supply company will charge a fee for their services and for providing the fresh adsorbent. An important limitation on this method of regeneration occurs in cases in which the adsorbate is considered to be a toxic material. In such cases the adsorbent loaded with adsorbate will be considered to be a hazardous waste, and transportation off-site to a regeneration facility may become prohibitively expensive.

Pressure-swing cycles

Pressure-swing adsorption (PSA) is a popular process for performing bulk separations of gases. Also, this process can sometimes be used on streams containing very low concentrations of adsorbates, such as can be found in the removal of small amounts of volatile organic compounds (VOCs) from vent streams. PSA uses both changes in total pressure and in gas composition to effect the adsorption and desorption steps. For example, reducing both of these variables during desorption, which is typical for a PSA process, causes desorption of the adsorbate. There is no PSA process for liquid systems, since total pressure changes, at least in the range of industrially realistic pressures, do not cause measurable changes in equilibrium loadings.

The feed pressure in PSA cycles is typically atmospheric pressure or higher, while the regeneration pressure can be above or below atmospheric pressure. In cases in which the feed pressure is near atmospheric and the regeneration pressure is below atmospheric, the process is often referred to as vacuum-swing adsorption (VSA). In those processes with feed pressures above atmospheric and regeneration pressures below atmospheric, the process is often referred to as vacuum-pressure-swing adsorption (VPSA). In our discussion here, however, we will in general refer to all such processes as simply PSA.

In the basic PSA process, feed is introduced to a bed at some total pressure P_f, and after a certain amount of adsorbate is adsorbed, the partial pressure of the adsorbate is reduced to a value of less than that in the feed, either by reducing the total pressure of the gas passing into the bed or by feeding an inert gas containing little or no adsorbate, or by both means. The original forms of the process were

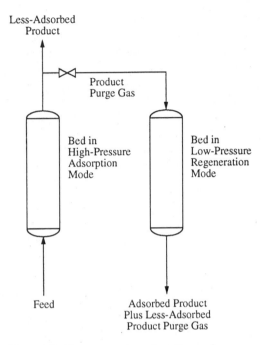

Figure 4.12   Pressure-swing adsorption cycle.

enunciated by Guerin de Montgareuil and Domine (1957) and Skarstrom (1958). The Skarstrom process is shown in Figure 4.12. In this process, the less-adsorbed gas passes through the bed to produce a more-or-less pure product. A fraction of this product stream is reduced in pressure and fed countercurrently to a second bed, operating at a lower total pressure of P_r, to help purge adsorbate from this bed. Thus, in this second bed, adsorbate is removed by a combination of reduced total pressure and the use of an essentially adsorbate-free purge stream. At the end of the feed period of the first bed, the two beds are reversed in function. Feed is introduced to the purged bed to increase its pressure and begin the adsorption step, while at the same time, the bed full of adsorbate is reduced in pressure and purged with part of the product from the other bed.

Note, however, that purging one bed with a fraction of the nonadsorbed product from the other has two important implications. First, the recovery of nonadsorbed product will be less than quantitative and will typically be in the order of 50–95%. In general, the greater the ratio of the feed pressure to the desorption pressure, the greater will be the percent recovery of the nonadsorbed product. And second, the product stream containing the adsorbate will be contaminated with nonadsorbed product. Thus, in general only one pure product can be recovered from a PSA process. These two limitations place a heavy restriction on the use of PSA and cause it to be limited to a relatively small number of separations such as air separation (where the raw material is free and on occasion only one pure product is desired) and hydrogen recovery (where some of the hydrogen can be lost to the purge stream, which is often used as a fuel gas). We will deal more with this in "Adsorption Processes."

It is often important to minimize the amount of nonadsorbed gas lost to the purge stream and to minimize the energy lost as the nonadsorbed gas used for purging is reduced in pressure and as the purged bed is increased in pressure for the feed step. This can be done by using more beds in parallel and by using more complex flows of gas between beds. A four-bed process for purifying hydrogen is shown in Figure 4.13, along with its sequence of flows. One sees in this

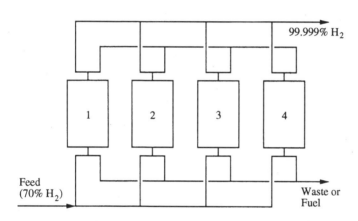

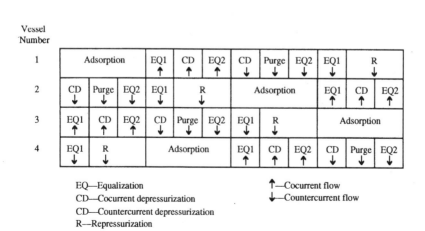

Figure 4.13   Four-bed PSA process.

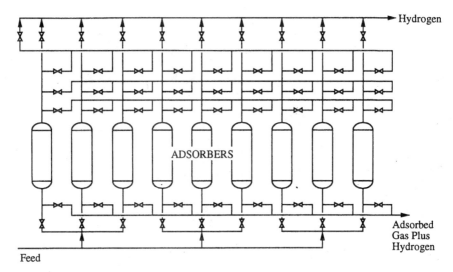

Figure 4.14 Polybed™ PSA process for hydrogen recovery.

process the use of pressure equalization steps in which the effluent from one bed at high pressure is used to repressurize another bed. Also, a complex arrangement of purge steps is used. An even more complex process flowsheet is shown in Figure 4.14, in which nine beds are employed. The whys and wherefores of multiple beds and complex flows seem far from clear at first glance. Table 4.8 can aid in understanding the functions of the various steps and aid in making sense of what seems to be a chaotic situation.

Obviously there is an economic tradeoff between the capital required to build in extra process complexity and the cost of losing more nonadsorbed product if that complexity is not present. Generally in air separation processes, 2–4 beds are used, while in hydrogen-recovery processes, 4 to perhaps 15 beds are used.

We learned that, in TSA processes, which involve trace separations, the heat of adsorption passes out with the nonadsorbed product, and that the bed temperature changes very little during the feed cycle. As a result, large amounts of adsorbate can be deposited in the bed before the bed is regenerated, if the adsorption isotherm permits. However, in PSA processes used for bulk separations, the heat of adsorption is stored in the bed during adsorption, which allows this heat to be available to effect desorption. To minimize the temperature rise, the delta loading is usually only a few percent of the difference in the equilibrium loadings between feed and regeneration conditions. Furthermore, significant adsorption and desorption occur usually in a narrow zone in the center of the bed, as shown in Figure 4.15. On

TABLE 4.8   Summary of the Elementary Steps Used in PSA Cycles

Elementary step	Mode of operation	Principal features
Pressurization	1. Pressurization with feed from the feed end	Enrichment of the less selectively adsorbed species in the gas phase at the product end
	2. Pressurization with raffinate product from the product end prior to feed pressurization	Sharpens the concentration front, which improves the purity and recovery of raffinate product
High-pressure adsorption	1. Product (raffinate) withdrawal at constant column pressure	Raffinate product is delivered at high pressure
	2. The column pressure is allowed to decrease while the raffinate product is drawn from the product end	Very high recovery of the less selectively adsorbed species may be achieved, but the product is delivered at low pressure
Blowdown	1. Countercurrent blowdown to a low pressure	Used when only raffinate product is required at high purity; prevents contamination of the product end with more strongly adsorbed species
	2. Countercurrent blowdown to an intermediate pressure prior to countercurrent blowdown	Used when extract product is also required in high purity; improves extract product purity and may also increase raffinate recovery
Desorption at low pressure	1. Countercurrent desorption with product purge	Improves raffinate product at the expense of decrease in recovery; purge at subatmospheric pressure reduces raffinate product loss but increases energy cost
	2. Countercurrent desorption without external purge	Recovery enhancement while maintaining high product purity is possible only in certain kinetic separation
	3. Evacuation	High purity of both extract and raffinate products; advantageous over product purge when the adsorbed phase is very strongly held
Pressure equalization	The high- and low-pressure beds are either connected through their product ends or the feed and product ends of the high-pressure beds are connected to the respective ends of the low-pressure bed	Conserves energy and separative work
Rinse	The bed is purged with the preferentially adsorbed species after high-pressure adsorption at feed pressure in the direction of the feed	Improves extract product purity when the lighter species are coadsorbed in large amount with heavier components

SOURCE: Ruthven et al, 1994. Used with permission.

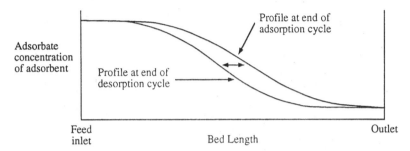

Figure 4.15 Movement of concentration profile in PSA operation.

either side of this zone, only very small amounts of adsorption and desorption take place. It is only because PSA cycle times can be made quite short (see Figure 4.10) that the process can be economical at all. This is because the amount of adsorbent required to remove a given amount of adsorbate per unit of time is equal to the amount removed per cycle times the number of cycles per unit time. Obviously the more rapid the cycle, the more often the adsorbent is used per unit of time, and less adsorbent is required.

Displacement purge cycles

Displacement-purge-adsorption (DPA) processes are processes based on concentration-swing regeneration cycles. They are used both with liquid feeds and with gas feeds, though such cycles are more common with the former. These processes are most often used for performing bulk separations. DPA processes, instead of using an inert material—something which does not adsorb to any significant extent—as the purging medium, a material called the desorbent, which adsorbs competitively with the adsorbate, is used. The desorbent thus displaces the adsorbate on the adsorbent. There are both positive and negative aspects to the use of a competitively adsorbing desorbent. On the positive side, the adsorbate can be more easily removed during the regeneration step since the desorbent can compete effectively for adsorption sites. This advantage is especially important in separating high-molecular-weight and other materials which bind relatively tenaciously to the adsorbent. In addition, the heat release during adsorption of the adsorbate (as well as the heat requirement during desorption) is essentially negligible, since as one material desorbs, another adsorbs. This means that large loadings can be effected in DPA processes without having to be concerned about large temperature changes in the bed. On the negative side, the desorbent becomes inevitably mixed in with both of the two product streams and must be separated from them in additional steps. Since distillation is often the process of choice for making these separations, this fact puts a requirement on the desorbent that it should have a relative volatility of at least two in both separations for the overall process to have a chance of being economically viable.

A schematic flowsheet for separating gas mixtures is shown in Figure 4.16. The bed undergoing adsorption is preloaded with desorbent *D*, which is displaced from the bed as feed is introduced. In a second bed, desorbent is added to remove the adsorbate. Two distillation columns are required to separate the products from the desorbent and return it to the process. This flowsheet should make it abundantly clear that, if two materials can be separated easily by distillation or another process, DPA processes will not be able to compete economically. There is just too much process complexity in these processes.

DPA processes for bulk liquid separations are even more complicated. Because of the much higher density of a liquid feed compared to a gas feed, in some cases no more than one bed volume of liquid feed can be passed through a bed before the adsorbent is completely saturated in adsorbate. Such a process, involving awkwardly short cycles, would be very difficult to actualize commercially. One way to deal with this situation might be to devise a process in which the adsorbent would move through an adsorbing zone, then to a regeneration zone and then back to the adsorbing zone. Such a process, shown in Figure 4.17, would resemble, for example, a countercur-

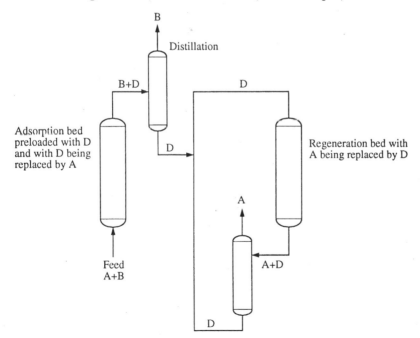

Figure 4.16 Displacement-purge cycle for gases.

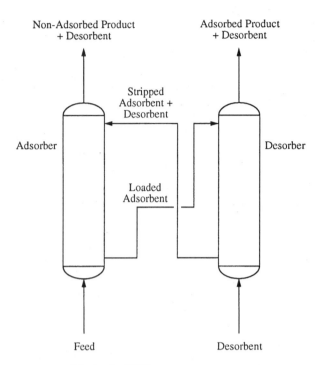

Figure 4.17   Moving-bed DPA process.

rent solvent-extraction process. Unfortunately, adsorption processes involving moving solids have had a very difficult and unsatisfactory history, and the number of commercially successful processes of this sort on liquid systems is vanishingly small. The problems include attrition of the adsorbent particles, erosion of equipment, and the difficulty of achieving plug flow of the solid and liquid phases. An alternative to moving solids is to maintain the solids in a fixed bed and move the feed and product removal points as a function of time. Such a scheme, shown in Figure 4.18, has been commercialized by UOP under the name Sorbex Simulated Moving-Bed technology. The heart of this process is a large, complex, rotary valve which shunts four streams to various points in the bed in a programmed manner.

Actually, a simulated moving-bed (SMB) adsorption process need not include a rotary valve, and other versions of the process include multiple, separate beds connected by lines, with additional lines between the beds leading to the feed, desorbent and product tanks. These lines have control valves on them so that the positions of various inlet and outlet flows can be programmed as a function of time. A schematic diagram of an SMB process is given in Figure 4.19. As a

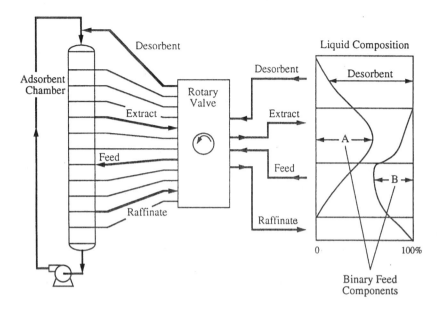

Figure 4.18   Sorbex Simulated Moving Bed for adsorptive separation.

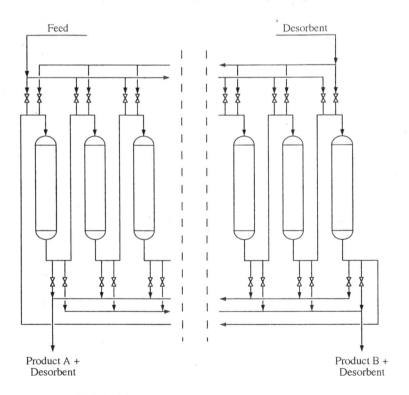

Figure 4.19   Multibed SMB process.

general rule, the rotary valve process of Figure 4.18 is favored economically when feed flows of over perhaps 50 million lb/yr are being
processed, while lesser feed flows—including those down to pharma-
ceutical-type volumes—are more economically handled in the multi-
ple-bed process of Figure 4.19.

Other cycles

What we have seen in the above sections are processes which use
changes in temperature, concentration, pressure, or the use of a
competitively adsorbing component to effect, first, adsorption and
then desorption. Virtually all adsorption processes are of this type.
But presumably any other variable which could effect changes in the
shape of an adsorption isotherm could also be used. One such variable is pH. The bonding between some adsorbents and adsorbates
such as amino acids in water can be changed remarkably as the pH
of stream is changed from above the isoelectric point (the pH at
which the molecule has zero charge) of the amino acid to below its
isoelectric point. Examples of materials other than amino acids
showing a pH effect on their isotherms have also been reported. The
economic problem with using pH swing as a means to drive a cyclic
process is the cost of the acid and base required to change the pH, as
well as the cost of disposal of the salt by-product. These considerations have limited the use of pH-swing processes to separating relatively small amounts of relatively high-priced materials such as
amino acids, pharmaceuticals, bioproducts, and the like. Therefore,
such processes are not likely candidates for separations in petro-
chemical and allied plants.

Another means for changing the shape of the adsorption isotherm
is the use of electric charge. Electrosorption involves adsorption when
the adsorbent is subjected to one voltage and desorption when the
voltage is changed. Typically the voltage change can be small, such as
1 V or less. An isotherm for ethylenediamine as a function of applied
voltage is shown in Figure 4.20. This process can only be accomplished in cases in which both the adsorbent and the feed stream are
highly conductive. Although process details have been reasonably
well documented (Eisinger and Keller, 1990), the economics do not
appear to be outstanding, and as a result there are no known cases of
commercialization of this technology.

It is possible that yet other means of effecting changes in adsorption isotherms are known or will be discovered. It is unlikely, however, that such means will be economically viable as bases for large,
adsorption-based separations.

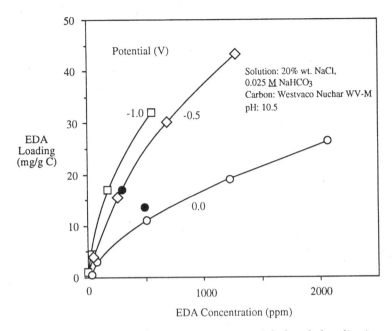

Figure 4.20 Adsorption isotherms at various potentials for ethylenediamine on carbon. (*Source: Eisinger and Keller, 1990.*)

Choosing a regeneration cycle

We have discussed three basic or archetypical regeneration schemes: TSA, PSA, and DPA. These cycles have been described in simple form, but, in fact, actual cycles can have substantial variations from these archetypes. (A number of these variations will be discussed later in "Adsorption Processes.") Nevertheless, in spite of a plethora of individual embodiments, it is possible to decide in a relatively straightforward manner which general type of process, or in some cases processes, should be considered for a given application. The scheme by which this is done is given in Figure 4.21.

It goes without saying that an adsorbent which will perform the separation must exist before an adsorption process can be considered at all. As a rough rule of thumb, the selectivity for the adsorbate of choice compared to the next most easily adsorbed material should be about 2 or greater. For those feeds for which such an adsorbent exists, the most important defining factors in determining the process of choice are the concentration of adsorbate in the feed and the condition of the feed, that is, whether it is a gas or a liquid. In some cases, if a liquid feed can be fully vaporized at a temperature below about

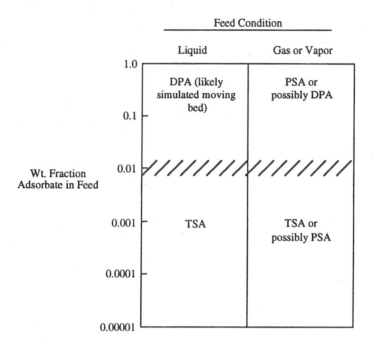

Figure 4.21 Regeneration cycle choices.

150°C, such a feed can be considered for either liquid-feed or gas-feed processes.

There are additional limitations which can be placed on adsorption systems, such as the following:

- The adsorbate should not undergo detrimental chemical reactions while on the adsorbent, especially if those reactions increase the difficulty of desorbing this material. This consideration may be particularly important in the case of TSA processes, since increased temperature can cause increasing problems with adverse reactions.

- Even if detrimental reactions occur at a rate small enough to be inconsequential to product purities, these reactions must not lead to fouling of the adsorbent. Except for very unusual cases, the adsorbent should have a lifetime of at least 3–6 months before having to be removed and externally regenerated to eliminate the fouling species.

- In the case of PSA-based adsorptions, one can produce only one relatively pure product—the nonadsorbed product. In addition, the recovery of that product is less than complete, since some of it is lost with the adsorbed product.

- In cases in which a desorbent is added (liquid bulk separations, primarily), the desorbent must be easily separable by distillation from all components of the feed.

Once a decision has been made regarding the general type of process to use, then studies can begin on the various process alternatives in this category. Almost always it will be necessary to begin interacting with process-equipment vendors at this point, since detailed process considerations and economics are almost never available in the open literature. In Appendix B a large number of adsorbent and adsorption equipment vendors are listed.

There is of course a further consideration with respect to the choice of a process, and that is whether adsorption is the right—that is, the cheapest and most workable—process for a particular separation. That question will be addressed in Chapter 7, after the reader has become familiar with the main process alternatives.

Design Considerations

A detailed discussion of the design considerations for adsorption processes is beyond the scope of this book, and as a result, only a qualitative treatment will be given here. For more detailed treatments, readers should refer to texts such as Cheremisinoff and Eleerbusch (1978), Gembicki et al. (1991), Keller et al. (1987), Ruthven (1984), Ruthven et al. (1984), Sherman and Yon (1991), Vermeulen and LeVan (1984), Wankat (1986), and Yang (1987). Beyond these sources, a large amount of information is available from individual articles and from vendors. A listing of vendors of adsorbents and adsorption processes is given later in Appendix B. From the standpoint of journals, the one of most interest is *Adsorption,* published by Kluwer Academic Publishers, 101 Philip Drive, Assinippi Park, Norwell, MA 02061. Other journals containing practical adsorption-process information include *Chemical Engineering* and *Hydrocarbon Processing.*

Among the design considerations for adsorption processes, the three most important aspects are (1) the adsorbent characteristics, (2) the way in which compositions change as a function of distance through the bed of adsorbent, and (3) the position and movement of the temperature front in the bed. These three aspects will be discussed later, along with other considerations such as pressure drop and bed geometry.

Adsorbent characteristics

The main factors in adsorbent selection are the adsorbent's isotherm characteristics, which are set by the type of adsorbent chosen, and the adsorbent geometry. Traditionally adsorbents have been fashioned almost exclusively in the form of particles, and most processes still use particles. Particle sizes typically range from 100×200 mesh (0.0029–0.0059 in. or 0.074–0.149 mm) up to about one-quarter inch (6.4 mm) or slightly larger in diameter. The choice of particle size is made by considering two sets of factors: (1) ease of mass transfer from the fluid to the surface, creation of as much interfacial surface area as possible and reduction of the intraparticle diffusion-path length, all of which favor smaller particles; and (2) maintenance of a low pressure drop, which favors larger particles. Typically the small particles are formed by crushing and screening larger particles, and as a result the smaller particles are of irregular shape. Small, polymeric adsorbent particles, which are normally spherical, are one exception to this generalization. Larger particles are most often either irregular granules, cylinders (made by extrusion) or spheres.

More recently adsorbents are being supplied as extended-surface shapes, as mats of fibers, and as monoliths. One of the most popular of the extended-surface particles is an extrudate with a cloverleaf or trilobe cross section called TRISIV (Keller et al., 1987), which is used for natural gas drying and for air purification. This shape creates more surface area than a cylinder of the same equivalent diameter and also reduces the pressure drop. Mats of fibers, in the order of a few hundred microns in diameter, although they have not yet achieved a significant number of commercializations, represent an interesting means of achieving high interfacial areas and low intraparticle diffusion paths and, at the same time, low pressure drops. This low pressure drop comes from a lower weight of adsorbent per unit volume than is typically found in packed beds. Monolithic adsorbents will be described in more detail later in this section.

Composition profiles

The movement of the adsorbate composition profile as a function of time inside a bed is critical to its performance for removing the adsorbate. In this section we will consider a bed operating essentially isothermally, and we will note how the adsorbate concentration wave moves through the bed. Figure 4.22 illustrates this movement as well as the outlet concentration of adsorbate as a function of time. We see three distinct zones within the bed: a zone nearest the inlet in which the adsorbent is fully saturated—in equilibrium—with the adsorbate concentration in the feed. Further down the bed is a zone in which the adsorbate concentration on the adsorbent roughly follows an S shape. It is in this zone, called the mass transfer zone (MTZ), that substantially all of the mass transfer from fluid to adsorbent occurs. Finally, there is a zone in which no adsorption is occurring since essentially all of the adsorbate in the feed has been depleted.

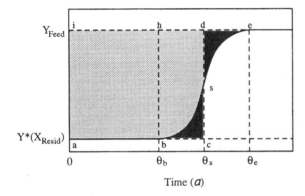

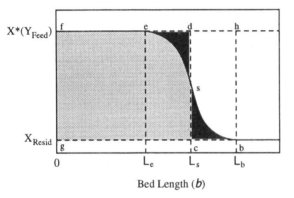

Figure 4.22 Time trace of adsorbate concentration in an adsorber effluent (a) and adsorbate loading along axis of an adsorber during adsorption (b).

At the beginning of the feed step of the cycle, the MTZ is positioned at the inlet of the bed. There will initially be an unsteady-state period in which the S-curve begins to take on its final shape. Provided that the adsorbate isotherm is concave downward (as it is with the Langmuir isotherm and in the large majority of commercial processes) and that the MTZ is reasonably isothermal (not more than about 10–20°C difference across it), the S-curve should maintain its shape until it reaches the end of the bed. As it passes through the end of the bed, then the adsorbate concentration in the nonadsorbed product will begin to rise, as shown in Figure 4.22, until the adsorbate concentration reaches its concentration in the feed.

The condition at which the leading point of the S-curve reaches the end of the bed is called breakthrough. It is at this point that the feed step must usually be stopped, even though a certain fraction of the adsorbent is less than fully loaded. The S-curve is roughly symmetrical, and its width can be defined as two times the distance from the point at which an unacceptably high adsorbate concentration in the nonadsorbed product stream is reached, to the point at which the adsorbate concentration is halfway between that in the feed and that resulting from any residual adsorbate loading from the previous cycle.

The width of the S-curve, or the MTZ, is obviously of great importance, as can be seen in Figure 4.23. This figure illustrates that, as the ratio of the width of the S-curve to the bed length increases, the bed utilization per cycle decreases, and more adsorbent will be necessary to process a given feed rate. There are a number of factors which influence the MTZ width, and these are listed in Table 4.9, along with strategies to minimize the effect of each.

Being influenced by a number of factors makes the prediction of the MTZ a difficult problem, however, especially for new systems. There are a number of references, for example, Keller et al. (1987), Ruthven (1984, 1991), Ruthven et al. (1994), Vermeulen and LeVan (1984), and Yang (1987) which can help along these lines, but it is often necessary to do some experimental work to get an accurate estimate. In general, for gas feeds, the MTZ for a well-designed system will be in the order

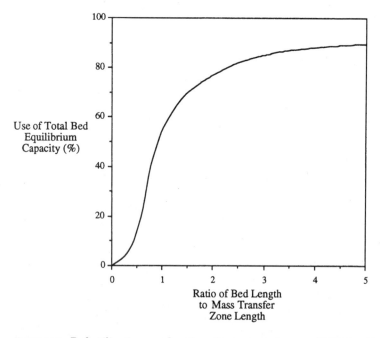

Figure 4.23 Bed utilization as a function of mass transfer zone (MTZ) length.

TABLE 4.9   Factors Affecting Mass Transfer Zone (MTZ) Width and How to Minimize Their Effects

Factor	How to minimize
Diffusion-rate limitation within the adsorbent particle	1. Reduce intraparticle diffusion path by reducing particle size.
	2. Use an adsorbent with a macropore network connecting the micropores to facilitate diffusion.
Adsorbent particle surface-area limitations	1. Reduce the particle size to create more area per unit weight of adsorbent.
	2. Use particles with higher surface areas per unit volume (e.g., special extrudates).
Uneven flow-rate distribution through the bed	1. Minimize void-fraction variation, which can create uneven flow in the bed.
	2. Provide uniform flow distribution at the inlet to and outlet from the bed.

of 1 to a very few feet (0.3–1 m), and in special cases the MTZ can be as small as 1 in. or so (25 mm). The MTZ will typically be longer for liquid feeds compared to gas feeds because of the much higher viscosities of liquid systems. These higher viscosities retard mass transfer to the particle surface as well as diffusion within the particle.

One popular way of specifying the dimensions of an adsorbent bed is to picture it as two zones. In the first zone, the adsorbent is at equilibrium with the adsorbate concentration in the feed, and its length is called the length of the equilibrium section (LES) or the weight of the equilibrium section (WES). This zone is followed by one in which the adsorbate loading is zero. The second zone is specified as the length of unused bed (LUB) or the weight of unused bed (WUB). Note from Figure 4.22 that the LUB is one-half the width of the MTZ. Also note that, except for extremely short beds in which the S-curve does not become fully developed, the LUB is essentially independent of the overall bed length.

In some cases it is important that as much of the bed be used as possible to minimize the use of an expensive adsorbent or to maintain the adsorber size at a minimum. This can be accomplished, with a modest increase in process complexity, by using a three-bed system employing lead-trim operation (Figure 4.24). The MTZ passes completely through the lead bed, assuring that this bed is fully loaded, and into the trim bed. At this point the lead bed is taken off-line and switched to the regeneration step, the trim bed becomes the lead bed, and the regenerated bed becomes the trim bed.

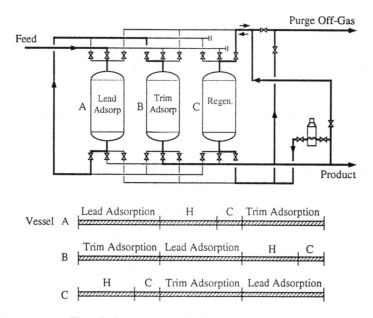

Figure 4.24 Three-bed system with lead-trim adsorption steps.

Temperature effects

The problem of heat release in a bed of adsorbent has already been stressed, and the processes which have been developed have had to take this highly important factor into account. What has been observed in non-DPA processes is that the heat of adsorption can be pushed out of the bed before the MTZ reaches the end of the bed, or the heat can be concentrated in the mass-transfer zone, or the heat can be concentrated behind the MTZ, depending on the feed composition, pressure, temperature, and the adsorbent isotherm. A quantity called the crossover ratio, R, is useful in determining the position of the thermal front vis-à-vis the MTZ (Gary and Ausikaitis, 1983):

$$R = \frac{C_{pg}/C_{ps}}{(Y_i - Y_0)/(X_i - X_0)} \tag{4.13}$$

where C_{pg} = heat capacity of the nonadsorbing fluid (Btu/lb fluid/°F or kJ/kg/°K)

C_{ps} = heat capacity of the adsorbent (Btu/lb/°F or kJ/kg /°K)

X_i = adsorbent loading in equilibrium with the concentration of adsorbate behind the mass transfer front (g adsorbate/g adsorbent)

X_0 = adsorbent residual loading ahead of the mass transfer front which is a result of a previous regeneration (g adsorbate/g adsorbent)

Y_i = inlet adsorbate concentration (g adsorbate/g nonadsorbing fluid)

Y_0 = outlet adsorbate concentration in equilibrium with X_0 (g adsorbate/g nonadsorbing fluid)

If X_0 is much smaller than X_i, and if Y_0 is much smaller than Y_i (both of these approximations are usually good for most TSA processes), the crossover ratio simplifies to:

$$R = \frac{C_{pg}/C_{ps}}{Y_i/X_i} \tag{4.14}$$

Furthermore, for gas adsorption systems at least, C_{pg} is slightly larger than C_{ps} (C_{pg} for nitrogen and air = 0.25 Btu/lb/°F, while C_{ps} for activated carbon = 0.165, for alumina = 0.186, and for silica = 0.169, for example), and therefore:

$$R = 1.4 X_i/Y_i \tag{4.15}$$

Depending on the values of these two concentrations in equilibrium with each other, R can be greater than, less than, or equal to 1.

Table 4.10 gives a summary of how the value of R affects the location of the thermal front compared to the mass transfer front, while Table 4.11 gives values of R for several different systems. As can be seen from this table, those separations which are typical for TSA processes all have values of R much greater than 1, indicating that the thermal front will stay well ahead of the mass transfer front. This means that the zones in which adsorption is taking place and has taken place will remain essentially at the inlet temperature. The

TABLE 4.10   Significance of the Crossover Ratio, R

Value of R	Location of thermal front
$\gg 1$	Well ahead of the MTZ and will normally leave the adsorbent bed before the leading edge of the MTZ reaches the end of the bed
~ 1	Located within the MTZ
$\ll 1$	Located behind the MTZ and will still be in the bed when the MTZ leaves the bed

TABLE 4.11   Typical Values of the Crossover Ratio, R, for Various Separations

Separation	Adsorbate Concentrations, kg/kg		R	Conclusion
	Y^*	X^*		
Acetone removal from air, 1 atm, 25°C, activated C	0.01	0.1	14	TSA
H_2 recovery from CH_4, 40 atm, 25°C, activated C	8.0 (CH_4)	<0.1	<0.02	PSA
O_2 recovery from air, 5 atm, 25°C, MSZ	3.3 (N_2)	~0.2*	<0.1	PSA

*Actually in this case considerable coadsorption of O_2 occurs.

implication of this conclusion is that the adsorption bed will attain the highest possible loading in the equilibrium section.

On the other hand, for air separation (making the assumption that only nitrogen is adsorbed and not oxygen, which is not absolutely correct) and for hydrogen-methane separation, then R is much less than 1, meaning that the heat of adsorption will be stored in the bed somewhat behind the MTZ. This fact has both a positive and a negative aspect. On the positive side, this heat, which is stored as sensible heat in the adsorbent, is subsequently available for desorption during the regeneration step, and therefore no additional energy is required for desorption. On the negative side, the fact that there will be a portion of the bed which is elevated in temperature means that the bed will be limited in capacity compared to an isothermal bed, and this exotherm will continue to rise as more feed is added until major adsorbate breakthrough occurs. Realistically, to keep the exotherm under control, the delta loading in a PSA process will be only a few percent of the equilibrium adsorbate loading at feed pressure and temperature.

It must be noted that the adsorption curves in PSA processes (see Figure 4.15) differ widely in shape and movement from those for TSA processes (see Figure 4.22). For this reason, the crossover ratio should not in any way be thought of as a quantitative design factor. Rather its value is only indicative of the general location of the heat front relative to the location of the MTZ, and in this sense the crossover ratio is very helpful in the selection of the correct process form for use in a particular situation.

If the value of R is close to 1, then the thermal and mass transfer fronts will be very close together, and the MTZ will be highly non-isothermal. The effect will be to widen the MTZ and in general degrade the separation, as pointed out in Table 4.9. This condition should be avoided.

It is worthwhile to make simple calculations for new systems, especially in cases in which the adsorbate concentration in the feed is in the gray area of about 2–10 wt%. The analysis should help in the decision as to whether PSA or TSA will be preferred. The detailed information to design many of the complex processes used today, however, remains mainly outside the domain of the published literature and is the property of the various vendors of their technologies. One must work with these vendors, a number of which are listed in Appendix B, to arrive at final designs and economics.

Standard adsorption vessel geometry

The overwhelming fraction of adsorption beds containing particles consists of vertically oriented, cylindrical vessels. The flow through these beds is also vertical. The main exceptions to this geometry arise from a need to reduce the pressure drop or a need to accommodate a nonparticulate adsorbent such as a monolith. These exceptions will be discussed in the next section.

The length-to-diameter ratio (L/D) of cylindrical, particulate adsorption beds is usually greater than about 1.5. Even higher values of L/D may be necessary for liquid feeds to minimize the problem of a large MTZ (see "Composition Profiles" in this section), which limits the effective bed capacity. Also, if the L/D is less than 1 even for gas feeds, then the MTZ problem may again be significant.

The adsorbent may be supported in the vessel by two means. The first involves the use of horizontally mounted I-beams to support a grating, which can support layers of finer-mesh screens. In some cases the grating can be supported from the vessel wall. If necessary, ceramic balls can be used on top of the screens to further inhibit loss of adsorbent through the support. The second method involves using large ceramic balls in the bottom of the vessel, onto which successively smaller and smaller layers of solids are placed, until the top layer will not allow passage of the adsorbent.

If the adsorbent must be dumped often from the vessel, then there may be no support system at all, that is, adsorbent completely fills the bottom of the vessel.

Especially in liquid-adsorption cases, it can be important to restrain the bed at the top, since failure to do so could lead to fluidization or flow maldistribution, which in turn leads to particle attrition and a severe lengthening of the MTZ. A typical hold-down strategy is to place a screen across the top of the bed and then add heavy support balls on top of the screen. To deal with the problem of settling of the bed over time, the hold-down device is allowed to "float" on top of the bed instead of being rigidly positioned.

Low-pressure-drop bed geometries

Gas streams which are vented to the atmosphere are often available at a pressure close to atmospheric. In such cases and in a number of others, the pressure drop through the adsorption bed becomes an important consideration. If that pressure drop is substantial, then a compressor or a blower may be necessary, which would be more costly from both investment and energy usage standpoints compared to a case in which the pressure drop would be much lower.

Two approaches are available to produce a lower pressure drop than would be experienced in a typical vertical, cylindrical packed bed:

- Reduce the gas velocity and/or bed depth

- Use an adsorbent geometry which has an inherently lower pressure drop

In the first approach, one strategy is to orient the bed in a vessel as shown in Figure 4.25. If the amount of adsorbent is kept constant and the cross-sectional area for flow is increased by, say, a factor of 10 (as could easily be the case in a horizontally oriented bed), then the bed depth will decrease by a factor of 10. The combination of these changes will result in a dramatic lowering of the pressure drop, as is illustrated in Table 4.12. Another means of effecting these same changes in cross-sectional area and bed depth is shown in Figure 4.26. In this example, taken from the heating, ventilation, and air conditioning industry, a number of thin panels packed with adsorbent particles are oriented in a slanted fashion with respect to the flow entering the adsorbent section. In this case the cross-sectional area for flow becomes the cross-sectional area of a panel times the number of panels. Typical bed thicknesses for slanted beds are in the range of 1 in. (25 mm) $-\frac{3}{4}$ in. (18 mm).

Changing the bed geometry obviously does not come without some cost. The horizontally oriented vessel will be larger than a vertically oriented one and will therefore be more costly. The slanted-bed panels will also be more costly per unit of adsorbent weight, and the volume

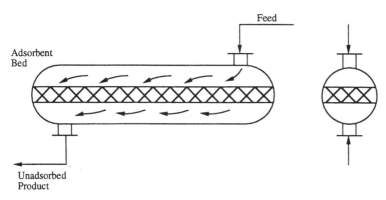

Figure 4.25 Horizontal bed.

TABLE 4.12   Pressure-Drop Implications of Unusual Bed Geometry

Regular-geometry packed bed vs. slanted packed bed:

- Slanted bed has 10 times the cross-sectional area and 1/10 the bed depth, i.e., the same amount of adsorbent.

- Air velocity = 20 ft/s (6 m/s) in the duct. This will give a superficial velocity of 20 ft/s in the regular-geometry bed and 2 ft/s in the slanted packed bed.

- Particles are 0.25-in. (6.4-mm) spheres.

- Void fraction = 0.35

- For these values, the pressure drop in the slanted packed bed will be about 1/1000 of that for the regular-geometry packed bed.

Regular-geometry packed bed vs. monolithic bed:

- Packed bed is the same as above.

- Monolith contains the same amount of adsorbent as the regular-geometry bed, has the same depth and has channels with equivalent diameter = 0.157 in (4 mm).

- Void fraction = 0.35.

- Pressure drop for the monolithic bed will be about 1/100 of that for the regular-geometry packed bed.

taken up by the beds will be several times that of a regular packed bed, because of the extra open volume between the panels. Also, one cannot simply reduce the bed depth indefinitely. Because mass transfer is never infinitely fast, an S-curve in adsorbate concentration in the fluid, as discussed earlier, will always exist in the bed. As the length of this S-curve begins to approach the bed depth (see Figure 4.23), the effective loading up to the point of breakthrough of the

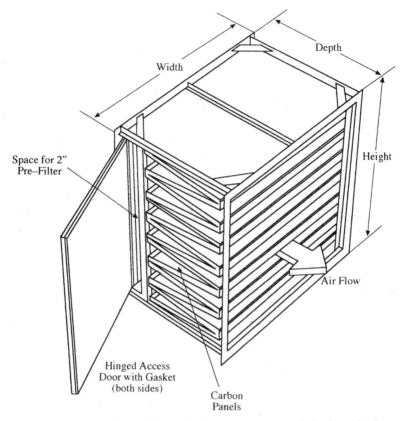

Figure 4.26   Barnebey & Sutcliffe panel adsorption system—side-loading adsorber.

adsorbate will decrease, and more adsorbent will have to be used per unit of gas processed.

To reduce the inherent pressure drop of the adsorbent, two strategies have been used. The first has to do with changing the geometry of the adsorbent particles. Trilobe extrudates, spheres, and cored cylinders, for example, produce less pressure drop per unit weight of adsorbent than do irregular particles. Recently, adsorbent fibers which can be woven into adsorbent mats have been developed.

Perhaps the most important commercialized development in the area of adsorbent geometry, however, has been the monolithic adsorbent. In monoliths, which so far are used exclusively in the area of gas adsorptions, the gas flows through very regular, parallel channels, and in doing so, the pressure drop associated with the fact that the gas must follow a tortuous, winding path through a particulate bed—called *form drag*—is dramatically reduced. These monoliths can be made entirely of adsorbent, and this is most often done when the

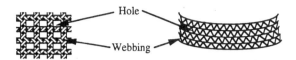

a. Extruded structure, consisting of square holes with webbing lined with adsorbent.

b. "Wrapped" structure, consisting of roughly triangular holes with webbing lined with adsorbent.

Note: In some cases, the entire webbing can consist of adsorbent.

Figure 4.27 Adsorption monolithic structures: typical cross sections.

adsorbent is activated carbon. Alternatively, monolithic adsorbents can consist of a coating of adsorbent on a nonadsorbing, ceramic, or, occasionally, metal monolith. This is usually the case when adsorbents such as silicalite are used. Examples of monolithic adsorbents are shown in Figure 4.27. The technology to produce adsorbent monoliths has progressed to the point that process streams from a few cubic feet per hour (0.1 cu m/h or so) to over 2 million cu ft/h (over 60,000 cu m/h) can be treated.

Table 4.12 also gives a summary of pressure-drop estimates for a monolithic adsorbent and compares them with those for regular-geometry and slanted beds. This table makes it clear that the magnitude of the reduction in pressure drop in the nonparticulate cases is remarkable compared to the particulate case.

Adsorption Processes

The purpose of this section is to describe a number of important commercial adsorption processes and adsorbents, as well as the situations in which they are used. This information should help the reader to choose which processes will be most beneficial in a given situation. The information is organized according to whether the feed is a liquid or a gas, and whether the adsorbate concentration in the feed is in the bulk separation or in the purification range. Since many of the newer separations which can be solved by adsorption are environmentally oriented, this aspect is emphasized.

Gas purification

Table 4.1 gave an overview of the main uses for adsorption in gas purification. One can see that quite a number of applications are in the environmental area, for example, the cleanup of vent streams, etc. and in the removal of small quantities of contaminants in feed gases.

In this latter case, the problem being addressed is often the removal of materials which will poison or inhibit catalytic activity in subsequent reaction steps.

Gas purification separations, once in a moribund state with regard to new adsorption technology directions, have recently become the recipients of a series of new-process options and new adsorbents. Still, the vast majority of these separations are being handled in a rather straightforward manner with fixed beds of traditional adsorbents such as activated carbon, activated alumina, silica gel, and zeolite molecular sieves. These beds operate in more or less traditional TSA modes, with the adsorption step operating very close in temperature to that of the feed. Since designs and adsorbents are generally widely available from vendors for these fixed beds, the emphasis in the sections below will be on some of the newer—but already commercialized—processes and situations in which new adsorbents and new process configurations are being used.

Continuous processes. Continuous adsorption processes potentially offer the advantage of process simplicity, which translates into lowered investment costs. The problem, of course, is that it is much more difficult to move particulate solids between adsorbing and regenerating zones than it is to move a liquid between adsorbing and regenerating zones. Problems include attrition of solids, erosion and corrosion of equipment, and difficulties in achieving plug flow of the solids. Plug flow is, of course, desired in both countercurrent and cocurrent operations to promote the greatest possible degree of mass and heat exchange between the phases. Attempts to develop continuous adsorption processes in the past have met with mixed success. For the most part, the approaches used have consisted of two basic types: physical movement of small particles from one zone to another, and, more recently, the use of large, circular, monolithic adsorbent shapes which rotate slowly between two zones.

Perhaps the first major attempt to develop a continuous gas adsorption process resulted in Union Oil's Hypersorption process (Berg, 1946, 1951) for separating light hydrocarbon gases. This separation was actually a bulk separation and used an activated carbon adsorbent. The process was scaled up to a rather large size but was soon shut down because of operational and economic problems. Later the Kureha Company in Japan developed a very hard, microspherical, activated carbon and used it in a fluidized bed/fixed-bed process for removal of organic materials from gaseous vent streams (Keller, 1983). This process, which was marketed in the United States under the name Purasiv HR, is shown in Figure 4.28. Though no longer sold under this name, Daikin Industries, Ltd. in Japan has continued to

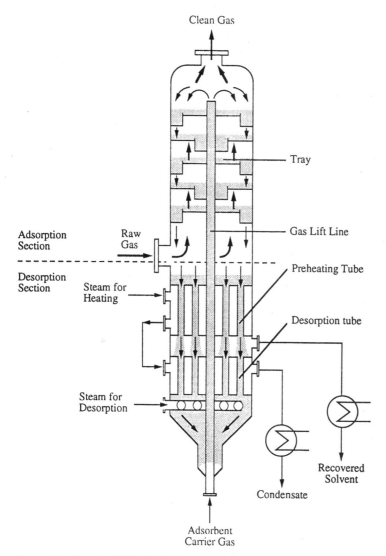

Clean Gas

Tray

Adsorption
Section

Raw
Gas

Gas Lift Line

Desorption
Section

Steam for
Heating

Preheating Tube

Desorption tube

Steam for
Desorption

Recovered
Solvent

Condensate

Adsorbent
Carrier Gas

Figure 4.28   Purasiv HR™ fluid-bed/moving-bed process.

market the process, now called SOLDACS, for various environmental
applications.

Styrene-divinylbenzene copolymer resins have for many years been
derivatized to form ion-exchange resins. More recently a number of
new uses have been and are being developed for the underivatized
resin as an adsorbent. The major plus for a resin-based adsorbent for
a continuous, moving-bed process is the fact that attrition, erosion,
and corrosion problems should be considerably less severe than what

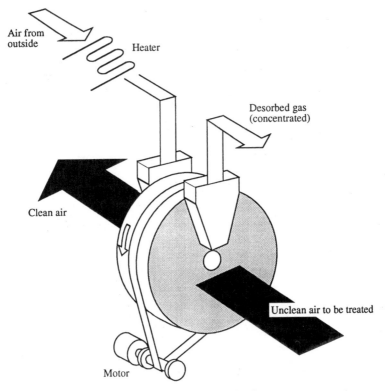

Figure 4.29 Adsorbent wheel with monolithic adsorbent.

would be encountered with a rigid, friable, solid adsorbent. Several companies are now offering moving-bed adsorption processes using resins as well as inorganic adsorbents (see Appendix B).

If experience has shown us that it is difficult to flow particles from one zone to another, a second tactic has met with considerably more success. Monolithic adsorbents in the form of rotors or wheels, as shown in Figure 4.29, are rapidly gaining favor as a means of removal of organic species from small-volume to very-large-volume vent-gas streams. This figure shows that, as the wheel rotates, the adsorbent is alternately contacted by the process gas and by a regeneration gas at an elevated temperature. The relative residence times of the adsorbent in each zone are set by the fractions of the circle covered by each of the zones, and the overall cycle time is set by the speed of rotation. Not only do these wheels offer a degree of process simplicity through the use of only one vessel to house the whole process, but also the cycle times can be considerably shorter for wheel-based, TSA processes than for fixed-bed, TSA processes. Instead of cycle times of several hours or more for typical fixed-bed processes, the wheel-based processes can have cycle times conceivably as short as about 5–10 min (Figure 4.10). The obvious result of this more rapid cycle is that the wheel-based processes will require considerably less adsorbent per unit of gas processed per unit of time. Counterbalancing the use of less adsorbent, however, is the fact that the adsorbent costs more per unit weight in monolithic form than it does as simple particles. Wheels can now be made with diameters in excess of 13 ft (4 m) and are available with silicalite, alumina, silica gel, and other adsorbents. These wheels can obviously treat truly massive flows, some in excess of 2 million standard cu ft/h (60,000 cu m/h).

A complication associated with adsorbent wheels is that there can be leakage of process gas into the regeneration section and of regeneration gas into the adsorption section. This cross-contamination leads to practical removals which are lower than can normally be achieved in fixed-bed TSA processes. In addition, the mass transfer rate per unit length of bed traversed is not as high as that in a packed bed of particles. (The monolithic bed mass transfer rate can be enhanced by using as small a dimension as possible for the hole size, and at hole sizes in the order of a very few millimeters or so, this mass transfer concern is effectively minimized.) As a result, instead of being able to achieve the 99% or higher removals which can often be accomplished by the fixed-bed processes, wheel-based processes will probably be limited to removals in the order of 95% or so.

It is also possible to construct an adsorbent wheel, mounted horizontally and shown in Figure 4.30, which can contain a particulate adsorbent. This approach represents a price-performance compromise. The adsorbent, being in a traditional, particulate form, will be

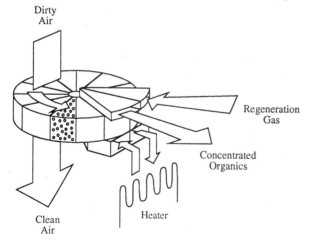

Figure 4.30 Adsorbent wheel with particulate adsorbent.

less expensive than a monolithic adsorbent per unit weight of adsorbent, but pressure drops per unit of bed thickness will be higher.

Irreversible adsorbents. Instead of forming reversible bonds—bonds which can be broken by mild changes in partial pressure or temperature—with adsorbates, these adsorbents actually promote the formation of chemical species with selected materials in a gas stream. Subsequent removal of the adsorbate is usually accomplished by effecting a second chemical reaction on the adsorbent surface, freeing up the adsorbate but usually in a new chemical form. An example of this is the reaction of hydrogen sulfide with iron sponge to form iron sulfide. This process has been known for a number of years and is sold by several companies and is used widely to remove relatively low concentrations of hydrogen sulfide from gas streams. Other reversible adsorbents include solid sodium hydroxide and other bases for removing carbon dioxide and hydrogen sulfide. MSZs preloaded with bromine can serve as effective traps for trace olefins by reacting them to the corresponding bromides, and activated carbon preloaded with cupric chloride can react with and remove trace mercaptans. Other specialized adsorbents for removing various radioactive gases and other dangerous trace pollutants have been developed.

Recently ICI Katalco technology (1993) for removing a number of different contaminants from process gas streams has been commercialized, as shown in Table 4.7. The regeneration conditions are so extreme that on-site regeneration is impossible. Thus a loaded bed must be dumped from the adsorber and returned to the manufacturer. To keep the costs associated with this method of regeneration reasonable, this technology should only be considered in cases in which the amount of material removed is about 220 lb/day (100 kg/day) or less. With such a removal rate, bed lifetimes can be in the order of several months or longer.

Parenthetically, off-site regeneration can also be carried out with more traditional, reversible adsorbents in cases in which the cycle time is in the order of several months or longer. In such cases, the reduction in investment which occurs from not having to install a regeneration system can effectively counterbalance the extra cost of "renting" the adsorbent.

In some cases an irreversible adsorbent, once loaded, is simply discarded. Examples of this are limited primarily to very small beds, such as those used to trap ozone, toners, and other emissions from copy machines, laser printers, and the like.

Biosorption. Biosorption, often called biofiltration, is potentially a strong contender for applications in which the organic components of

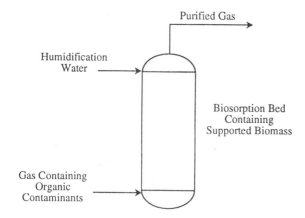

Purified Gas

Humidification Water

Biosorption Bed Containing Supported Biomass

Gas Containing Organic Contaminants

Figure 4.31 Biosorption process.

a stream are to be destroyed and not recovered (Fouhy, 1992; Leson and Winer, 1991; Togna et al., 1993). Activated sludge, such as that from a waste treatment facility, is fixed on a porous wood or another substrate to create a significant area for mass transfer. The material is then typically mounted in a vertical vessel. This active biomass is maintained viable by adding water and perhaps some nutrients at the top of the vessel. As shown in Figure 4.31, gas is passed upward through the vessel, and the contained organics are first dissolved into the wet biomass and then transformed by the biomass into carbon dioxide, water, and other simple molecules if heteroatoms such as chlorine are present in the organics. This process is thus the gas-phase analogue to the well-known biological waste treatment facility for aqueous waste cleanup.

At first glance, it is a bit surprising that biosorption has not found a wide number of applications, given the tremendous number of aqueous waste treatment facilities. And in fact, applications overseas, and especially in Europe, are far more numerous than in the United States. The perceived problems are twofold. First, the rate of metabolism of the organics by the biomass can be slow enough that removal rates per unit volume of equipment can be much lower than typical values for traditional adsorption processes. This problem also can limit the percentage removal of organics from a gas stream to values lower than can be reached by the traditional processes. And in some cases, organics which are very difficult to metabolize are hardly removed at all. This problem is exacerbated in biosorption compared to the liquid process by virtue of the fact that gas residence times in the biosorber are orders of magnitude lower than the water in a biotreatment facility, making the metabolism kinetics problem more serious. The second problem biosorption faces is maintenance of the stability of the biomass. Close control of the wetness of the biomass represents a type of problem not present in other processes, creating perhaps a psychological barrier to the acceptance of biosorption.

In spite of these problems, biosorption represents a technology which will undoubtedly undergo improvements in the future and become more competitive for applications in which removal but not recovery is desired.

PSA processes for purifications. Until relatively recently, the use of PSA for the removal of relatively small amounts of organics from vent streams was virtually nonexistent. Dow Environmental, a subsidiary of Dow Chemical Corp., now sells a two-bed PSA-based process, called Sorbathene (Pezolt et al., 1995), which can remove a variety of organics from these streams. Some examples of materials which can be removed by such a process are given in Table 4.13. This process, which is shown in Figure 4.32, often operates with a feed pressure close to atmospheric, and the regeneration step operates under vacuum. The adsorbent is activated carbon or a polymer adsorbent, depending on the separation to be made. The regeneration stream, which is richer than the feed in the organics, is compressed to near-atmospheric pressure, a large fraction of the organics is condensed out, and the remaining noncondensed gas is added to the feed stream going to the other bed.

For the cycle to operate in such a manner, it is necessary for the concentration of organics in the feed to be somewhat close to their saturation concentration. Only in this case will significant condensation occur in the regeneration stream. It is also possible, however, to send the regeneration stream to either a flare or an incinerator or use it as a low-grade fuel. In such cases the organics concentration in the

TABLE 4.13   Examples of VOC Removal by the Sorbathene PSA Process

- Aromatic hydrocarbons such as benzene and styrene
- Aliphatic hydrocarbons such as pentane and isopentane
- Oxygenated organics such as methanol, acetone, propylene oxide, methyl ethyl ketone, and butylene oxide
- Chlorinated organics such as methylene chloride, carbon tetrachloride, ethylene dichloride, 1,1,1-trichloroethane, trichloroethylene, perchloroethylene, chloropropane, and chlorobenzene
- Other materials such as trichlorofluoromethane (R-11), epichlorohydrin, and ethylenediamine

Figure 4.32 Pressure-swing adsorption unit using carbon adsorbent. (*Photo provided by Dow Environmental.*)

feed can be well below the saturation concentration. The effect of the cycle, in this case, is to concentrate the organics to the point that little or no additional fuel would have to be added to the flare or incinerator to support combustion.

Selection of a regeneration cycle. We have seen three primary types of regeneration cycles for removing small concentrations (generally below about 2 wt%) of various constituents from gas streams: fixed-bed TSA, moving-bed TSA (including both particulate-adsorbent-based and wheel-based processes) and PSA. Among the moving-bed processes, it seems likely that there will be a growing number of applications for wheel-based processes, compared to the moving-particulate-bed processes. Biosorption is a possibility in some cases, although there remain questions regarding the achievement of high (95–99%) removals, the inability to react certain organic species and the long-term stability of the biomass and its adaptability to process upsets. For these reasons biosorption should be considered for those cases in which problems exist with the three primary processes.

There seem to be three factors which can help in selecting which process is preferred among the three primary processes:

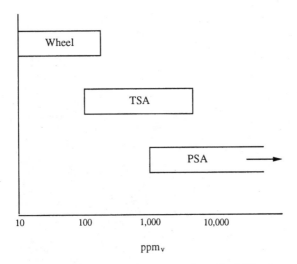

Figure 4.33 Approximate ranges of applicability for various adsorption processes.

- The composition of adsorbate in the feed. Our speculative feed concentration range for application of the three processes is given in Figure 4.33. Here it can be seen that, the higher the concentration (and the nearer the adsorbate concentration is to saturation), the more PSA will be favored. Wheel-type processes are probably best suited for much lower concentrations, while fixed-bed processes will be able to cover a wide range of concentrations.

- The percent removal of adsorbate from the feed. The two fixed-bed processes can in most cases remove at least 99% and in some cases can reach 99.99% removal. Even higher-percent removals may be possible with irreversible adsorbents. On the other hand, wheel-based processes seem destined to be limited to somewhat lower removals—perhaps 95% or so.

- The flow rate of the feed stream. The smaller the stream, the more will process simplicity become important. For streams of only 100–1000 cu ft/h (3–30 cu m/h), PSA processes will probably begin to lose favor. With respect to wheel-based processes, clever packaging of the technology should make them economically competitive with fixed-bed TSA for small flows.

We shall see in Chapter 5 that membranes can also be used to remove organic contaminants from vent streams. In general, if both an acceptable membrane and an acceptable adsorbent exist for a given separation, adsorption will be favored if high (greater than 95–99%) recoveries are important, while membranes will be favored if lower recoveries can be tolerated. Also, the higher the concentration of contaminants in the vent stream, the more will membranes be favored. Process improvements are being made in both areas, however, and it is important to investigate both processes for most separations of this type.

Flowsheets for combined processes. There are some significant instances in which adsorption can be combined with other processes to successfully resolve a separation problem. One example is the combination of air stripping of wastewater and adsorption for subsequently capturing the stripped organic constituents from the air. A flowsheet is given in Figure 4.34. Air stripping can be a very effective means of removing relatively hydrophobic species from wastewater. These species exit from the stripper in quite low concentrations in the air, but without their removal from this stream, one can turn a water pollution problem into an air pollution problem. Adding an adsorption process to the stripper effluent can produce a clean air stream for venting. The organic species can subsequently be recovered or burned. A variant on this concept can be used in situations in which

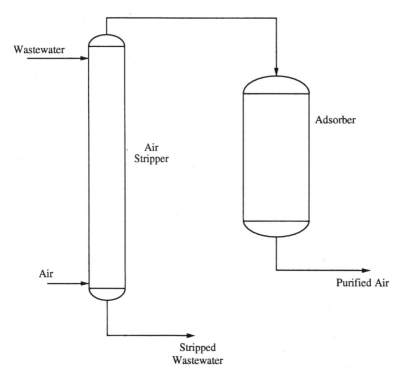

Figure 4.34 Air stripping/adsorption hybrid process for VOC removal from wastewater.

contaminated soil is air stripped to remove volatile contaminants, and these contaminants are then adsorbed.

In situations in which dilute contaminants in gas streams are to be destroyed and not recovered, adsorption can play an important role in concentrating the contaminants so that they can be more economically burned. The problems with burning a large stream containing less than a few percent organics are the cost of the fuel required to raise the stream to combustion temperatures and the investment in a large burner. Even when catalysts are used to lower the combustion temperature, there can still be significant fuel and capital costs. Adsorption processes can be used to concentrate the organics to the point that little or no supplemental fuel is needed. Investment can also be reduced by substantially reducing the volume of the stream to be treated.

Bulk gas separation

Bulk gas separations via adsorption are not very common, as was portrayed in Table 4.1. The vast majority of these separations are carried out by PSA, since it can deal with the problem of heat liberation and consumption during the cycle much better than can TSA. But with this advantage comes the severe disadvantage mentioned earlier: less than complete recovery of the nonadsorbed product and an impure adsorbed product. It can be seen from this table that the uses for PSA are primarily in the following situations:

- The raw material is relatively inexpensive (e.g., air, in oxygen or nitrogen production). This situation makes high recovery of the nonadsorbed product less important.

- An impure adsorbed product is not critically important (some hydrogen can be lost to the fuel-gas stream in hydrogen upgrading).

- The feed flow is relatively small. We can see from Figure 4.35, which compares cryogenic, PSA, and membrane processes for nitrogen production from air that PSA, compared to cryogenic separation, becomes more of an option for smaller flow rates. This is also true of membrane separations, which we shall see in the next chapter.

There are a few instances in which other cycles are used. One notable example is the IsoSiv Process, which was pictured earlier in Figure 4.16 and is based on a DPA cycle. There are almost no other commercial examples of DPA-based processes for gas separations, however, and there appear to be some strong reasons for this situation. As has been noted earlier, DPA-based processes are inherently complex and

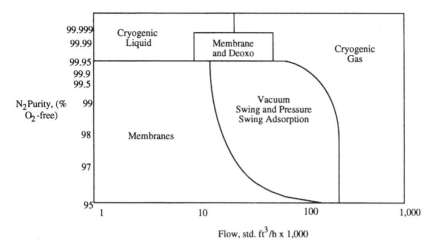

Figure 4.35   Areas of dominance for various technologies for nitrogen production. (*Courtesy of Praxair.*)

require recovery of the desorbent from the two product streams. Thus the investment tends to be relatively high. This complexity problem is further compounded for gas separations in that the use of a desorbent dilutes the product streams and makes quantitative recovery of the desorbent especially difficult. In this respect it is interesting that the IsoSiv Process separates materials which are liquids at room temperature and uses a desorbent which is also a liquid at room temperature. Thus in this case the product-desorbent separations are not as difficult and costly as they would be if products and desorbent were low-boiling gases such as methane, ethane, nitrogen, and oxygen.

There is a variation of PSA which is also used for bulk gas separations. This variation involves the use of an inert gas as a purging medium for desorbing the adsorbate. The total pressure is kept essentially constant. In this sense the process can be thought of as a partial pressure-swing adsorption process. An illustration is shown in Figure 4.36, for separating the ethanol-water azeotrope. Nitrogen or ethanol is used as the purging medium, and the heat of adsorption is stored in the bed (the same as it would be if a PSA process were used) to supply the heat for desorption of the adsorbate. Note again, however, that this process operates in excess of 100°C and separates materials which are liquids at room temperature, and there is little evidence of use of this process for separation of low-boiling gases. The IsoSiv Process can also be fashioned to use an inert gas such as hydrogen to purge the adsorbate instead of a desorbent.

The carbon molecular sieve (CMS) technology for air separation was championed early on by Bergbau Forschung (Knoblauch, 1978)

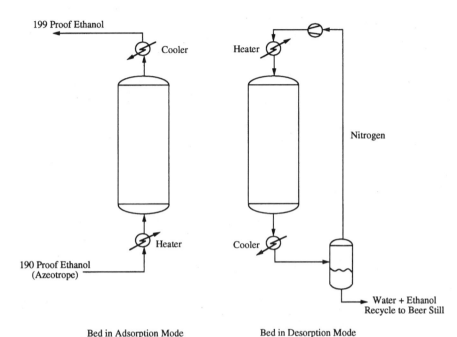

Bed in Adsorption Mode   Bed in Desorption Mode

Figure 4.36 Ethanol drying process. (*Adapted from Garg, 1983.*)

and more recently has been sold by a number of companies including Air Products in the United States. This technology makes use of widely differing pore diffusion rates to effect the separation. Typical time versus loading curves for oxygen and nitrogen are shown in Figure 4.37. The dramatic difference shown in the two curves is caused by the fact that the pore diameter of the carbon is just very slightly larger than the diameters of the two gases, and nitrogen has a much more difficult time traversing a pore than does oxygen. Thus, even though there is virtually no equilibrium selectivity between the two on the carbon, a vacuum pressure-swing process can be used to produce nearly pure nitrogen (Figure 4.38). Unfortunately this method of producing selectivity has not been successfully applied to other separations. This is probably because the pore diameter is very critical to producing the selective diffusion-retarding phenomenon which is the heart of the separation. Micromanipulation of pore diameters is apparently not simple.

There are a number of other PSA-based flowsheets which have been proposed for various separations. These are covered extensively in Ruthven et al. (1994).

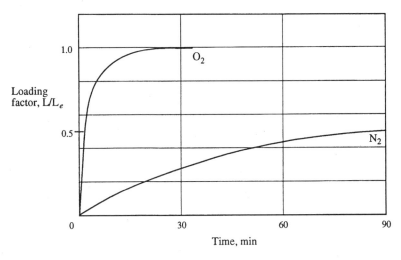

Figure 4.37   Carbon molecular sieve uptake rates.

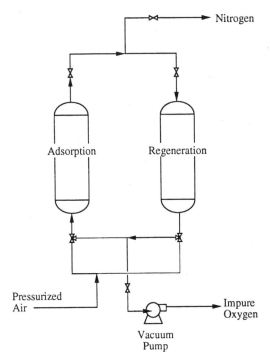

Figure 4.38   Bergbau-Forschung CMS process for nitrogen production.

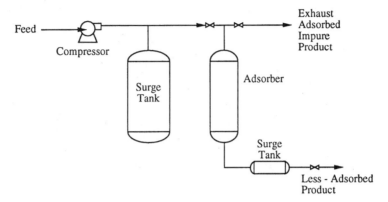

Figure 4.39 Schematic of rapid pressure-swing adsorption process.

The three main areas for improvement of PSA are:

- Reduction of cycle times (this promotes smaller equipment and, thus, lower investment)
- Reduction of energy costs
- Improved percentage recovery of the nonadsorbed product

Unfortunately an improvement in one of these three areas may actually be detrimental to the other two, and so progress has been painfully slow. One attempt to deal with the issue of process simplicity is shown in Figure 4.39 and goes by the name Rapid Pressure-Swing Adsorption (RPSA). This technology has been commercialized for the production of oxygen from air in the homes of patients with lung deficiencies caused by cancer, emphysema, black lung, etc., as well as for the recovery of chlorocarbons from a vent stream (Keller, 1983). The time cycle is short enough (generally less than 10 s) that one bed can produce a constant flow of nonadsorbed product while being fed a constant flow of feed. Surge tanks smooth out these flows. To facilitate the very high rate of mass transfer to and from the adsorbent, the particle size is considerably smaller (less than 1 mm in diameter) than typical particles in PSA beds (one-quarter in.—6 mm—or larger). These smaller particles, coupled with the higher gas velocities in RPSA, make it mandatory that the bed be constrained against fluidization. Thus, even though the adsorbent volume is smaller and the process flowsheet simpler for RPSA compared to PSA, the bed constraint problem prevents RPSA from achieving large-scale usage.

One advance which should be possible to achieve in the near future is the application of monolithic-adsorbent technology to PSA processes. Monoliths should reduce pressure drop and fluidization in the bed, which will be growing problems with particulate adsorbents as cycle times are decreased.

Liquid purifications

A number of typical liquid purifications were shown in Table 4.1. Note that many of these purifications have to do with cleanup of wastewaters from various sources. These separations are performed for the most part using three different process configurations:

- Rather standard, vertically oriented, fixed-bed vessels filled with particulate adsorbent. The adsorbent is periodically regenerated. In some cases, this regeneration is carried out by dumping the adsorbent from the bed, hauling it to a remote regeneration facility and then returning it to the adsorbent bed.

- Mixer-settler arrangements, using powdered adsorbent which is recovered but may or may not be regenerated after being loaded.

- Waste treatment facilities, into which powdered activated carbon is added and subsequently incorporated into the activated sludge.

In processes in which regeneration is carried out, it is almost always done by raising the temperature and supplying a purge fluid. In liquid-feed systems, however, it is often preferable to operate the regeneration part of the cycle in the gas phase using either steam or a hot gas to supply the heat. To carry out a gas-phase regeneration requires several extra steps to be added to the cycle. First the vessel must be drained and then heated and purged with a gas stream. Following regeneration, the vessel must be cooled and filled with liquid. Many times this liquid comes from the product stream, so that no adsorbate can exit from the bed during the cooldown. Then the feed is started. At the end of the adsorption cycle, the liquid is drained from the bed as thoroughly as possible, and then the hot gas flow is again commenced.

Upflow is most often used during the feed-flow part of the cycle, and the bed can be allowed to expand slightly, especially if the feed contains some suspended solids, so that these solids will not become trapped in the bed. It is necessary, however, to keep the bed from fluidizing in any gross sense, since not only can adsorbent attrition become a problem, but also premature breakthrough will occur by virtue of the adsorbent particles' becoming backmixed.

Bulk liquid separations

Bulk liquid separations via adsorption are constrained for the most part to DPA approaches involving simulated moving beds (SMBs), basically because of the difficulty of dealing with the problems of the heat of adsorption and the relatively high density of the liquid feed. The liquid phase is actually so dense that, in a number of cases, less than one bed volume of liquid could fully load the adsorbent even if there were no heat effects. In addition, there is no simple pressure-swing option for desorbing the adsorbate in a liquid separation. So engineering problems become immense for anything but a DPA/SMB process, which essentially eliminates the heat effects and also dilutes the feed stream with desorbent.

However, as Figures 4.18 and 4.19 clearly point out, SMB processes are far from simple mechanically, and they nearly always lead not only to one or more adsorption beds but also to two distillation columns. As a result, bulk liquid separations via adsorption are relatively rare and are limited to separations which cannot be handled easily by distillation. Table 4.14 lists all of the SMB processes which have been commercialized by UOP, while Table 4.15 lists separations which have been demonstrated in the laboratory or in pilot-plant scale. (Other companies such as U.S. Filter have also commercialized SMB processes, and their applications are similar in nature to UOPs.) A study of these tables and the figures is instructive in helping to decide which separations are candidates for SMB. First, it ought to be clear that, if a bulk separation can be accomplished easily by distillation (say, a relative volatility of about 1.2 or greater), then it is highly unlikely that an SMB process could begin to compete economically. Beyond this criterion, we can characterize bulk separations which are possibilities for SMB processes in the following manner:

TABLE 4.14   Commercial Separations with UOP Sorbex Technology*

Process name	Separation	No. of units
PAREX	p-Xylene/C_8 aromatics	53
MOLEX	n-Paraffins/i-paraffins, C_{10}–C_{14}	24
	C_4–C_6	9
OLEX	Olefins/paraffins	6
CYMEX	p- or m-Cymene	1
CRESEX	p- or m-Cresol	1
SAREX	Fructose/glucose	5
CITREX	Citric acid/fermentation broth	1
		100

* Adapted from UOP information, 1992.

TABLE 4.15   Separations Demonstrated with UOP Sorbex Technology*

Hydrocarbon Separations

m-Xylene/C_8 aromatics

Ethylbenzene/C_8 aromatics

o-Xylene/C_8 aromatics

i-C_6 olefins/n-C_6 aromatics

1,3-Butadiene/C_4 hydrocarbons

Indene/alkyl aromatics

β-Pinene/α-pinene

3,5-Diethyltoluene/diethyltoluene isomers

2,6-Diethyltoluene/diethyltoluene isomers

2,6-Diisopropylnaphthalene/diisopropylnaphthalene isomers

2,7-Diisopropylnaphthalene/diisopropylnaphthalene isomers

Isoprene/C_5 naphtha and gas oil cracking fractions

1,3-Butadiene/butadiene isomers

p-Ethyltoluene/ethyltoluene isomers

o-Diethylbenzene/diethylbenzene isomers

Industrial Chemical Separations

p-Chloronitrobenzene/other chloronitrobenzene isomers

o-Chloronitrobenzene/other chloronitrobenzene isomers

2,6-Toluene diisocyanate/toluene diisocyanate isomers

2,4-Toluene diisocyanate/toluene diisocyanate isomers

4,4-Dichlorodiphenylsulfone/dichlorodiphenyl sulfone isomers

Hydroxyparaffinic dicarboxylic acids/olefinic dicarboxylic acids

Dihydroxybenzene isomers

Coumarone/indene

o-Nitrotoluene/p-nitrotoluene

p-Toluidine/toluidine

Picoline isomers

Nitrobenzaldehyde isomers

2,5-Dichlorotoluene/dichlorotoluene isomers

m-Dichlorobenzene/dichlorobenzene isomers

Toluenediamine isomers

Methyl p-hydroxybenzoate/methyl o-hydroxybenzoate

Thiophene, pyridine, phenol/naphtha

TABLE 4.15   Separations Demonstrated with UOP Sorbex Technology*
(Continued)

Fatty Chemical Separations

Unsaturated fatty acid methyl esters/saturated fatty acid methyl esters

Saturated fatty acids/unsaturated fatty acids

Fatty acids/unsaponifiables

Stearic acid/palmitic acid

Oleic acid/linoleic acid

Fatty acids/rosin acids

Triglycerides by degree of unsaturation

Diglycerides/triglycerides

Monoglycerides/triglycerides

Monoglycerides/diglycerides

Trans-olefins/*cis*-olefins

Fatty acid esters/rosin acid esters

Biochemical Separations

Lactic acid/fermentation broth

Phenylalanine/fermentation broth

Ethanol/water

Carbohydrate Separations

Glucose/mannose

Glucose/polysaccharides

Sucrose/molasses

Arabinose/pentose plus hexose

Maltose/glucose

Psicose/saccharides

Monosaccharides/disaccharides

* Adapted from UOP information, 1992.

- Separation of close-boiling isomers. Such separations can be made on the basis of differences in molecular geometry (the MOLEX process, for example, can separate straight-chain paraffins from others which need to have no more than a methyl branch) or polarity (the PAREX process, for example).

- Separation of materials which differ substantially in polarity. This can be seen in the OLEX process, which takes advantage of the extra polarity added by one double bond to discriminate between long-chain paraffins and olefins.

- Separation of very-high-boiling materials. The SAREX process separates essentially nonvolatile sugars, while the CITREX process separates high-boiling, fermentation-derived citric acid from other high-boiling components of the broth.

- Separation of relatively expensive materials. It is not likely that SMB processes will be commonly used to separate, for example, azeotropic mixtures of water and low-cost organics, or waste streams from which various species must be removed. The investment and operating-cost implications are just too great for such separations to bear. On the other hand, as a product's value increases, then SMB technology becomes more of a possibility. Separations of products in the flavors and fragrances area, the personal care area, the pharmaceutical and biochemical areas, the electronic chemicals area and a number of others should be investigated for resolution by SMB technology if the separations are inherently difficult by other means.

There are other and quite different approaches to making the adsorbent bed "think" that it is moving between adsorption and regeneration zones. These approaches actually transport the containers in which the adsorbent is housed between the two zones. One such approach is illustrated by the ISEP Continuous Contactor, sold by Advanced Separation Technologies, Inc. In this process, separate beds of adsorbent actually move in a circular manner on a carousel, encountering fresh feed, feed from other beds and desorbent streams as the carousel rotates. Note that, in this process, the adsorbent remains fixed in beds which physically move. As did the SMB technology the ISEP process also illustrates well the problem one encounters in attempting to fashion a continuous adsorption process. The ISEP process is also used for ion exchange operations.

Another option for moving the adsorbent between adsorption and regeneration zones has been developed over the years at Oak Ridge National Laboratory (Figure 4.40). Called the Continuous Annular Chromatograph (CAC), an annular bed of adsorbent rotates slowly

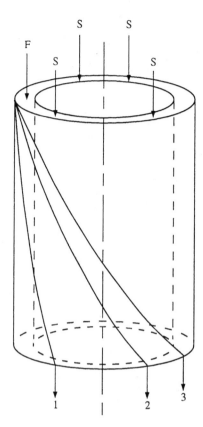

Figure 4.40 Schematic of continuous annular chromatograph.

past stationary feed and desorbent (or eluant) points. The feed constituents travel down the bed at different velocities, forming what appear to be stationary patterns to one watching the bed rotate. Stationary drawoff points collect the various products. Strictly speaking, this process belongs more in the chromatography family that in the adsorption process family, but in cases such as this, the line of demarcation becomes a bit fuzzy. In spite of a tremendous amount of research and publicity over a long period, the CAC process has seen virtually no significant commercializations, and the likelihood of commercializations in the future appears to be small. One major drawback to the process is the fact that it uses substantially more adsorbent to process a certain feed rate than alternative processes.

The overall complexity of SMB and moving-bed processes means that, to investigate these approaches for a given separation, one must work closely with vendor companies which supply the technology almost from the very start of the investigation. Simple screening experiments to determine if there is any likelihood that a selective adsorbent can be found can be done, but beyond this point, contact with these companies will be necessary. In some instances, small pilot plants can be bought or rented to help develop a process, and in other instances, the vendor companies can carry out pilot-plant studies.

New applications and technology directions

In the previous sections we have discussed adsorbents and adsorption technology, as well as their commercial applications. In this section we will extrapolate this information to provide a look into the future directions which adsorption technology and applications will take.

The strongest driver for expanded use of adsorption processes is clearly environmental concerns. These processes are particularly suited for "end-of-pipe" applications, for which TSA processes are often particularly well suited. Wheel-based processes represent possibly a major breakthrough from a technological standpoint, because they can:

- Easily be adapted to a very wide range of feed flow rates in a straightforward manner

- Incorporate adsorption and regeneration steps in one unit, that is, there is no need for two or more beds

- Have considerably shorter cycle times, thus reducing the volume of adsorbent and housing required

In addition, a wide range of adsorbents can now be used, eliminating that possible limitation.

The major negatives for wheel-based processes are the high cost of the adsorbent per unit weight and the inability to reach per-pass removals of adsorbate much over 95%. The adsorbent cost becomes particularly important in cases in which it becomes fouled quickly and must be replaced every few months or so. It is not known at this point whether this fouling can be burned off in place. If it can be in a simple manner, then activated-carbon wheels may be particularly disfavored because of the difficulty of regenerating the carbon without destroying it once it becomes fouled. Silicalite, as a result, should come into greater favor with rapidly fouling applications, despite its higher cost, since it can be burned off with no detrimental effects to it.

It is less clear how adsorption can be used for in-process environmental applications. As adsorbate concentrations become higher, then TSA cycles become less of an option, leaving PSA and SMB cycles as the only options. The shortcomings of these cycles, which have to do primarily with the lack of complete separation (PSA) and the extreme complexity of the process (SMB), militate against using them on a wide scale in processes. Also, the fact that PSA and SMB processes are difficult to develop and design militates against the vendor companies being willing to supply one-of-a-kind processes in this area.

There will be a few new applications for bulk liquid separations, but such separations, to be viable for adsorption processes, must have two characteristics:

- They cannot be carried out by distillation in any reasonable way, that is, the relative volatility is below 1.2, or there must be some other major impediment to the use of distillation.

- The products being separated must be able to bear the cost of an SMB process, which will be in the range of 5–10 cents/lb of useful product for 50-plus million lb/yr products and considerably more per pound for lesser-volume products.

As a result of these restrictions, it is likely that most of the applications for new SMB processes will be for specialty chemicals, pharmaceuticals, bioproducts and the like. Table 4.15 can serve as a rough guide for future applications.

References

Berg, C., *Trans. Inst. Chem. Engrs.*, **42**, 665, 1946.
Berg, C., *Chem. Eng. Prog.*, **47**, 11, 585, 1951.
Breck, D. W., *Zeolite Molecular Sieves*, Wiley Interscience, New York, 1974.
Breck, D. W. and R. A. Anderson, "Molecular Sieves," in *Kirk-Othmer Encyclopedia of Chemical Technology*, 3rd ed., Vol. 15, Wiley, New York, 1982, p. 638.
Eisinger, R. S. and G. E. Keller, *Environmental Progress*, **9**, 4, 235, 1990.
Flanigen, E. M. et al., *Nature*, **271**, February 9, 1978, p. 512.
Fouhy, K., *Chemical Engineering*, **99**, December 1992, p. 41.
Garg, D. R. and J. P. Ausikaitis, *Chem. Eng. Progress*, **79**, 4, April 1983, p. 60.
Gembicki, S. A., A. R. Oroskar, and J. A. Johnson, "Adsorption, Liquid Separations," in *Kirk-Othmer Encyclopedia of Chemical Technology*, 4th edition, Wiley, New York, 1991, p. 573.
Geurin de Montgareuil, P., and D. Domine, French patent 1,223,261 to Air Liquide, 1957.
Humphrey, J. L., personal communication to G. E. Keller, September 1996.
ICI Katalco, Puraspec Purification Process, promotional literature, 1993.
Keller G. E., "Gas-Adsorption Processes: State of the Art," in Whyte, Jr., T. E., Yon, C. M., and E. H. Wagener, eds., *Industrial Gas Separations*, ACS Symposium Series 223, American Chemical Society, Washington, DC, 1983.
Keller, G. E. et al., "Adsorption," in Rousseau, R. W., ed., *Handbook of Separation Process Technology*, John Wiley, New York, 1987, p. 644.
Keller, G. E., *Chem. Eng. Progress*, **91**(10), 56, October 1995.
Knoblauch, K., *Chem. Eng.*, November 6, 1978, p. 87.
Leson, G. and A. M. Winer, J., *Air Waste Manage. Assoc.* **41**(8), 1045, 1991.
Pezolt, D. J. et al., "Gasoline Vapor Recovery Using Vacuum Swing Adsorption Technology," presented at the AIChE Spring Meeting, Houston, TX, March 1995.
Ruthven, D. M., *Principles of Adsorption and Adsorption Processes*, Wiley, New York, 1984.

Ruthven, D. M., "Adsorption," in *Kirk-Othmer Encyclopedia of Chemical Technology,* 4th ed., Vol. 1, Wiley, New York, 1991, p. 493.

Ruthven, D. M. et al., *Pressure-Swing Adsorption,* VCH, New York, 1994.

Sherman, J. D. and C. M. Yon, "Adsorption Gas Separations," in *Kirk-Othmer Encyclopedia of Chemical Technology,* 4th ed., Vol. 1, Wiley, New York, 1991, p. 529.

Shultz-Sibbel, G. M. W. et al., *Talanta,* **29,** 447, 1982.

Skarstrom, C. W., U. S. patent 2,944,627 to Esso Research and Engineering Co., 1958.

Streat, M. and F. L. D. Cloete, "Ion Exchange," in R. W. Rousseau, ed., *Handbook of Separation Process Technology,* Wiley, 1987, p. 697.

Togna, A. P. et al., "Field-Pilot Results of Styrene Biodegradation Using Biofiltration: A Case Study," Envirogen, Inc., presented at the 48th Annual Purdue University Industrial Waste Conference, West Lafayette, IN, May 11, 1993.

Vermeulen, T. and M. D. LeVan, "Adsorption and Ion Exchange," in *Perry's Chemical Engineers Handbook,* 6th ed., McGraw-Hill, New York, 1984.

Wankat, P. C., *Large-Scale Adsorption and Chromatography,* vols. 1 and 2, CRC Press, Boca Raton, FL, 1986.

Yang, R. T., *Gas Separation by Adsorption Processes,* Butterworths, London, 1987.

Membrane Processes

Introduction

Membrane-based separation processes create separations by selectively passing (permeating) one or more components of a stream through a membrane while retarding the passage of one or more other components. Thus the composition of the material passing through the membrane is different from that not passing through. This type of operation is shown schematically in Figure 5.1. In practice the membranes are contained in a vessel to form what is called a module. In this figure the stream passing through the membrane is called the *permeate,* and the stream retained by the membrane is called the *retentate*. In the vast fraction of membrane processes, a pressure difference between the feed and permeate streams creates the driving force for the separation. The percentage of the feed stream which passes through the membrane is called the stage cut. For a certain component which is being held back or retained by the mem-

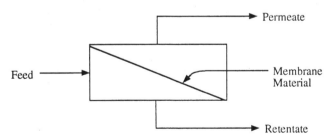

Note: The absolute pressure on the permeate side of the membrane is lower than that on the feed/retentate side.

Figure 5.1 Simple membrane process.

brane, the fractional removal, or rejection, of this component from the permeate is shown in this figure.

Actually there is a great deal of confusion regarding the names of the retentate and permeate streams, and a large number of names has grown up. To help bring some uniformity in this area, the International Union of Pure and Applied Chemistry (IUPAC) has developed a list of preferred names for these streams as well as for other parameters describing the technology. These names are given in Appendix C at the end of this book, and the recommendations given there are used throughout the chapter.

The reader can rapidly conclude that a single chapter of this length can hardly do justice to the extremely broad and deep information base which comprises membrane technology. For that reason we recommend a short library of additional references, including Cheryan (1986), Eykamp and Steen (1987), Ho and Sirkar (1992), Humphrey et al. (1991), Hwang and Kammermeyer (1975), Kesting and Fritzsche (1993), Klein et al. (1987), Koros (1995), Lee (1987), Noble and Stern (1995), Parekh (1988), Porter (1990), Sirkar (1995), Sirkar and Lloyd (1988), and Turner (1991).

Membrane processes are astonishingly varied and extend from the separation of relatively large, visible particles from a liquid or a gas, to the separation of molecules and ions of molecular weight less than 100 from each other. In the former category are a wide variety of filters made of paper, sintered metal, fiber mats, and many other substrates. The technologies using these filters are well developed and have been a part of the chemical industry scene for many years. Because these tech-

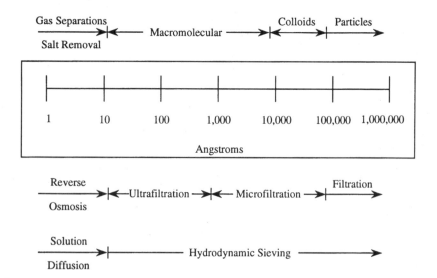

Figure 5.2 Classification of membrane separations.

TABLE 5.1   Membrane Characteristics

Membrane process	Separation mechanism	Pore or inter-molecular size, Å	Transport regime
Particle filtration	Size exclusion	>50,000	Macropores
Microfiltration	Size exclusion	500–50,000	Macropores
Ultrafiltration	Size exclusion	20–50	Mesopores
Nanofiltration, reverse osmosis	Size exclusion Solution/diffusion	<20 <5	Micropores Molecular
Gas separation	Solution/diffusion	<5	Molecular
Pervaporation/vapor permeation	Solution/diffusion	<5	Molecular

nologies are so widespread, relatively established, and well known, they will not be discussed here. Rather, we will confine ourselves here to the newer technologies which are used to remove particles of only a few microns or less from various streams, as well as technologies for molecular separations. It is in these areas that membrane technology is evolving most rapidly and that the most new uses are likely to be found. This is also the area in which membranes so far have made only a small penetration into the chemical, petroleum refining, and allied industries.

Membrane processes are classified by the size of particles or molecules being separated. This is shown in Figure 5.2 and in Table 5.1. The category called nanofiltration (NF) is a relatively new one and includes the upper-molecular-weight range of the reverse osmosis (RO) domain and the lower-molecular-weight range of the ultrafiltration (UF) domain. It is important to note that there are three different mechanisms by which membranes can perform separations:

- By having holes or pores which are of such a size that certain species can pass through and others cannot. This mechanism is called size exclusion.

- By selective retardation by the pores when pore diameters are close to molecular sizes. In gas separations, for example, membranes with very small pores can separate by Knudsen diffusion, in which the rate of diffusion of molecules under the same partial-pressure driving force will vary as the inverse of the square root of their molecular weight.

- By dissolution into the membrane, migration by molecular diffusion across the membrane and reemergence from the other side. This means of passage is called solution diffusion.

The markets for membrane-based processes are far from small, as shown in Table 5.2, and are growing rapidly. The dollar values listed

TABLE 5.2   Worldwide Membrane Markets*

Process	$ Million		
	1986	1991	1996
Microfiltration	550	885	1525
Ultrafiltration	120	230	530
Hemodialysis	500	500	510
Reverse osmosis	350	500	1050
Gas separation	20	50	100
Electrodialysis	70	105	150
Total	1610	2270	3865

*Amounts include membranes, modules and peripheral equipment.

here include not only the membranes themselves but also the modules or housings for them and peripheral equipment. Included in this list are electrodialysis and dialysis/hemodialysis, which will not be discussed here. The implications for the use of these two processes in the industries covered here, though not zero, are not considered to be very large. Pervaporation, which is not listed in this table because its commercial sales are still very small, will be discussed. Table 5.3 gives representative commercial uses for membrane processes. As can be seen from this list, except in the area of gas separations, most of today's uses for membranes lie in other industries.

There is a major difference in the flow patterns which are used most often when removing relatively large particles compared to removing very small particles or making molecular separations, as shown in Figure 5.3. For normal filtration operations, dead-end flow is normally used, in which the flow is perpendicular to the filter surface, and essentially all of the flow which can permeate passes through the filter. As more and more fluid passes through the filter, more and more material builds up on the surface of the filter until the flow resistance becomes too large for the desired flow rate to be maintained. At that point, the operation must be shut down, the solids removed, and the flow restarted. In a few filtration processes, a continuous solids removal is used which can facilitate a continuous operation.

For small-particle or molecular separations, crossflow is used, in which the feed flow is parallel to the membrane. In this case, not all of the fluid which could permeate through the membrane does, the nonpermeating species are for the most part concentrated and carried along in retentate stream, and nonpermeating species do not build up

TABLE 5.3   Representative Commercial Uses for Membrane Processes

Gas separations

  Nitrogen recovery from air (mild concentration of oxygen)
  Hydrogen upgrading from fuel gas, ammonia process blowoff streams, etc.
  Synthesis gas (carbon monoxide/hydrogen) ratio adjustment
  Carbon dioxide removal from various gas streams
  Drying of gas streams
  Removal of organic compounds from vent streams and other streams

Reverse osmosis (RO) and nanofiltration (NF)

  Recovery of fresh water from seawater and brackish water
  Reduction of chemical oxygen demand (COD) of wastewater and groundwater
    streams
  Color removal from wastewater streams
  Removal of various ions from wastewater streams
  Cleanup of wastewaters from electroplating baths
  Concentration of spent sulfite liquor from paper plant effluents
  Pretreatment of boiler feed water
  Recovery of sugars in food-processing steps
  Concentration of milk and whey for cheese production

Pervaporation

  Removal of small amounts of water from organic solutions, e.g., water
    from isopropanol
  Removal of small amounts of organics from water, e.g., in wastewater cleanup

Ultrafiltration (UF)

  Concentration of latex particles in water and recovery of latex particles
    from wastewaters
  Concentration and fractionation of proteins
  Partial dewatering of clay slurries
  Separation of wax components from lower-molecular-weight hydrocarbons
  Removal of polymer constituents from wastewaters
  Clarification of wine
  Separation of oil-water emulsions
  Pretreatment step for RO and NF feed streams

Electrodialysis (ED)

  Desalination of brackish water
  First step in the production of table salt (Japan)
  Deionization of boiler feed water
  Production of ultrapure water
  Processing of rinse waters in the electroplating industry
  Electrocoating of automobile bodies
  Resolution of salts into acids and bases (bipolar membranes)

Microfiltration (MF)

  Removal of micron-size particles from a wide variety of liquid streams
  Concentration of fine solids
  Pretreatment step for RO, NF, and UF feed streams
  Sterilization of various streams in the pharmaceutical industry
  Clarification of beverages
  Purification of fluids in the semiconductor industry

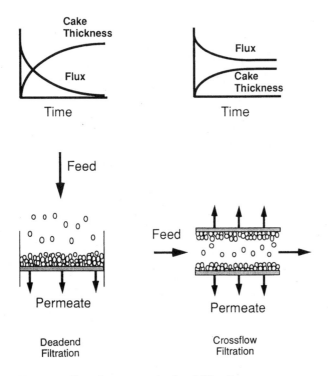

Figure 5.3 Crossflow versus dead-end filtration.

continuously on the membrane surface. Rather, they tend to be swept along by the velocity of the retentate stream. Thus the process can be truly continuous instead of discontinuous, as it is in most dead-end filtrations.

Membrane processes have a number of pluses and minuses compared to alternative means of performing separations. The pluses include the following:

- Because membrane processes can separate at the molecular scale up to a scale at which particles can actually be seen (see Table 5.1 and Figure 5.2), this implies that a very large number of separation needs might actually be met by membrane processes.

- Membrane processes generally do not require a phase change to make a separation (pervaporation, as we will see later, is the major exception). As a result, energy requirements will be low unless a great deal of energy must be expended to increase the pressure of the feed stream in order to drive the permeating component(s) across the membrane. Especially in liquid separations, membrane processes often show very low energy usages compared to other separation processes.

- Membrane processes present basically a very simple flowsheet. There are no moving parts (except for pumps or compressors), no complex control schemes, and little ancillary equipment compared to many other processes.

- Membranes can be produced which, in many cases, can produce extremely high selectivities for components to be separated. In general, the values of these selectivities (which will be derived later but can be thought of here as a measure of the relative tendencies for species to cross the membrane under equal driving forces) are much higher than typical values for relative volatility for distillation operations.

- Because of the fact that a very large number of polymers and inorganic media can be used as membranes, there can be a great deal of control over separation selectivities.

With these pluses, however, membrane processes bring with them some heavy negatives which limit their usage as a viable choice for many separations. These include the following:

- Membrane processes seldom produce two pure products, that is, one of the two streams is almost always contaminated with a minor amount of a second component. In some cases a product can only be concentrated as a retentate because of osmotic pressure problems (this will be discussed later), while in other cases the permeate stream, because the membrane selectivity is not infinite, contains significant amounts of materials which one is trying to concentrate in the retentate.

- Membrane processes cannot easily be staged compared to such processes such as distillation and solvent extraction, and most often membrane processes have only one or sometimes two or three stages. This means that the membranes being used for a given separation must have much higher selectivities than would be necessary for relative volatilities in distillation. Thus the trade-off is often high-selectivity/few stages for membrane processes versus low-selectivity/many stages for other processes.

- Membranes can have chemical incompatibilities with process solutions. This is especially the case in typical chemical-industry solutions which can contain high concentrations of various organic compounds. Against such solutions, many polymer-based membranes, which comprise the majority of membrane materials used today, can dissolve, swell, or weaken to the extent that their lifetimes become unacceptably short or their selectivities become unacceptably low. Many polymers, for example, swell in the presence of aromatics and various oxygenated solvents.

- Membrane modules often cannot operate at much above room temperature. This is again related to the fact that most membranes are polymer-based, and that a large fraction of these polymers, as well as the glues, O-rings, gaskets, etc. which are used in typical modules, do not maintain their physical integrity at much above 100°C (373 K). This temperature limitation means that membrane processes in a number of cases cannot be made compatible with chemical process conditions very easily.

- Membrane processes often do not scale up very well to accept massive stream sizes. (There are some notable exceptions to this rule, however, including sea water desalination and hydrogen recovery.) Membrane processes typically consist of a number of membrane modules in parallel, which must be replicated over and over to scale to larger feed rates. This necessity of module replication, instead of simply being able to install bigger modules, means that the investment will rise more nearly linearly with throughput than would, for example, distillation. Thus, the economy of scale for membrane processes is not as pronounced as it is in some other separation processes. We will see this later when we study the production of nitrogen, for which membrane processes are economical for flows of up to nearly 40,000 cu ft/h (1200 normal cu m/h) of product, while much larger flow rates are the domain of cryogenic distillation and, to a lesser extent, adsorption.

- Membrane processes can be saddled with major problems of fouling of the membranes while processing some types of feed streams. This fouling, especially if it is difficult to remove, can greatly restrict the rate of permeation through the membranes and make them essentially unsuitable for such applications.

In the sections to follow, we will discuss:

- The theory behind membrane processes
- The nature of membranes, the forms in which they come, and the ways in which they are housed in modules
- Design considerations
- Examples of membrane processes of interest to the chemical industry
- Some speculations on new applications for this industry

Theory

In this section we will discuss the factors which determine the rate of transfer of species across a membrane, as well as other factors which are determinative in the performance of a membrane system.

Flux and driving force

For pressure-driven separations, the flux of a component A through a membrane can generally be described by Equation (5.1):

$$N_A = (P_A/L)\Delta\Phi_A \qquad (5.1)$$

where N_A = flux of component A through the membrane, mass/time-area

P_A = permeability of A, mass-length/time-force

L = thickness of the separating layer of the membrane, length

$\Delta\Phi_A$ = driving force of A across the membrane

For gas-phase separations, $\Delta\Phi_A$ is the partial pressure of component A on the feed side of the membrane minus the partial pressure of A on the permeate side, and Equation (5.1) becomes, for a pure component:

$$N_A = (P_A/L)\Delta\Pi \qquad (5.2)$$

where $\Delta\Pi$ = difference in total pressure across the membrane

The more common case is that in which the gas feed contains a mixture of components. In this case:

$$N_A = (P_A/L)\Delta p_A \qquad (5.3)$$

where Δp_A = difference in partial pressure of component A across the membrane

For liquid-phase separations, the picture is somewhat more complex. For a pure liquid, $\Delta\Pi_A$ is equal to the pressure difference across the membrane. If a small amount of solute is added to this liquid, and if this solute is held back—that is, partially or totally rejected—by the membrane, then the flux of the original liquid can be reduced by an osmotic pressure effect. The osmotic pressure difference between the feed and the permeate streams works to reduce the effect of the full pressure difference between these two streams. Thus the driving force for a solvent to cross the membrane is equal to the total pressure difference minus the osmotic pressure:

$$N_A = (P_A/L)(\Pi - \pi_A) \qquad (5.4)$$

where π_A = osmotic pressure of component A

Osmotic-pressure effects can be completely neglected in gas-phase separations and pervaporations, as well as in separations of species with molecular weights of greater than a few thousand daltons. For liquid systems containing solutes with molecular weights of less than about 1000 daltons, the effect can be surprisingly large—large enough to dramatically limit the ability to separate low-molecular-weight solutes from, for example, water. For a relatively low concentration of a solute A, the osmotic pressure of a solution is given by the van't Hoff equation (Eykamp and Steen, 1987):

$$\pi_A = C_A RT \tag{5.5}$$

where C_A = concentration of solute in the solution, moles/volume
      R = ideal gas constant, pressure-volume/moles-temperature
      T = absolute temperature, °R or °K

Qualitatively, osmotic pressure can be visualized as follows. Consider a system, shown in Figure 5.4, consisting of a membrane which is permeable to water but not to a solute contained in the water. If a solute-containing solution is added to one reservoir and pure water to the other, there will be a tendency for water to flow from the pure water side to the solute-containing side to minimize the difference in concentrations. This flow, which is called osmosis, will continue until a certain difference in the heights of liquid in the two reservoirs occurs. At that point no further permeation will take place. The pressure difference exerted by this difference in height is the osmotic pressure of the diluted solution. If more solute is then added to the

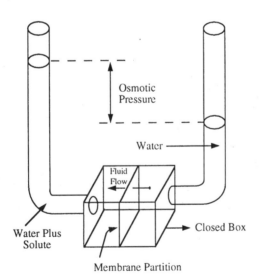

Figure 5.4 Effects of osmotic pressure.

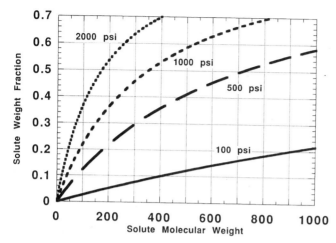

Figure 5.5 Osmotic pressure limitations on maximum solute weight fraction.

solute-containing side, more liquid will be drawn across the membrane, and the height will increase until it creates a pressure difference which represents the osmotic pressure of the new solution. If there is more than one solute retained by the membrane in the feed, then the osmotic pressure of the solute-containing stream would be equal to the sum of the osmotic pressures of all of the solutes.

As mentioned earlier, the osmotic pressure effect opposes increasing the concentration of the solute or producing a pure water stream. Figure 5.5, which has been derived from Equation (5.5), depicts the magnitude of this effect. As can be seen, for low-molecular-weight materials, the osmotic pressure at a given concentration is very large and clearly limits the ability to reach high concentrations of nonpermeating species. For example, in the molecular weight range of many typical oxygenated solvents (molecular weights of 100 daltons or less), if the pressure difference is 500 psi (3.5 MPa), then the maximum concentration to which an aqueous retentate stream could be raised would be in the order of only 22 wt% or less. At such a concentration, the osmotic pressure would equal the pressure difference across the membrane, and the flux would be zero. This would occur if the membrane were completely refractory to permeation of these solvents. Somewhat higher-concentration retentate streams could be produced if the membrane allowed some passage of the solvents.

Although this discussion has used water as the solvent, any solution will exhibit an osmotic-pressure effect against the pure solvent. Figure 5.5 will qualitatively apply to these cases also.

Separating capability

In gas-phase separations, Equation (5.1) can be used to compare the separating capability of a membrane for two or more species. If we write this equation for species A and for species B and then divide one by the other, we arrive at Equation (5.6):

$$N_A/N_B = (P_A/L)/(P_B/L)(\Delta p_A/\Delta p_B) = (P_A/P_B)(\Delta p_A/\Delta p_B) \qquad (5.6)$$

If we now consider a situation in which the driving forces are equal, Equation (5.3) simplifies to:

$$N_A/N_B = (P_A/P_B) = \alpha \qquad (5.7)$$

where α = selectivity

Equation (5.7) states that the ratio of fluxes through a membrane is equal to the ratio of permeabilities when driving forces are equal. This ratio is often called the *selectivity* of the membrane. It is usually defined such that component A has the highest rate of permeation of any component of interest, so that selectivities will always be greater than 1.

Membrane selectivity can be thought of as a close analogy with relative volatility in distillation and other vapor-liquid separations. Clearly the higher the selectivity (or relative volatility), the easier the separation. The range of acceptable values for membrane selectivity is considerably different, however, than the range for distillation. For the latter, relative volatilities as low as about 1.2–1.3 can be accommodated and are done so on massive industrial scales. Such low numbers are acceptable because in distillation it is extremely easy to stage the operation; 100 or more stages can be achieved in some columns. For membranes, staging is much more difficult. Most often, only one stage is used, and very seldom will one find more than three.

In liquid phase separations, it is common to use the term rejection when describing a membrane's ability to retain a solute and permeate a solvent.

Rejection is defined as

$$R = [1 - (C_{Ap}/C_{Af})]100 \qquad (5.8)$$

where R = rejection, %
C_{Ap} = solute concentration in the permeate
C_{Af} = solute concentration in the feed

Equation (5.8), then, is a measure of degree to which the membrane holds back or rejects the solute.

Operating characteristics

From these basic equations above and flow characteristics of the feed and permeate streams in the vicinity of the membranes we can ascertain some important operating characteristics of membrane processes. The flow patterns in various types of modules can vary widely, and three archetypical patterns are shown schematically in Figure 5.6. Cocurrent and countercurrent patterns are similar in concept to those encountered in typical shell-and-tube heat exchangers, except that the feed/retentate and permeate flow volumes change as a function of distance along the flow path. In countercurrent flow, the permeate flow at the point at which the retentate exits will be zero unless a purge or sweep stream is added to the permeate side of the membrane. Addition of a sweep stream is a relatively occasionally used option, and, as we shall see later, to do so requires a specially designed module and a special type of membrane. This will be discussed later.

Crossflow is less common than the other two flow patterns in heat exchange and other separation processes, but it is very common—indeed the most common—in membrane processes. In crossflow, the permeate crossing the membrane at any point is immediately removed from any further contact with any other part of the membrane. Thus the permeate fluid, by virtue of its composition, has no chance to influence the transfer of fluid at any other point.

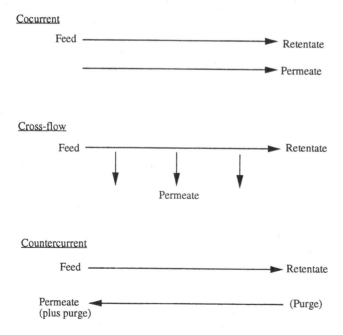

Figure 5.6 Archetype flow patterns.

In addition to the flow patterns of the feed and permeate streams, the ratio of the feed to permeate pressures, $P_{feed}/P_{perm} = R$, the stage cut, S, and of course the membrane selectivity, α, are also important in determining the degree of separation which can be attained in a membrane process. These effects are investigated in Figures 5.7a–d. These figures show exit stream compositions for retentates and permeates as a function of pressure ratio R, flow pattern, stage cut S, and selectivity α for a gas separation. The feed to the membrane unit is 11.8 mol% of an organic contaminant in nitrogen. The higher the value of R, the more efficiently the membrane will act as a separator. This can be seen in Figure 5.7b, which shows that about 60% of the retentate can be recovered as virtually pure nitrogen with $R = 15$ and countercurrent flow (stage cut = 0.4), while only about 25% can be recovered as pure nitrogen when $R = 1.5$ (stage cut = 0.75). These

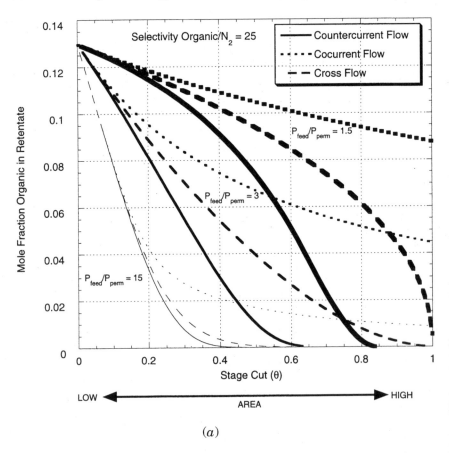

(a)

Figure 5.7a Retentate and permeate concentrations as functions of selectivity, stage cut, pressure ratio, and flow pattern. Feed = 11.8 mole percent organic in nitrogen.

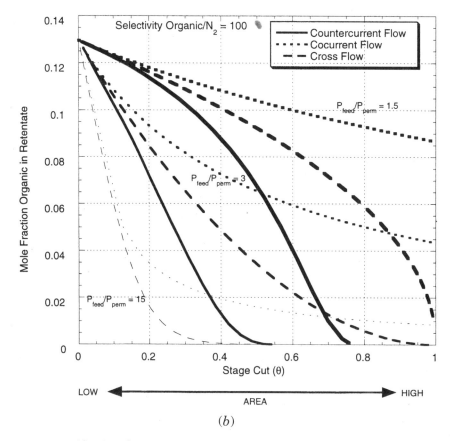

Figure 5.7b (*Continued*)

percent recoveries are only slightly smaller as a function of pressure if the selectivity is reduced from 100 to 25. Note that the permeate cannot be recovered in high purity under any set of conditions analyzed (Figure 5.7c and d), although higher purity is promoted by reducing the stage cut and increasing R. This helps to explain, for example, why high-purity oxygen cannot be recovered as a permeate from commercially available membranes, which have selectivities for oxygen over nitrogen of less than 10.

These figures also show that major differences in membrane performance, especially of the retentate concentration, can be caused by the nature of the flow patterns in the membrane module. Countercurrent flow is always preferred if overall separation is the chief criterion. Next in desirability is cross flow. At high values of R such as 15 and with an α of 100, countercurrent flow and crossflow are essentially equivalent in their separation performance, and when $\alpha = 25$, the two

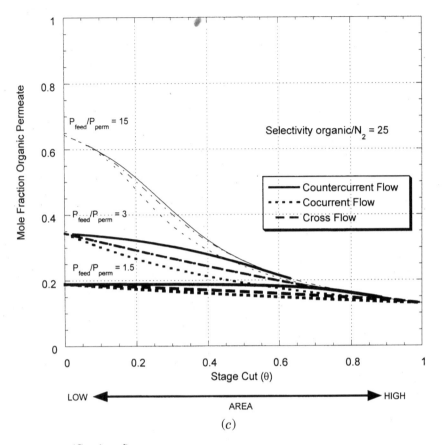

Figure 5.7c *(Continued)*

flow regimes are nearly equal in their performance. At lower values of *R,* cross flow becomes distinctly inferior. Almost always cocurrent flow should be avoided, as separation performance deteriorates dramatically compared to crossflow at all conditions covered in the figures. The greatest deterioration occurs at low values of *R.*

In the range of selectivities covered by these graphs, the separation performances of the membranes are not dramatically different but are clear enough to establish the directions of performances of membranes with both lower and higher selectivities. For example, with lower values of α lower than 25, a lower recovery of a given purity retentate would result, while the reverse would be true for values of α greater than 100. It also appears to be the case that flow patterns will produce the same dramatic differences in separation performance no matter what the membrane selectivity is.

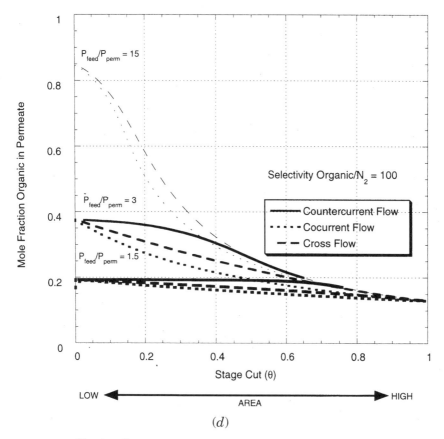

(d)

Figure 5.7d (*Continued*)

Not shown in these figures is the effect of temperature on separation performance. A seldom-violated rule is that, as temperature is raised, permeabilities will increase but selectivities will decrease. If a temperature change causes a polymeric membrane to cross its glass transition temperature, that is, the temperature at which it changes from a glassy to a crystalline polymer or vice versa, then the changes in both permeability and selectivity can be quite dramatic. Nearly always permeabilities rise sharply as the polymer changes from a crystalline to a glassy state.

It was mentioned earlier that it is possible to introduce a purge gas or liquid consisting of a component not present in the feed to a module to dilute the permeate flow. Characteristics of the module design will be discussed later, but the effect of this addition can be thought of as increasing the effective pressure ratio R for the separation. This comes about by virtue of the fact that the partial pressure of permeating species is reduced in the permeate, giving a higher effective pressure ratio between the feed/retentate and the permeate.

Sources of membrane selectivity

Materials permeate through membranes by passing through holes and/or by dissolving into the membrane at one surface, diffusing through the membrane to the other surface and then emerging from that surface (the solution-diffusion mechanism). The first mechanism is operative in the case of regular filtration, microfiltration, ultrafiltration, and some cases of nanofiltration. At dimensions of about 0.5–1.0 nm (5–10 Å), the second mechanism begins to become operative, and below 0.5 nm, there seems to be little disagreement that the solution diffusion mechanism predominates.

If a membrane had holes or pores which were all exactly the same diameter, then those molecules or particles whose diameters were smaller than the pore diameter would pass through the membrane, and those molecules or particles with larger diameters than the pore diameters would be totally rejected. Such a membrane would show an infinite selectivity, such as might exist with a molecular sieve zeolite. Several factors can complicate this simple picture, however, and cause the selectivity to be less than infinite. First, seldom will all pores in a membrane be exactly the same size. Thus smaller pores might exclude one component and larger holes permit it to pass. In such a case, the selectivity would be a function of the relative populations of various pore sizes. Second, molecules may be able to deform to some extent and may actually enter pores slightly smaller than their original diameter. In other cases, molecules of one type may adsorb on the walls of the pores and reduce the effective diameters of these pores. In this case a pore's effective diameter might vary with the feed/retentate and permeate compositions, depending on how much of the adsorbing component is in each of these streams. A third factor which can reduce a membrane's apparent selectivity is associated with its packaging: any leakage of the feed stream through tears, breakages, and incompletely sealed joints and connections will clearly reduce the apparent selectivity of the assembled membrane system compared to the inherent selectivity of the membrane alone. This factor becomes especially important in cases in which very high selectivities—in the range of 100–1000 or higher—are sought. Very small leakages of feed into the permeate can produce disastrous effects in these cases.

Actually membranes having pores can still show acceptable selectivities in cases in which the pore diameters are not small enough to exclude several components of the stream. For example, if a membrane is given an electrical charge by attaching ionic species to it, then in liquid systems the diffusion rates of ionic components can be drastically reduced compared to their diffusion rates through neutral membranes with similar pore diameters. In such cases, like-charged ions will be repelled by the membrane charge, and the counter-ions will also be retarded because the solutions must maintain charge neutrality. This mechanism is often used in desalination of seawater and brackish water to produce potable water.

In gas systems, a membrane having pore diameters slightly larger than the diameters of the various gas molecules can exhibit Knudsen diffusion characteristics. In such cases, the selectivity between two components will be equal to the inverse of the square root of the ratio of their molecular weights. Knudsen diffusion occurs when the mean free path of a molecule is greater than the pore diameter. Seldom, however, will this type of diffusion produce an acceptable selectivity. For example, the selectivity for separating hydrogen and methane is only 2.83 [$= (16/2)^{1/2}$], and the selectivity for separating hydrogen and carbon monoxide is only 3.74—values which, as we will see later, are hardly large enough to produce a substantially pure stream of either component. Actually there is evidence that higher selectivities than those predicted by the Knudsen-diffusion equation can be attained in some cases. These cases seem to involve surface diffusion through the pores rather than through the gas phase. Other cases may involve adsorption on the pore walls of one component which can make easier the diffusion of another.

The solution-diffusion mechanism seems to be operative for reverse osmosis, most gas separations, pervaporation, and vapor permeation. This mechanism is almost entirely limited to polymer membranes and to the case of hydrogen diffusion in various metal membranes containing palladium and a few other elements. The permeability, P_A, of component A is equal to:

$$P_A = S_A D_A \qquad (5.9)$$

where S_A = solubility of component A in the membrane, and
D_A = molecular diffusion coefficient of component A in the membrane.

This mechanism thus allows for two factors to control the selectivity. As a general rule, for liquid systems, since molecular diffusion coefficients of various components are usually fairly much the same, then solubility

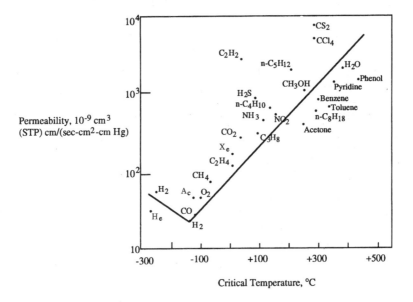

Figure 5.8 Vapor permeability as a function of critical temperature. Membrane material is polydimethylsiloxane. (*Adapted from Baker et al., 1987*).

in the membrane—usually a polymer—is the key to producing industrially important selectivities. In the case of separating ionic or organic components from water, chemical and ionic differences between water and the other components can be used to create highly different solubilities in various polymers. In gas systems, both factors can be important, as shown in Figure 5.8. For light, low-boiling gases such as hydrogen and the atmospheric gases, the permeability goes down as the molecular weight goes up. This suggests that molecular diffusion coefficients are quite important for these species, since the diffusion coefficient decreases as the molecular weight increases. As the normal boiling points of the species increase further, permeabilities then begin to increase and actually take on orders of magnitude larger than those of the light gases. Since molecular diffusion coefficients would still be decreasing with increasing molecular weight, it must be that solubility in the polymer becomes the controlling mechanism.

Membranes and Modules

Membranes have been in existence at least as long as life has existed on this planet. Living cells nearly always have membranes which allow for the selective passage of nutrients into the cells and the passage of wastes out of the cells. Such membranes are not forced to have fluxes of the magnitudes which are required by industrial processes, however, and so the development of membranes for industrial processes have actually involved some major technological breakthroughs before economical process embodiments could be arrived at. We will first discuss the nature of membranes which are used commercially and then move to the ways in which they are "packaged" for use in processes.

Membranes

Early synthetic membranes consisted of dense, thin, polymer films. Permeation fluxes were too low to be of commercial interest on any scale at all, and if a greater pressure difference were used to drive the permeate across the film, then the film had to be made thicker to resist being torn or deformed. The conceptual breakthrough came about in the 1950s when researchers began making asymmetrical membranes (Sourirajan, 1963). Several types of these membranes are shown in Figure 5.9. A membrane must simultaneously produce a flux as high as possible and resist deformation from the pressure difference, but these two functions are cleverly separated in the membranes shown here. Most of the membrane width consists of a very open structure which provides mechanical strength against the compressive force but does virtually no separating. Flow through this layer, if the membrane is well designed, is almost completely unimpeded. On top of this support layer is a dense film which is the separating medium and across which almost all of the pressure drop occurs. The thickness of the dense film is a very small percentage—perhaps 1–10%—of the total thickness of the membrane. Thus the membrane can be made thick enough to withstand the compressive forces but has a thin enough separating layer to provide acceptable fluxes. The dense film is always on the high-pressure side of the

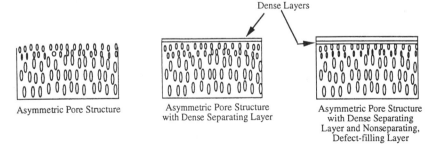

Figure 5.9 Asymmetric membranes.

membrane, that is, the feed side, since in this way maximum use of the support layer is made in stabilizing this film.

The separating, or dense, layer can be of two types. In the first case, the membrane is made in such a way that the polymer is densified on one surface, and thus the dense layer is the same polymer as the support layer. (For some membranes of this type, a second polymer is coated onto the dense surface to plug any holes remaining after the densification process, but this coating does not affect the basic selectivity of the dense-layer polymer.) In the second case, the support layer is formed, and then a second polymer is coated on one side to form the dense layer. This second method of forming an asymmetric membrane seems to be growing in popularity, since, at least in principle, the best polymer for forming the support layer and the best polymer for doing the separation can be used simultaneously to form the membrane.

Membranes formed from ceramic or other inorganic materials are typically made using two or more types of particles. The support layer in these cases consists of relatively large particles which form large channels for flow between the particles. On top of this layer is deposited a very thin network of much smaller particles which form the separating layer. Thus in an alumina-based membrane, the support layer consists of relatively large alpha-alumina particles, while the surface layer consists of much smaller gamma-alumina particles. The interstices between the particles are the holes through which the permeate passes.

The number of polymers, ceramics and other materials which have been either used in or proposed for commercial membranes is very large; a subset of this number is shown in Table 5.4. Many more of these polymers are in the category of "proposed" than those which have actually been commercialized. The polymers which are used most often in commercial installations are highlighted in this table. It can be seen that this number is relatively small. It should be noted that, in the case of coated membranes, the amount of the coating used is very small, since this layer is usually in the order of only a few micrometers to about 0.1 μm thick, and therefore expensive polymers can be used to create interesting selectivities with little impact on the overall cost of the membrane.

Membranes are configured for the most part in three ways: (1) as long cylinders (hollow fibers, capillaries, and tubes), (2) as sheets which are either rolled up or maintained in a flat condition, and (3) as various monolithic designs. Hollow fibers can have diameters from about 25 to a few hundred μm, capillary diameters are in the order of a very few millimeters, and tube diameters are in the order of a few centimeters. In the case of cylindrical configurations, the dense layer

TABLE 5.4 Examples of Membrane Materials for Reverse
Osmosis, Ultrafiltration, and Gas-Separation Applications

Polymers
 Cellulose derivatives, including acetates, other esters, and nitrate*
 Regenerated cellulose*
 Aromatic polyamides*
 Polyimides*
 Polybenzimidazole and azolone
 Polyacrylonitrile
 Polyacrylonitrile and derivatives
 Polysulfone*
 Poly(dimethyl phenylene oxide)
 Poly(vinylidene fluoride)
 Poly(methyl methacrylate)
 Polydimethylsiloxane*

Ceramics
 Alumina*
 Zirconia

Metals
 Palladium and palladium alloys

*Materials used most often.

can be either on the inside or the outside, since it is possible to feed a
process through the bore or outside. A comparison of various aspects
of these configurations is shown in Table 5.5. Here it can be seen that,
if membrane costs were the only issue, hollow fibers would be used
almost universally. Other factors related to suitability of polymers,
fouling, temperature of operation, flow patterns, etc. have made it
necessary to develop other forms as well, and commercial practice
seems to favor spiral-wound and monolithic forms as well as hollow-
fiber and capillary forms.

Modules

From an overall cost standpoint, not only is the cost of membranes
per unit area important, but also the cost of the containment vessel
into which they are mounted. Basically the problem is how one can
pack the most area of membranes into the least volume, to minimize
the cost of the containment vessel, consistent with providing acceptable flow hydrodynamics in the vessel. A module, then, consists of the
combination of the membranes and the housing. In this section we
will describe the basic module designs.

TABLE 5.5   Module Design Characteristics

Item	Hollow fibers	Spiral-wound	Plate-and-frame	Ceramic
Manufacturing cost ($/m²)*	50–200	300–1000	1000–3000	3000–12,000
Packing density (m²/m³)	~15,000	~900	~250	300+
Resistance to fouling	Poor	Moderate	Good	Good†
Parasitic pressure drops	Can be a problem	Moderate	Low	Low
Suitable for transmembrane ΔΠ>3.5 MPa?	Yes	Yes	Yes	No
Limited to specific types of membranes?	Yes	No	No	Yes

SOURCE: Baker, 1990, modified and updated.

*Includes the cost of the membranes and the module.

†In addition, ceramic membranes can be outfitted with on-line back-flushing to facilitate removal of fouling components.

Hollow-fiber modules.   The development of hollow-fiber modules to a commercial reality was accomplished by DuPont in the 70s and Monsanto in the 80s. The Monsanto Prism hollow-fiber separator technology provided the economic breakthrough which led to processes for hydrogen recovery and air separation. For both separations, the competition consisted of cryogenic distillation and pressure-swing adsorption. A schematic diagram of a prism module is shown in Figure 5.10. In this module the feed gas contacts the outside of the fibers, part of the gas permeates to the bores of the fibers and passes out of the module. The retentate flows along the outside of the fibers and then out of the other end of the module. The high-pressure feed side is separated from the low-pressure permeate side by a tube sheet somewhat similar to that found in a shell-and-tube heat exchanger. This tube sheet is typically a polymer into which the ends of the fibers are "potted," as shown in Figure 5.11. The fiber ends are placed into a reacting polymer system, such as an epoxy resin, while it is still a liquid. After the polymer becomes solid, one end is sawed off, opening the ends of all of the fibers so that gas can then flow into the permeate plenum. Fibers may either be sealed off at their other ends, as shown in Figure 5.10, or they may be looped and both ends potted

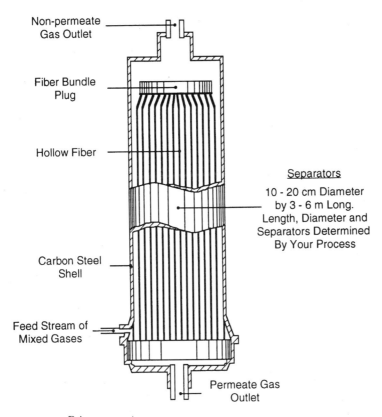

Figure 5.10 Prism separator.

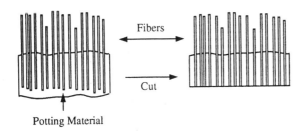

Step 1. Fiber ends are embedded into a polymerizing material such as an epoxy resin.

Step 2. After plug hardens, the bottom is sawed off, opening all fiber bores.

Figure 5.11 Potting of hollow fibers.

in the tube sheet, or occasionally tube sheets can be placed at either end of the containment vessel and permeate allowed to exit from both ends. If feed is added to the fiber bores, then there must be tube sheets at either end to allow the feed to enter and the retentate to flow out. Hollow-fiber modules are characteristically 4–8 in. (10–20 cm) in diameter and 3–5 ft (1–1.6 m) long. As time goes on and the technology improves, diameters and lengths of modules will increase over these values. On a volume basis, hollow-fiber modules can contain more surface area for permeation than any other type of module.

Spiral-wound modules. Spiral-wound modules represent a very different approach to creating high amounts of surface area per unit volume. A schematic diagram is shown in Figure 5.12. Two rectangular sheets of membrane material, with the dense layers facing away from each other, are sealed together on three sides. A spacer material is added inside the envelope, the open side of the envelope is connected to a porous tube, and then the envelope is wrapped up in jellyroll fashion around the tube. A second spacer material is added outside of the envelope as it is being wrapped up. This open area becomes the channel for the feed and retentate flow, while the open area inside the envelope becomes the channel for the permeate flow on its way to the central collection tube. It is also possible, as shown in Figure 5.13, to have more than one envelope wrapped around the central tube. At this time, modules containing spiral-wound membranes are used more than any other type of module for reverse osmosis, nanofiltration, ultrafiltration, and gas separations.

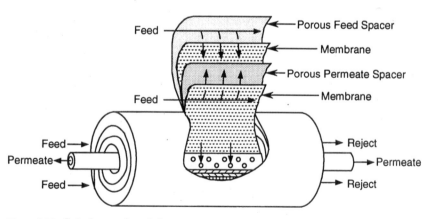

Figure 5.12 Spiral-wound module.

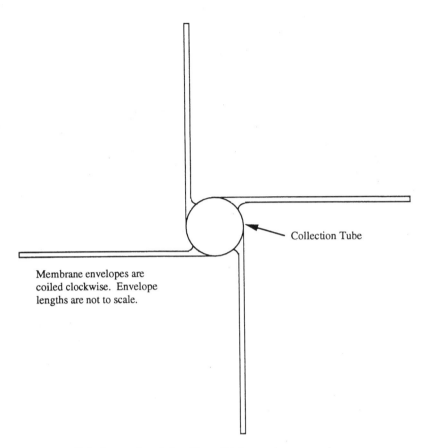

Collection Tube

Membrane envelopes are
coiled clockwise. Envelope
lengths are not to scale.

Figure 5.13   Spiral-wound module with multiple membrane envelopes.

Plate-and-frame modules.   Plate-and-frame membrane configurations
had their genesis in plate-and-frame filters, which have been in existence for well over a hundred years. Nevertheless, they have also
been subjected to modern engineering development, and today the
new plate-and-frame configurations represent an impressive and efficient way of contacting feed flows with the membranes. As in spiral-
wound configurations, an envelope of two membranes is formed with
a spacer in between. This envelope is roughly circular with a circular
hole in the center, These envelopes are packed onto a porous tube, as
shown in Figure 5.14. O-rings are placed between the envelopes and
next to the tube to complete the seal between the feed and the permeate sides. By the use of baffles to control the direction of the flow
across these envelopes, it is possible to maintain the feed/retentate
flow velocity nearly constant throughout the entire module by varying

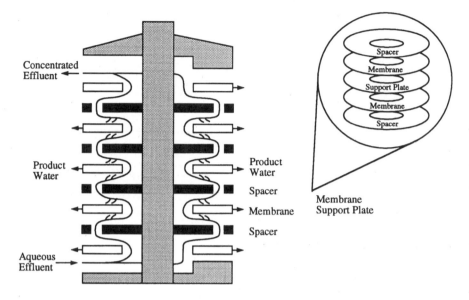

Figure 5.14   Assembled plate-and-frame module.

the number of envelopes between baffles. This flow-setting ability can be important in minimizing the buildup of permeating species near the membrane surface, and it can only be achieved in the plate-and-frame configuration in a single module. Plate-and-frame modules contain less membrane surface area than the two configurations discussed earlier, but the ability to engineer the high-pressure flow pattern better makes this module economical to use in a number of applications. A possible major advantage of plate-and-frame configurations is that they can apparently be designed and successfully operated at much higher transmembrane pressure drops than can other configurations. Currently hollow-fiber and spiral-wound modules seldom operate with transmembrane pressure drops of much above 67 bar (1000 psi), while plate-and-frame modules, designed and built by Rochem (Stanford and Vail, 1992), are commercially operating at twice that pressure drop and in the future will be operating at three to five times that pressure difference. High transmembrane pressure drops can become important in performing separations in which osmotic pressures are high (see Figure 5.5).

Nonpolymer, monolithic modules.   Monolithic configurations, two forms of which are shown in Figure 5.15, are quite different from the above-mentioned three. Monoliths become important when ceramic materials are used, since hollow-fiber, capillary, spiral-wound, and plate-and-frame configurations are usually not practical configurations for friable materials. The monoliths shown in the figure consist of a

a.   b.

• Feed flow is normal to the drawings and through the holes.

• The membrane is a thin layer on the surfaces of the holes.

• The permeate flows into the porous, solid structure, out the edges
  and into the shell surrounding the monolith.

Figure 5.15   Two ceramic monolith geometries.

porous ceramic material through which parallel holes for flow of the feed have been formed. This material has no separating ability and is porous enough that the permeate, which flows through it, will encounter very little pressure drop. On the surface of each hole is deposited a very thin layer of much smaller particles than exist in the monolith, and this layer performs the separation. The separating layer is most often gamma alumina or zirconia. The monolith can be made of alpha alumina or several different mixed-metal oxides. The size of the pores is controlled by the size of the particles on the surface. In practice, pore-size distributions can be controlled quite closely, and pore sizes can be varied from about 40 Å (4 nm) up to several microns. Thus these membranes are not capable of performing RO, nanofiltrations and gas separations, and in fact their use in ultrafiltrations is limited to the upper size range of this category. There is some indication, however, that classified work at the Oak Ridge National Laboratory (Fain, 1995) has succeeded in producing ceramic membranes with holes about an order of magnitude smaller than what is available commercially today. In the future these membranes may be available for use in the civilian world.

In a monolithic structure, the feed flows through the parallel holes, and the permeate flows through the separating layer and then flows primarily by gravity to the bottom of the monolith and into an annulus between the monolith and the shell which contains it. Once it reaches the shell, the permeate flows out through one or more exit lines.

A major problem with ceramic monoliths is that the area per unit volume of ceramic is much smaller than the area per unit volume of polymer membranes. This problem is a major factor in the higher costs of ceramic membranes versus polymer-membranes per unit area, as we shall see later. The design of Figure 5.15b, however, does represent a creative means of partially rectifying this problem. This design is sold by the CeraMem Co. However, the U.S. Filter technology is used in a very large fraction of the existing applications. The CeraMem design can also be modified to create a microfilter for removal of particles from gases, by blocking the opposite ends of adjacent channels and forcing the gas to cross at least one separating surface before leaving the monolith.

Nonpolymeric (primarily ceramic) membranes present some notable pluses and minuses compared to polymeric membranes. First, the temperature range of ceramic membranes is much broader than that of polymeric membranes. For example, the latter seldom can operate at temperatures much above 100°C, and in most cases 50°C is more likely a practical limit. Not only can higher-temperature operation become important in some cases (high-viscosity streams are a good example), but also, if a ceramic membrane becomes heavily fouled, the fouling material can often be burned off without damage to the membrane. Such an operation, which can be important in systems in which highly sterile conditions must be maintained, obviously could not be carried out with a polymeric membrane, for which chemical cleaning may be the only option.

Second, it is possible to backflush a ceramic membrane by increasing the permeate pressure to a value above the feed pressure to cause a reverse flow through the membrane. Usually this reverse-flow period is in the order of only about 1 s, and it can be performed on a time cycle which maintains an acceptably high permeation rate. Backflushing of a polymeric membrane can lead to separation of the dense layer from the porous layer, which will destroy the membrane. Third, ceramic membranes are less subject than polymeric membranes to chemical attack, swelling, and other types of damage by organic liquids or harsh liquids such as acids and bases. And fourth, in some instances ceramic membranes with the same molecular weight cutoff can have higher permeation rates than polymeric membranes.

On the other hand, as shown in Table 5.5, ceramic membranes, per unit area, are considerably more expensive than polymeric membranes. In addition, since the area for permeation per unit volume is much lower for ceramic membranes compared to polymeric membranes, the containment vessel will be much larger for the former. This fact and ceramic membranes' limited ability to separate low-molecular-weight molecules restrict the use of these membranes to rather special situations, and their usage in the chemical, petroleum, and allied industries is and will remain a small fraction of the usage of polymeric membranes.

Metallic, dense membranes have not achieved commercial significance to date. The most widely researched membranes are palladium and palladium alloys, which are used for selective passage of hydrogen over other gases. Such membranes were commercialized in the 1950s but could not compete economically with alternative processes and were taken out of service. It remains to be seen whether improvements in the technology will make them economically competitive in the future. This technology is used today in small units to generate pure hydrogen for use in laboratories.

Modules with purge streams. In all of the module designs discussed so far, there has been only an exit point for the permeate and no means of adding a stream to the permeate side. In some cases, the ability to do so would facilitate transfer across the membrane by reducing the partial pressure (or the concentration) of a permeating material. In practice, the use of a purge stream is not simple. In cases in which cross flow is used, as it is in spiral-wound, plate-and-frame and monolithic modules, purging the permeate side creates little or no benefit. The permeate passes virtually perpendicular to the membrane surface, and it is very difficult to get a purge stream next to the membrane surface to dilute the permeate. Only in the case of hollow-fiber and other tubular membranes does it seem practical to use a purge stream. With these membrane geometries and a flow pattern which resembles that of a shell-and-tube heat exchanger, the full beneficial effect of diluting the permeate side can be realized. A schematic diagram is shown in Figure 5.16. Such modules are sometimes called four-port modules, referring to the extra port on the permeate side for introducing the purge stream.

Design Considerations

The design of modules, that is, the type and configuring of membranes in modules, is virtually entirely the prerogative of the vendor compa-

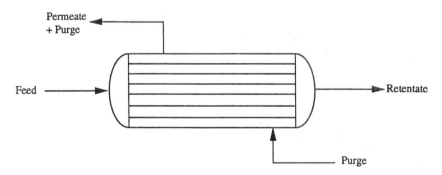

Figure 5.16 Four-port module.

nies. These companies have invested in polymer technology and have studied how membranes are configured to such an extent that user companies can almost never come up with superior membranes and modules. This being said, there are still design problems to which user companies can supply valuable inputs. These problems include fouling, concentration polarization, achieving high-stage cuts, and attaining high degrees of separation. In general, the first three of these problems are more prevalent and more severe in liquid separations compared to vapor separations, while the fourth can be difficult for both. These four problems and how to minimize them are discussed below.

Fouling

Fouling is probably the most frequently encountered problem with membrane processes, and especially with processes involving liquid feeds. Fouling is the coating of the membrane surface or blocking of the pores with a solid or gelatinous material which creates a barrier through which the permeating species must pass. The net effect of this blockage is to reduce the flux passing through the membrane. This blockage may also take the form of a second but nonselective resistance and thus decrease the overall selectivity of the membrane. Fouling materials can enter the module in the feed as particulates, gels or soluble, high-molecular-weight species, or they may precipitate from solution as part of the feed permeates. Because there is a flux toward the surface of the membrane, caused by the flow of material through the membrane, these fouling substances tend to migrate toward the membrane surface. This situation is shown in Figure 5.17.

There are several strategies which can be employed to ameliorate the effects of fouling. First is the reduction of the concentration of particles and gels in the feed entering a module. Prefiltration is nearly always a prerequisite for processes involving RO, NF, and UF. Even in many gas separations it is necessary to prefilter the feed stream. Relatively simple dead-end microfilters are often used in this service.

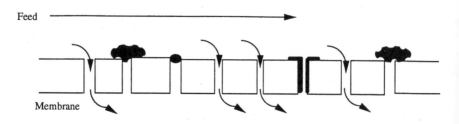

Figure 5.17   Membrane surface fouling.

Even in these cases, though, solids buildup of solids can seldom be totally eliminated.

Next, the proper form of membranes must be chosen. It can be seen from Table 5.5 that different types of membrane configurations have different susceptibilities to fouling: hollow fibers are the most susceptible and plate-and-frame configurations are the least. In that regard, note that there seems to be an inverse relationship between the surface area per unit volume of membranes and the tendency for these membranes to foul. Also, it appears that the more fouling-resistant membrane configurations cost more per unit of surface area. This fact contributes in a major way to explaining why we have several fundamentally dissimilar membrane configurations instead of just one.

A third strategy involves maintaining the velocities on the feed side of the membranes as high as possible. The higher the velocity, the more shearing action is exerted near the membrane surface, and the greater will be the tendency to re-suspend surface-coating in the feed/retentate stream. Modules obviously have their velocity limits, however, which must not be surpassed. Ceramic membranes typically have the highest velocities, which can range up to about 10 ft/s (3 m/s) or more.

A fourth strategy involves the use of a cleaning cycle. In many such cases, a relatively small fraction of the total number of modules is valved out of the product-producing system and attached to the cleaning system. A cleaning system of this type is shown schematically in Figure 5.18. Periodically the modules being cleaned are switched back to being

Membrane Module in Production Mode

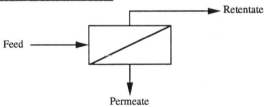

Membrane Module in Cleaning Mode

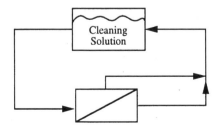

Figure 5.18 Cleaning cycle for membrane process.

product producers, and other modules are switched to the cleaning cycle. With time, all modules are cleaned. A second means of cleaning involves stopping the feed and taking all modules out of service at the same time. This strategy is only practical in cases in which the feed flow can be interrupted for a period without adverse consequences. Cleaning solutions either dissolve the solids or reduce their physical bond with the membrane surface so that they can be swept away. These solutions can consist of acids or bases, which are used in cases in which a change in pH will dissolve or loosen the solids, or special solvents.

A fifth strategy involves the use of *blowback,* or a reverse flow from the permeate side of the membrane to the feed/retentate side. Blowback is effected by increasing the pressure of the permeate stream to a value greater than the feed/permeate pressure. With polymer membranes, blowback is seldom possible because, as permeate is forced back through the membrane, it is possible for the bond between the dense layer and the support layer to rupture, causing destruction of the separating capability of the membrane. Ceramic membranes are another story, and indeed one of their main areas of application is to rapidly fouling feed streams. Blowback can be rather easily accomplished as shown in Figure 5.19. Some experimental results from a laboratory study using a ceramic membrane for microfiltration of very fine solids from a coal tar stream are shown in Figure 5.20. Note that fouling occurred rapidly in spite of the fact that a velocity of over 1 m/s was used across the membrane surface. This figure also shows that the fouling layer apparently does not grow indefinitely but rather grows to some thickness and stops, that is, the

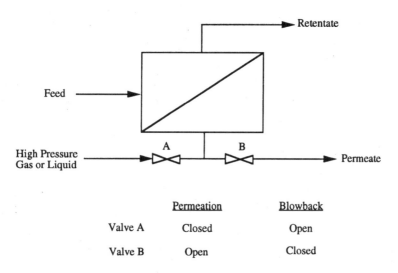

	Permeation	Blowback
Valve A	Closed	Open
Valve B	Open	Closed

Figure 5.19 Blowback scheme for membrane process.

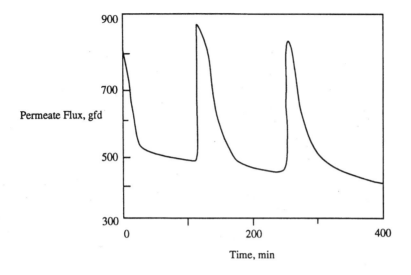

Feed: Coal tar containing 0.2% solids < 1 micron diameter

Figure 5.20 Results from blowback study.

permeation rate remains at a low but nearly constant rate. This thickness is the value at which the rate of solids deposition equals the rate of resuspension by the shearing action of the feed/retentate stream. Blowback in this case lasted only a very few seconds, and therefore that part of the cycle was an insignificant fraction of the overall cycle. On the other hand, dramatic increases in permeation rate were obtained immediately following blowback. Obviously a higher average permeation rate could be reached in this case by increasing the frequency of blowback, and in a real case, blowback would almost always be used before the steady-state flux is reached.

There is still another approach, which is different from the ones listed above, for minimizing the effects of fouling, and that is to create, by mechanical means, intense shearing right at the surface of the membrane. This approach has been commercialized in an innovative design by New Logic Research. In this design, circular plates containing membranes on either side are oscillated at a very rapid rate, creating a shearing action which disrupts the buildup of foulants at the membrane surface and keeps the membrane surfaces relatively clean. New Logic membrane units have found major uses in microfiltration of aqueous clay streams, resulting in retentates with the consistency of pastes. A second major use is the nanofiltration of various aqueous waste streams which contain substantial amounts of solids. One example is the concentration of pulp mill wastes. Although the number of mechanically enhanced units such as those of New Logic is still very small in the chemical and allied industries, the inherent advantages of this approach in treating high-solids-concentration feeds may stimulate increasing interest in this approach in the future.

Concentration polarization

Concentration polarization is somewhat akin to fouling in that a layer is formed near the surface of the membrane which is different in concentration from the bulk fluid. With concentration polarization, however, the restraining layer consists of a buildup in the concentrations of nonpermeating or slowly permeating components in the feed as the more permeable components pass through the membrane. The net effect of concentration polarization is to reduce the permeation rates of the more rapidly permeating components of the feed over what would be calculated using the bulk fluid concentration, since their concentrations next to the membrane surface will be lower than those in the bulk of the fluid. On the other hand, for those components which permeate relatively slowly, their permeation rates can actually increase, since their concentrations next to the membrane will be greater than those in the bulk of the fluid. The major detrimental effect of concentration polarization is thus a decrease in membrane selectivity.

As with fouling, the negative effect of concentration polarization is minimized by the use of feed/retentate velocities as high as possible, consistent with membrane hydrodynamic limitations. Of considerable interest in this respect is the fact that feed velocities in small-scale devices, discussed later, for evaluating membranes in the laboratory seldom approach those used in full-scale modules. As a result, the selectivities determined by these devices are usually lower than those obtained on the plant scale.

Promotion of high-stage cuts

Consider a module in which permeation is occurring. As more and more material is permeated, less and less material is present as retentate. The hydrodynamic effect of this fact is that, if the feed flow velocity is fixed near the feed end to be optimal for mass transfer, then this velocity will become suboptimal as the feed/retentate stream approaches the product end of the module. With most modules, this fact makes it impossible to reach stage cuts of, for example, 90% or so without suffering a major reduction in the efficiency of permeation. Since concentration polarization becomes worse as the velocity is reduced, both the apparent selectivity and the actual permeation rate will fall below their predicted values with no concentration polarization present.

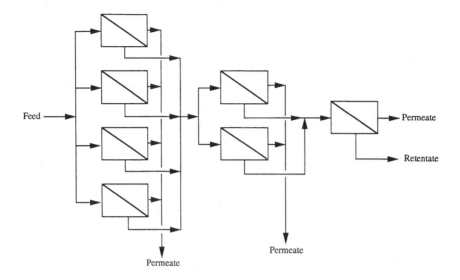

Figure 5.21   Membrane module cascade for high-stage cut operation.

A common way to overcome the detrimental effects of reduced velocities is to use a cascade of several stages of parallel membrane modules in which the number of modules in each stage decreases as the feed/retentate passes toward the outlet from the cascade. A typical arrangement is shown in Figure 5.21. If we set a criterion that a stage cut of 50% is achieved in each stage, then the number of modules in the next stage will be one-half of the previous stage. Thus in the three stages shown in this figure, nearly a 90% rejection could be achieved, with the velocity profiles remaining nearly the same in each stage.

Promotion of improved separations through staging

A major drawback of membrane processes is their inability to attain high degrees of separation in some cases. This problem is especially acute in situations in which solution-diffusion is the primary mechanism for attaining selectivity. Thus, for microfiltration and ultrafiltration, relatively high degrees of separation can often be attained, but for RO, gas separations and pervaporation, the problem of simultaneously attaining high-purity permeate and retentate streams can be daunting. This is in spite of some separation factors which can be dramatically higher than those normally encountered in distillation, as was discussed earlier. The problem is that membrane processes cannot be easily staged, such as distillation can. Fifty or more theoretical stages are easily attained in distillation, while the separation in a membrane module can be thought of as one stage. And even the separation in that stage is usually less than could be achieved if true countercurrent flow patterns on both sides of the membrane could be realized.

There are, of course, membrane-process flowsheets in which multiple stages can be realized by the use of multiple modules in series. Figure 5.22a can be considered a generalized flowsheet for simultaneously producing high-purity permeate and retentate streams. Note that the permeate flows progress to the left, while retentate flows progress to the right. In principle, as many stages as are necessary

a. Generalized Cascade

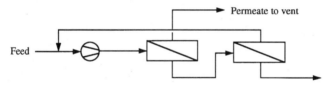

b. Two-stage process for retentate purification

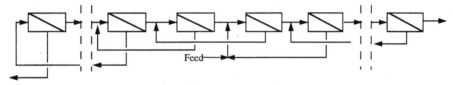

c. Three-stage process for nitrogen (retentate) purification

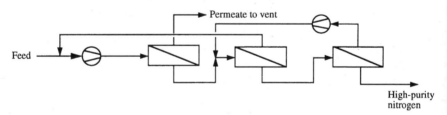

d. Two-stage process for high recovery and concentration of material in retentate

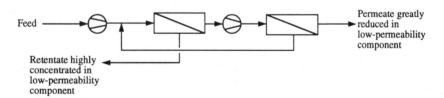

Figure 5.22 Multiple-stage membrane configurations.

can be configured in such a manner, but in practice, the story is not that simple. First and most important, any time that a permeate stream becomes a feed stream to another stage, either a pump or a compressor, if the stream is a gas, it is necessary to increase the pressure. (These pumps or compressors are not shown in Figure 5.22a; they simply add to the complexity already shown in this figure.) Thus a significant investment cost over and above the cost of the modules results. This additional cost is especially important in gas separations, since compression equipment can be expensive. In addition, the energy cost to recompress gas streams several times can be a major economic detriment. A second factor which militates against staging is that greater flows of feed to each module result, so that the membrane area required is generally increased. As a result, processes involving more than two or three stages are extremely rare. Some exceptions to this rule are the separation of uranium isotopes as the hexafluoride gas, and the recovery of very costly pharmaceuticals and biochemicals. In cases of this sort, the very high value of the product can compensate for the costly membrane flowsheet. Such cases are virtually unknown in the more conventional chemical and allied industries.

There are other multistage flowsheets which can be used to create one relatively high-purity or highly concentrated stream. Two such flowsheets for producing high-purity retentates are shown in Figure 5.22b and c. The first is a standard two-stage flowsheet for producing medium-purity nitrogen from air. In this case, note that the retentate does not have to be recompressed. On the other hand, the second-stage permeate, which is richer in nitrogen than air itself, is compressed and returned as part of the feed to the first stage. This recycle of enriched nitrogen, though it comes at the cost of a compressor and its associated energy requirement, reduces the air feed rate to the cascade and increases the product nitrogen purity which can be attained. Typical nitrogen purities are in the range of 95–99% in such processes. If a higher-purity nitrogen is required, hydrogen can be added to the stream to "burn" the remaining oxygen to produce water, which can then be easily removed. As an alternative to chemically reacting out the oxygen, a three-stage process (see Figure 5.22c) can produce nitrogen in the range of 99.9% purity.

The flowsheet of Figure 5.22d can be used to produce a high-purity permeate when the lower-permeability component in the feed is in small concentration, for example, a very few percent down to parts per million. Such might be important in concentrating a soluble catalyst as a retentate from a reactor-effluent feed stream, or in the recovery of any valuable, low-concentration, low-permeability material in a feed stream. The recycle of the retentate stream from the second stage is important to minimize losses of the high-value material in this stream. This flowsheet could also be used in concentrating a small amount of a contaminant as a retentate in a wastewater stream. In this case the second-stage retentate is recycled to maintain as high a recovery of the contaminant as possible.

It is interesting that neither of the flowsheets of Figure 5.22b and c nor any other membrane flowsheet has been able to produce oxygen (95+%) as a permeate at a price competitive with that from cryogenic distillation and pressure-swing adsorption. The reason apparently lies in the high investment and operating costs associated with the multiple stages and multiple recompressions required to produce a reasonable purity of oxygen. It should be remembered that to concentrate oxygen from 21 to 95% is clearly much more difficult than to concentrate nitrogen from 79 to 95%. We also learned earlier from Figure 5.7a–d how naturally difficult it is to produce high purities in permeate streams. To make high-purity oxygen via membranes will apparently require membranes with one to two orders of magnitude higher selectivities for oxygen to nitrogen than the values of present-day membranes, which typically are less than 10. In the case of the concentration of hydrogen in a permeate stream, recompression of the hydrogen to purify it further is seldom if ever used, and if hydrogen purer than can be produced in one membrane stage is required, a nonmembrane process is apparently more economical.

Other multistage flowsheets are of course possible, and we will see an example in the discussion of the recovery of organic contaminants from gas streams. But even with specialized flowsheets, the rule of no more than two or three stages seems to hold well.

Experimental programs

Because of the problems listed above, as well as others, it is almost always necessary to conduct experimental studies before a membrane process can be safely specified. Only in cases such as desalination and certain gas separations can a membrane process be specified without undergoing a test program. In the organic-chemical and petroleum-refining industries, chemical compatibility of polymer membranes with a feed stream—especially a liquid feed stream—constitutes a major problem. A second problem to be faced is the determination of the selectivity and permeation rates of membranes. Fouling must also be evaluated on samples of actual feeds because of the possible presence of small amounts of contaminants.

Round one testing can often be done cheaply in small cells from companies such as Osmonics. With these cells one can quickly determine which membranes will have the required chemical stability, and semiquantitative estimates can be developed for permeabilities and selectivities. A qualitative feel can also be developed for whether there is likely to be a serious fouling problem. Simple cleaning cycles can often be worked out. This small-scale screening work can be carried out in the user company's laboratories, and it is usually economical to do this work in-house.

To become more quantitative, however, it is necessary to test at least one full-scale module. Often vendor companies will have pilot-scale facilities for doing such testing, and if there are no problems such as highly toxic, odiferous, or chemically unstable feeds, it often makes economic sense to use their facilities. Companies which expect to develop a number of commercial membrane processes should install at least primitive, flexible pilot facilities.

Some Current Uses

The current uses for membrane-based processes in the chemical, petroleum, and allied industries are small in number and are mostly confined to rather small streams and are not usually used in large separation steps. Nevertheless, because of the important positive factors of membrane processes and the changing demands for separations, the uses for these processes seem destined to grow. We will study the opportunities for growth, first by discussing the current uses and then, by projecting from these uses and the nature of the research being carried out, the opportunities for expanded and larger uses. Below we will discuss current uses for gas separations, liquid separations, and pervaporation.

Gas separations

Gas separations represent the largest area of growth for membranes in the last 20 or so years. As was pointed out earlier, the revolution was begun by the Monsanto Company's Permea Division with the successful commercialization of hydrogen recovery units for purge streams from ammonia plants. (The Permea organization is now a part of Air Products and Chemicals Co.) A typical flowsheet is shown in Figure 5.23. An ammonia plant incorporates a high-pressure gas loop containing primarily nitrogen and hydrogen. Over time, argon, which enters with the nitrogen feed, builds up in the cycle and must be purged. Prior to 1980, this purge stream was usually subjected either to cryogenic distillation or to a PSA process to recover the valuable hydrogen from this stream. Then Permea's Prism process was commercialized, and within a few years, most large ammonia plants in the world were retrofitted with this technology. The Prism process

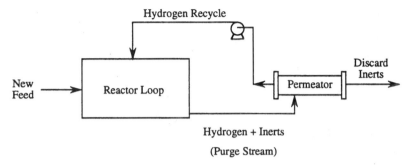

Figure 5.23 Ammonia plant flowsheet.

involved the use of hollow-fiber membranes (see Figure 5.11), based on polysulfone fibers, for the first time in a gas separation. So important was this breakthrough that the Prism technology won the 1981 Kirkpatrick Award (Rosenzweig, 1981), given by *Chemical Engineering* magazine, for the top commercialized development in the world.

It was in some ways fortunate both that the Prism process worked so well (for this established the viability of the technology) and that the market for ammonia plant purge streams was relatively small. The technology having been proven for high-pressure hydrogen recovery, companies selling membrane processes began to search for new applications for separating hydrogen. These applications began to present themselves rather quickly. By far the most prevalent of these uses were the recovery of hydrogen from mixtures of hydrocarbon gases and the adjustment of the hydrogen to carbon monoxide ratio in a synthesis-gas stream by the selective permeation of part of the hydrogen.

It is clear that membrane processes are not the only possibilities for separation of hydrogen in these streams. Cryogenic distillation and PSA are both viable technologies with a long history of successful applications for hydrogen recovery. In addition, PSA especially can produce hydrogen at a purity which is considerably greater than that which a membrane process can attain. Nevertheless, the simple, non-cyclic and often low-investment nature of the membrane flowsheet provides a major incentive for a number of applications. Process capacities have continued to rise over the last decade so that feed streams of over one million standard cubic feet per hour are now being processed. Table 5.6 compares operating conditions and other characteristics of the three systems. These data are over a decade old, and it is very likely that improvements in the intervening years in the membrane process make it even more competitive against its competitors.

TABLE 5.6   Product Purity, Recovery, and Relative Economics of Treating a Catalytic Reformer Off-Gas Stream

Process	Product purity, %	Product recovery, %	Relative capital cost	Relative operating cost	Relative product cost
Cryogenic	97.5	96.0	1.44	1.22	1.06
Membrane	96.9	89.4	1.00	1.17	1.09
PSA	99.9	86.0	1.40	1.00	1.00

SOURCE: Spillman, 1989.

Feed stream contains 75–85% hydrogen at 1.72 MPa (250 psig).

Polymer membranes were actually not the first membranes used to recover hydrogen commercially. In the late 1950s and early 1960s Union Carbide Corporation developed and installed several large hydrogen purification processes, based on palladium-silver alloy membranes, in its own and in a few other companies' facilities. Palladium and a small number of other metals can provide a virtually infinite selectivity for hydrogen over all other gases. The diffusion process involves (1) the dissolution of hydrogen into the membrane accompanied by the dissociation of the hydrogen into atomic hydrogen on the high-pressure side of the membrane, (2) the diffusion of the atomic hydrogen across the membrane, and (3) the recombining and release of hydrogen gas from the low-pressure side. Temperatures considerably above ambient and usually above 600 K are necessary to produce this activated-diffusion process at a reasonable rate. Silver was added to the palladium to minimize the brittleness and physical-cracking problems associated with the pure metal.

Despite the commercialization of the palladium technology, the improving economics of competing hydrogen-recovery processes and the maintenance costs and poisoning tendencies of the palladium-silver membranes eventually made them uneconomical, and by the late 1960s all of the commercial units were shut down. In the intervening years, research has been carried out around the world on such membranes, but no breakthroughs have resulted. Today palladium membranes are used in laboratory hydrogen generators but are no longer found in process plants.

Another major separation was spawned by the Prism technology: the production of nitrogen from air. Flowsheets were given in Figure 5.22b and c. In this process, air is compressed, usually to less than about 90 psig (0.7 MPa), cooled and then sent to the membrane section. Nitrogen purities are generally in the 90–95% purity range for the two-stage process and can be 99% or higher for the three-stage process. Originally nitrogen production units were quite small, with flows of 300–3000 ft³/h (10–100 m³/h) of nitrogen being common. Cryogenic distillation and PSA were powerful competitors. But membranes, being a less mature technology, improved in economics with additional research and development to the point that nitrogen flows an order of magnitude or greater are now in operation.

Cryogenic distillation and PSA are still strong competitors for nitrogen production, but membrane processes have carved out an important niche in the overall supply picture. A recent breakdown of the market was shown in the chapter on adsorption. The gains made by membrane processes have been primarily at the expense of PSA, while PSA has eroded part of the market earlier held by cryogenic distillation. If massive nitrogen flows are required, however, the economics produced by cryogenic distillation will never be rivaled by either of the competing processes. This is because cryogenic-distillation columns scale up at close to the 0.6 power, that is, a doubling of output would produce an investment of $2^{0.6}$ or only about 1.52 times that of the smaller unit. PSA and membrane processes have higher scale-up factors—probably in the order of 0.9 for the latter—so that they become increasingly disadvantaged as production needs increase.

With the increasing need for reducing harmful gas emissions from plants, new processes have been developed and installed for removing various organic materials from gaseous vent streams. Some sources of these vent streams are shown in Table 5.7, and various materials which can be recovered are given in Table 5.8. Membrane-based processes have found a definite niche for this task. This niche seems to be with the smaller-flow streams—streams whose flow rates are generally less than about 60,000 ft³/h (2000 m³/h). The membranes perform the seemingly unlikely feat of selectively permeating the much larger organic molecules compared to permanent gases such as nitrogen, which usually makes up the bulk of vent streams. That this can happen is a triumph for the solution-diffusion model's explanation of permeation through dense films (see the earlier discussion of this mechanism). In the case of these membranes, the separation layers, which are usually siloxane-based polymers, even though the diffusion coefficients of the large organic molecules are smaller than those of the permanent gases, the solubilities of the former are dramatically larger, and this effect is ultimately determinative in setting the selectivity of these membranes. Figure 5.8 showed a graph of the permeabilities of various materials through an organic-selective membrane. On the left-hand side of the graph, as molecular weight increases, the fixed gases show a declining tendency to permeate through the membranes. In this region molecular diffusion coefficients are the primary determinant of selectivity. This is the region

TABLE 5.7   Sources of Chemical Plant Gaseous Emissions

Storage tank vents

Storage tank filling operations

Process vents

Process leaks

Tank car and tank truck filling operations

Polymer devolatilization bin vent off-gas

Air-stripping process vents

Biotreatment facility off-gas

In addition to these sources, there are large numbers of sources in other types of facilities, such as spray painting lines, gasoline filling stations, dry cleaning plants, microchip fabrication operations, refrigerant change-out operations, etc.

TABLE 5.8   VOCs Which Can Be Removed from Vent Streams by Membrane Permeation

Acetone	Isopropanol
Benzene	Methanol
Butane	Methylene chloride
Carbon tetrachloride	Methyl ethyl ketone
CFC-11	Octane
CFC-12	Perchloroethylene
CFC-113	1,1,1-Trichloroethane
HCFC-22	Trichloroethylene
HCFC-123	Toluene
HFC-134a	Vinyl chloride

SOURCE: Baker, R. W., et. al. (1987), and promotional literature by MTR Corp.

which is operative for both hydrogen separation and air separation processes. But on the right-hand side of the graph, solubility effects begin to dominate, and permeabilities increase as the molecular weight increases in spite of the fact that molecular-diffusion coefficients for the higher-molecular-weight materials are larger.

A typical flowsheet for membrane processes is shown in Figure 5.24. In this process, note that the organic material is removed by condensation. This condensation is accomplished by compressing the feed, as shown in the figure, or by pulling a vacuum on the permeate

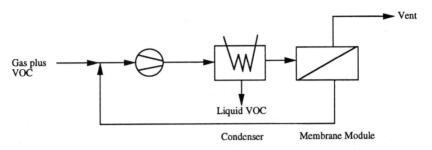

Figure 5.24 Flowsheet for Membrane Technology and Research (MTR) process for VOC removal.

and then recompressing the permeate, or by both strategies. Thus these processes work best on streams which are nearly saturated in the material to be removed. In ideal cases, the percentage removals will be in the range of 90–95% or slightly higher. Considerably higher removals can be effected, but the unit costs tend to rise substantially. One way to effect higher removals is to create a greater pressure ratio between the feed and the permeate. This will permit the permeate stream to be enriched more in the organic material. A second way is to reduce the temperature of the condenser in which the organic material is removed. In general, this will require refrigeration, which will add to the investment cost of the unit. If a relatively light constituent, such as propylene, is to be removed, a hydrocarbon-depleted retentate and a hydrocarbon-rich permeate stream can still be produced, but the separation will not be as dramatic as in cases in which condensation can be effected.

Hybrid processes involving membranes for treating gas streams are not yet numerous at all. There are a few installations of membranes followed by PSA for reducing pollutants in gas streams vented to the atmosphere. A schematic flowsheet is shown in Figure 5.25. Other proposed hybrids will be discussed later.

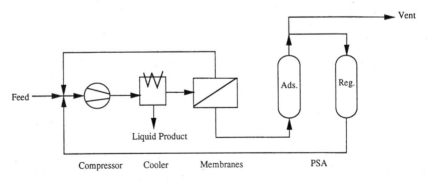

Figure 5.25 Membrane/PSA unit for VOC reduction.

Liquid separations

The use of membranes for separation of liquid mixtures in the chemical industry is truly minuscule, outside of the multitude of dead-end micro-filters used for particle removal. There are several reasons for this situation. First, for many streams, polymer membranes, glues, and other module components simply do not possess the chemical or dimensional stability required to stand up to a wide range of organic chemicals and other products. Second, often the desired processing temperature is higher than can be handled. Third, other conditions such as a very high or a very low pH can result in poor stability. And fourth, the less than complete separation which is typical of many membrane processes can be a major detriment to their use. The first three of these problems could be dealt with by ceramic membranes and improved polymer systems, but these technologies are still too new to have been used broadly.

The main uses which have developed have been confined to the treatment of water streams. Membrane use as one step in the production of boiler feed water is one example—one which is ubiquitous across industry. In a few cases, process water for a plant may be produced from brackish water via membranes. The concentration of various small-particle-containing, aqueous waste streams has been commercialized. Ultrafilters, for example, are being used to concentrate dilute latex (less than 1 μm) streams and thereby make recovery of polymer easier. Also, the New Logic membrane technology is beginning to be used in the concentration of inorganic pigments and clays.

Beyond these rather modest uses and others akin to them, membrane technology is very seldom used. Fortunately, as we shall see in the next section, new membrane technologies seem poised for a number of new applications.

Pervaporation and vapor permeation

Pervaporation involves the use of a liquid feed to produce a vapor permeate and a liquid retentate. Vaporization occurs as the permeating species pass through the membrane. The feed is normally supplied at an elevated temperature—usually slightly above 100°C—and at a pressure just above atmospheric. The permeate pressure is maintained at a vacuum well below atmospheric. The separation capability of the membrane can be amplified in some cases if the relative volatility of the primary permeating material is greater than one. The permeate side normally includes a condenser before the vacuum pump to recover the permeate as a liquid. A schematic diagram is shown in Figure 5.26. Figure 5.27 is a vapor-liquid diagram showing a normal equilibrium diagram and the separation which can be achieved in pervaporation.

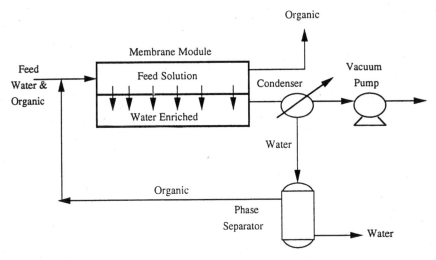

Figure 5.26 Pervaporation process.

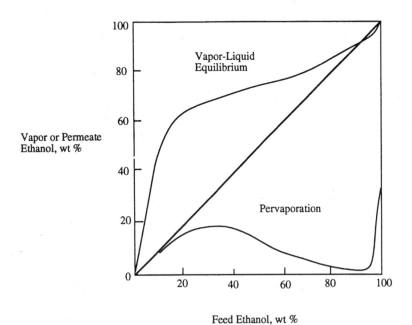

Figure 5.27 Pervaporation performance.

Considering all of the ballyhoo which attended the introduction and original development of pervaporation over a decade ago, one might have thought that this technology would have taken over a large fraction of the overall membrane business. This has hardly happened, and there are two main reasons:

- The heat to vaporize the permeate can only realistically be supplied by a reduction in the sensible heat, and therefore the temperature, of the feed. Reducing the feed temperature in turn reduces the driving force for permeation through the membrane. Thus only small amounts of permeate can be produced before the feed must be reheated, and this problem causes the need for several heat exchangers to reheat the feed as it passes through the permeator.

- Driving forces for permeation are low. Whereas in gas separations, trans-membrane pressure drops are usually from about 3 atm (0.3 MPa) to about 68 atm (6.8 MPa), pressure drops in pervaporation are seldom much over one atmosphere (0.1 MPa). The net effect is to produce relatively low permeation rates across the membranes and thus to create a need for more membrane surface area than is economically desirable in some cases.

As a result of these disadvantages, the market for pervaporation has been limited to two general areas: the removal of small amounts of water from organic materials and the removal of small amounts of organic materials from water. In such separations, the number of stages of reheat required can be manageably low, and the amount of material permeated is not excessive. In the first case, the most common application is in breaking azeotropes and producing a relatively pure organic material. In the second case, the usual objective is to remove small amounts of organic species from wastewater. Examples of uses of both types are given in Table 5.9. By far the most common systems employ hydrophilic membranes and selectively permeate water. Drying of isopropanol appears to be the most common use, and units processing up to 3300 lb/hr have been built. Pervaporation of the ethanol/water azeotrope is less common, although units processing up to about 200 L/h have been built. How large pervaporation units will be built in the future is unclear. This process, like almost all other membrane processes, does not scale up as well as competing processes, such as azeotropic distillation, from an investment standpoint. Therefore, like nitrogen production via membranes, pervaporation may always be relegated to processing relatively small flow rates, leaving the large streams to competing processes.

Pervaporation processes to remove small amounts of organic materials from water are far less common than those discussed above. As

TABLE 5.9   Pervaporation Applications

Production of relatively pure products from feeds containing azeotropes (selective permeation of water)

- Drying of isopropanol/water azeotrope

- Drying of ethanol/water azeotrope (liquid feed)

- Drying of ethanol/water azeotrope (vapor feed, i.e., vapor permeation)

Drying of organic liquids (selective permeation of water)

- Removal of low concentrations of water (up to a very few percent) from nearly pure organics

Aqueous-waste-stream cleanup (selective permeation of organics)

- Removal of low concentrations (up to a very few percent) of relatively hydrophobic species

---

mentioned above, the chief use for such processes appears to be the removal of various relatively low-boiling species from wastewaters. The more hydrophobic the species, the more easily they can be removed by pervaporation, but the same trend also occurs with adsorption, extraction, and steam stripping. The economic picture for this application versus other means such as adsorption and steam stripping is tenuous.

Another limitation for pervaporation is the purities of the permeate and the retentate. Like most other membrane processes, the separation is basically a one-stage operation. For systems such as isopropanol/water and ethanol/water, there will be alcohol losses in the permeate and water left in the retentate. If the highest-quality ethanol is to be produced, then the typical water specification for this material of about 100 ppm cannot be met in any reasonable way. Furthermore, the greater the percentage removal of water, the more stringent the conditions on the permeate side of the membrane. Thus, the vacuum would have to be greater than before and the condenser temperature lower than before—even below the freezing point of the condensing mixture in an extreme case. An additional problem accompanying high removals of the high-permeability component is that losses to the permeate of the low-permeability component will increase.

Vapor permeation resembles pervaporation except that the feed is a vapor instead of a liquid. Vapor permeation differs from the gas separations discussed earlier in that all components are near to their condensation points. Also, transmembrane pressure drops are usually considerably less than those used in fixed-gas separations and usually less than those used in the removal of organic materials from vent gases. Vapor permeation has two advantages over pervaporation for accomplishing the same separation. First, the heaters to reheat the feed are eliminated because there is no latent heat of vaporization which must be supplied. Second, mass-transfer on the feed side of the membrane is easier to accomplish through the vapor phase than through the liquid phase. This advantage should lead to a somewhat smaller membrane area to accomplish the same separation performance.

Future Uses for Membranes

As was pointed out earlier, the present impact of membrane technology on processes in the chemical, petroleum, and allied industries is exceedingly small. There is reason to believe, however, that this impact will grow in the future, driven for the most part by improvements in membrane performance (both in selectivity and in permeability) as well as by the need to reduce both gaseous and liquid emissions from these plants. It is our contention that most of the new and expanded uses will be in the environmental area.

As was also pointed out earlier, membrane technology is generally bedeviled by scaleup economics, in that distillation and adsorption generally improve in costs per unit of product more than membranes do as the feed-stream rate increases. Thus most applications will be relatively small in size, at least initially. Two major exceptions to this generalization are water desalination and hydrogen recovery. For water desalination, distillation has relatively high energy costs and materials-of-construction problems, and for adsorption there is no economical adsorbent- or ion-exchange-resin-based process to remove the large amount of salts from the feed. For hydrogen recovery, cryogenic distillation represents a relatively high energy-cost alternative, while PSA presents a highly complex flowsheet.

In the sections below we present speculations on new and growing uses for membranes.

Gas separations

The vast fraction of the present-day uses of membranes—nitrogen from air, hydrogen recovery and upgrading, synthesis gas ratio adjustment, carbon dioxide removal, water removal, and vent stream cleanup—will remain the workhorse applications in the future. But as the technology improves, it can easily be predicted that the maximum rates of production or removal for most of these applications will rise.

Air and hydrogen separations.   Membrane-based nitrogen generation will be increasingly installed to supply gas for blanketing storage tanks and process vessels, for instrument air and for purging of process vessels. Hydrogen separations have already found wide uses, with volumes presently reaching 1 million cu ft/h (70,000 m³/h). An earlier economic study, summarized in Table 5.6, shows that membranes compete well with cryogenic and PSA-based separations. As time goes on, this comparison ought to become even more favorable for membranes, since this technology is not as highly optimized as are the other two. There may also be possibilities for interfacing membrane and PSA units for simultaneously producing improved hydrogen recovery and purity. Two flowsheets are shown in Figure 5.28.

Metal membranes, which usually contain palladium either pure or as an alloy, for hydrogen purification still do not appear to be economical. Future applications will depend on whether problems of brittleness, coking, poisoning, and maintenance costs can be successfully

High-Purity Hydrogen with High Recovery

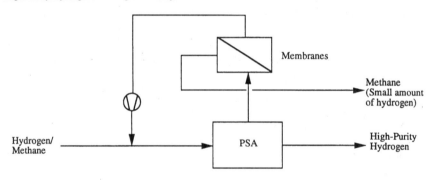

Synthesis Gas Ratio Adjustment Plus High-Purity Hydrogen

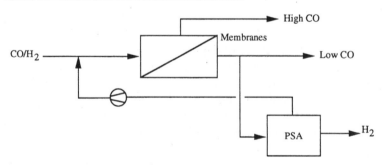

Figure 5.28 Membrane/PSA hybrids for hydrogen recovery.

dealt with. There has been a considerable amount of metal membrane research lately, and there is some hope that, in the future, a new generation of metal membranes will overcome these problem areas.

There is classified, ceramic-membrane technology at the Martin Marietta K-25 Site at Oak Ridge (Fain, 1995; Fain and Roettger, 1994) which purportedly can produce hole sizes in the order of 5 Å or smaller, which will allow permeation of hydrogen but not many larger hydrocarbons. If this technology is declassified, then it may compete well against polymer membranes and alternative processes for hydrogen purification, as well as for other gas separations involving molecules of quite different molecular diameters.

Oxygen production. Membrane-based, high-purity oxygen production is not a likelihood in the next 5–10 years, since selectivities with nitrogen many times higher than can presently be reached will be necessary. There has been a considerable amount of work done in the past on ceramic, oxygen-permeable membranes, but no development so far appears to be economical for the production of high-purity oxygen. Markets may develop for lower volumes of enriched oxygen (25–40%) for combustion purposes. Such enrichment increases the thermal efficiency of engines since not as much nitrogen must be heated and subsequently lost at high temperature from the engine. If such markets develop, membranes could be in a favored position to supply this need.

Carbon dioxide removal. Carbon dioxide removal will remain some-what problematical for use in cases in which a very high retention is required of the gas from which the carbon dioxide is being removed. Typical commercial selectivities between carbon dioxide and nitrogen or low-molecular-weight hydrocarbons are in the order of 20, and at such a value, there will be a significant transfer of other gases across the membrane along with the carbon dioxide. This can be seen by inspecting Figure 5.7a–d. Permeates with such a selectivity are hardly pure, and often such losses of process gas in the permeate are economically intolerable. Selectivities will probably have to reach close to 100 before carbon dioxide removal via membranes from large gas cycles such as would be found in hydrocarbon-oxidation processes will become economical. Even in these cases, a two-stage process (see Figure 5.22), in which the permeate is recompressed and repermeated, with the retentate being recycled to the first stage feed, may be necessary to minimize the loss of other gases.

Recovery of pollutants from gas streams. The largest growth application for membranes for gas separations is likely to be in the recovery of organic and other relatively high-boiling species from various streams of fixed gases. Examples of such streams were given earlier in Table 5.7. One important factor which favors the use of membranes is a partial pressure of the compound to be removed which is close to its vapor pressure. This near-condensation condition facilitates its removal, either from the permeate or by pressurizing the feed. A second important factor is the degree of removal required. Membrane processes appear to be best suited for feed streams from which 90–95% or so removal is acceptable. If removals of 99% or higher are required, compression costs can increase rapidly and other processes will compete more favorably. It is possible, however, that even in high-removal situations, membranes may fit in well with other processes to effect the most economical solutions. Two possibilities are shown in Figure 5.29. In these flowsheets, membranes remove the bulk of the organics—say 90+%—and the second process removes the balance up to the point of acceptability.

There are many other processes, shown in Table 5.10, which compete with membranes for removal of organic contaminants from gas streams. In Chapter 7 we will put these processes into an economic context which should be helpful in choosing between them.

Membranes plus PSA

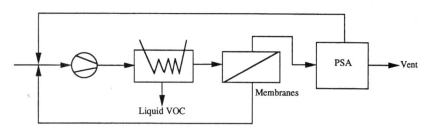

Membranes plus Rotary Wheel

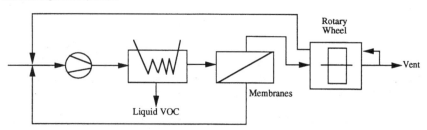

Figure 5.29 Membrane-containing hybrids for VOC removal.

TABLE 5.10   Competing Processes for Vent Stream Cleanup

Absorption

Adsorption

  Pressure-swing adsorption (PSA)
  Circulating-bed temperature-swing adsorption (TSA)
  Fixed-bed TSA
  Wheel-based TSA

Biosorption/biofiltration

Condensation—cooling of the stream using cooling water or refrigeration

Freezing—indirect cooling with liquid nitrogen to freeze all organic constituents in the stream onto solid surfaces

Incineration

Membrane processes

Membrane reactors.   Membrane reactors have had no significant applications up until now. The gas-phase reactions most often being investigated are dehydrogenations. Such reactions are characteristically equilibrium-limited at typical operating temperatures, and since in many cases the dehydrogenated product is difficult to separate from the reactant, there is a major incentive to increase the conversion as much as possible. This can be done by selectively permeating hydrogen. The common problems encountered in membrane-reactor research are (1) the inability to produce an essentially pure hydrogen stream when dehydrogenating light paraffins such as ethane and propane, and (2) the inability of the membranes to operate at typical dehydrogenation temperatures (generally greater than 400°C). The first problem arises from the fact that, if hydrocarbons must be recovered from the permeate, the savings in overall recovery and purification costs will suffer. With respect to the second problem, it is possible to separate physically the dehydrogenation reaction from the membrane-separation function (Figure 5.30) and therefore use heat exchange to reduce the gas temperature to the point that polymer membranes can be used. The gas as it exits from the membrane step must then be reheated through a second heat exchanger to achieve the desired dehydrogenation temperature. However, the investment cost of the heat exchangers, especially since they will be gas-gas exchangers and therefore of considerable area, will excessively add to the investment of the process. Recent work with relatively high-operating-temperature, ceramic-plus-polymer combinations (Rezac et al., 1995) give indications of easing but probably not completely eliminating this problem.

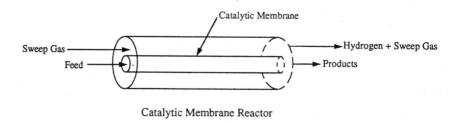

Catalytic Membrane Reactor

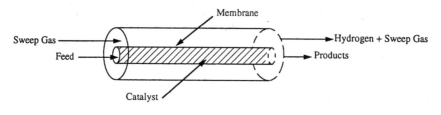

Inert Membrane Reactor

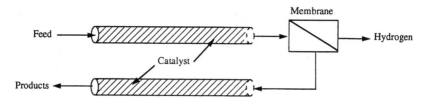

Reactor Plus Separate Membrane Unit

Figure 5.30   Reactor plus membrane combinations for dehydrogenation reactions.

Liquid separations

Whereas for gas separations, most of the new uses will be rather mild extrapolations from what is being done now, this will not be the case for liquid separations. At least two new uses should burgeon in the future: waste treatment and the recovery of soluble catalysts from products. The former category, that is, the removal of various species from wastewater, will provide a much greater number of uses. In time, it is likely that an approach to wastewater treatment involving separation processes will become a competitor economically for biotreatment. Separation processes also recover organic materials from the wastewater rather than destroying them as does biotreatment. If the recovered materials have economic value, separation approaches will take on an even more important advantage. This case will be studied in Chapter 7. Various ions are already being removed from waste streams. For example, over 100 RO units are in operation for concentration and recovery in nickel plating baths. The recovery and concentration of dilute latexes has also been commercialized on small streams. Oil/water emulsion-breaking to produce a clean water stream has also been demonstrated. The concentration of soluble polymers and surfactants is technologically feasible and should find considerable usage in the future.

Wastewater applications.   One of the most intriguing uses is the separation of relatively hydrophilic or polar species, such as glycols, acids, and the like, from wastewater. In many such cases a small concentration of a material cannot be separated by distillation at a reasonable cost, especially as the boiling point of the material increases. If water is the lower-boiling component, then large amounts of water would have to be boiled up to concentrate the waste materials, and the resulting energy cost would be excessive. Also, with respect to extraction, hydrophilic organics will also be difficult to separate from water because solvents which exclude water will also tend to exclude the organics. Adsorption will also suffer by having to remove polar constituents from water; adsorbent capacities will be low because water will compete strongly for the adsorption sites.

A number of RO and NF membranes show a tendency to selectively reject organic pollutants and permeate water. Figure 5.31 shows a band of rejections, as a function of molecular weight, for various organics with various membranes. This figure shows that, for a very large fraction of even hydrophilic organics, rejections of 90–99+% are possible. Therefore membranes would appear to be a very reasonable technology to investigate for this service.

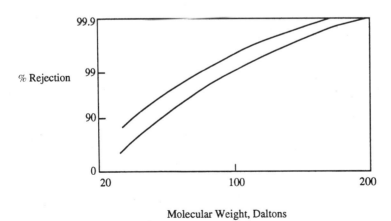

Molecular Weight, Daltons

Figure 5.31 Approximate rejections for nonionized organic species by RO membranes.

One important consideration, discussed earlier, for concentration of low-molecular-weight materials is the problem of osmotic pressure, which becomes worse as the molecular weight of the material being retained decreases. Since, in the presence of osmotic pressure, higher transmembrane pressure drops must be used, these pressure drops can become excessive for many types of membranes and modules. Recently Rochem (Stanford and Vail, 1993) has commercialized membrane systems for cleanup of landfill leachate which operate at pressure drops of 2000 psi (13 MPa), and there is some indication that eventually achievable pressure drops could increase to 4000 psi (27 MPa)–5000 psi (33 MPa). Such pressure drops can result in water recoveries of 80%, which results in substantial concentration of low-molecular-weight materials in water (see Figure 5.5).

Membrane-process investment costs will prove to be important in their being accepted into waste-treatment service. In all likelihood, rates of water permeation will have to exceed 10 gal/ft² day, and assembled modules will have to cost no more than about \$10–20/sq ft (\$100–200/sq m) of membranes plus modules to be competitive with biotreatment for large-scale uses. This comparison will be made in Chapter 7.

Catalyst/product separations. Soluble-catalyst reaction systems involving close-coupled membrane systems for return of the catalyst to the reactor while permeating the products represent a useful area for research and development. A schematic diagram is shown in Figure 5.32. Today a few bio-catalyzed reactions are operated in this fashion, but no larger-scale reactions are known to have been commercialized. Table 5.11 gives some examples of soluble-catalyst reactions. A number of these are carried out on very large scale and could be improved in some cases by the use of membranes for recycling of the catalyst. Especially when catalysts, reactants, or products are thermally labile, the ability to make the separation at low temperatures, compared to distillation, for example, is a major plus for membranes. The major minus, so far, for membranes is the lack of chemical and/or dimensional stability of most polymer membranes in the presence of major concentrations of organic species.

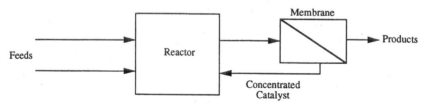

Figure 5.32 Reactor with soluble catalyst recycle.

TABLE 5.11   Examples of Reaction Products Involving Soluble
Liquid-Phase Catalysts

- Esterifications to give formates, acetates, propionates, butyrates, and many others
- Transesterifications or ester interchanges
- Hydroformylations to give methanol, butyraldehyde, propionaldehyde, and others
- Polymerizations to give polyethylene glycols and polypropylene glycols, as well as a variety of solution-phase polyolefin processes
- Ethoxylations to form nonyl phenol-based and nonionic surfactants
- Glycol ethers via an alcohol plus ethylene oxide or propylene oxide

A variation on the catalyst/product separation theme is that of selectively permeating one product from a reacting mixture to displace the equilibrium toward the formation of more products. Esterifications, transesterifications, and liquid-phase dehydrogenations are prime examples of this type of situation. For example, the selective permeation of water—perhaps by pervaporation—in an esterification reaction might be accomplished by pervaporation.

Low-molecular-weight organic separations.   Recent information from Exxon (Ho et al., 1995) reveals that liquid-phase organic separations can be performed between species of much the same molecular weight of different chemical types. In their work aromatics are selectively permeated at the expense of nonaromatics. This allows the recovery of a relatively-high-octane stream along with a relatively low one which can be used for other purposes than standard engines.

Hybrid systems.   A number of hybrid systems—membranes close-coupled with another process like distillation to perform a given separation more cheaply—have been proposed. BP America has developed the most widely publicized of these: the separation of light paraffins from olefins. Propane from propylene is their prime example. This separation is quite difficult to make by distillation. A process (Figure 5.33) has been proposed for debottlenecking a propane-propylene splitter column by removing part of a vapor stream from the column, contacting it with membranes and returning the propane-rich retentate to the column and the propylene-rich permeate to the product stream. Unfortunately no commercial membrane has a selectivity high enough to separate similar-carbon-number olefins from paraffins, and so in the BP process, an aqueous solution containing silver ions contacts the permeate side of the membrane. The olefin permeates and is selectively complexed by the silver, while the paraffin is

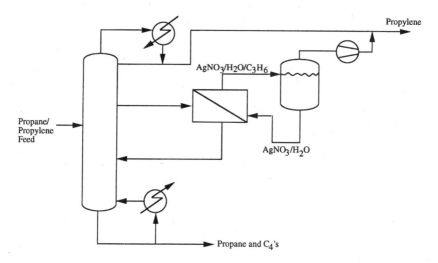

Figure 5.33 BP hybrid distillation/membrane process for propane/propylene separation.

rejected because, not only will it not complex with silver, but in addition it has almost no solubility in water. The olefin is subsequently stripped from the aqueous silver solution, which is then returned to the membrane. This process was operated in a pilot plant, but to date no commercializations have been announced. This process's economics would undoubtedly improve substantially if a highly olefin-selective membrane were a commercial reality.

As discussed in Chapter 2, pervaporation membranes are used in conjunction with distillation to dehydrate azeotropes. Such processes are used commercially in both Europe and Japan. Whether or not hybrid processes involving membranes will achieve widespread commercialization is still unclear. There are some possibilities in waste treatment, and these will be discussed in the chapter on process selection.

New membrane materials. New membrane materials are indeed one of the most exciting possibilities for the future. Table 5.12 gives examples of what may emerge in the next decade. We have not seen the end of new polymers being proposed, although performance gains over existing systems will in most cases be small. The addition of materials to polymers to facilitate the permeation of one component at the expense of another have been studied for some time. Materials added to the polymer include molecular sieves, ions such as $Ag+$ for facilitating olefin permeation and various hemoglobin mimics and similar compounds for facilitating gas-phase oxygen permeation. No commercializations are known, and the future of such membranes is difficult to predict.

TABLE 5.12 **Membrane Materials Which Could Become Available in the Future**

■ New polymers with improved selectivity

■ Ceramic and/or metal membranes with uniform-diameter pores of 5–20 Å

　- Molecular sieving of gas streams

　- RO and NF separations on the basis of molecular size

■ Polymer-based membranes with dramatically improved resistance to a wide range of organic solvents

■ Facilitated-transport membranes

　- Improved oxygen/nitrogen selectivity for air separation

　- Separation of olefins from paraffins

■ Improved-water-flux RO membranes and modules for use in large-volume wastewater applications

■ Improved module designs for minimization of concentration-polarization and flow-maldistribution effects

■ Improved module designs for successful separation of high-viscosity and high-solids-content feeds

---

Ceramic membranes seem destined for major expansions of use, especially if pore sizes can be reduced to 5 Å or so. With such a pore diameter, ceramic membranes could become aggressive competitors for various RO, NF, and UF applications. Ceramic membranes have the advantages of (1) a high degree of stability in the face of organic solvents compared to most polymer membranes, (2) an improved capability for removal of fouling layers on the membrane by backflushing, (3) the ability to operate at much higher temperatures, and (4) the possibility that considerably higher fluxes can be achieved compared to those of polymer membranes. These higher fluxes will definitely be needed in many applications because it is unlikely that ceramic-membrane modules will ever be able to achieve price parity with hollow-fiber and spiral-wound modules per unit area. Another drawback of ceramic membranes is their brittleness, although this seems to be a manageable problem for the most part.

Metal membranes neatly solve the brittleness problem and many of the module fabrication problems associated with ceramic membranes. But metal membranes have not found nearly the number of applications that ceramic membranes have when pores well below one micron are required. It is apparently much more difficult to get uniform and stable pore sizes in these pore diameters. Metal membranes will still remain prime contenders for separations in the MF range.

Pervaporation

As mentioned earlier, pervaporation uses are for the most part limited to removing relatively small amounts of one material from large amounts of another. So far, the primary use has been the removal of water from the ethanol/water and isopropanol/water azeotropes. In addition, the volumes of feed streams have been relatively small—up to about 6000 L/day of feed ("Membrane News Column," 1994). Adsorption processes compete directly with pervaporation for feed flows of this size and somewhat larger. For even larger feed flows, azeotropic and extractive distillation appear to be the processes of choice.

With time, pervaporation technology will improve and costs will be lowered. For situations in which water is to be permeated at the expense of organics, it is easy to predict that pervaporation will be used in larger and larger applications in azeotrope-breaking. In fact, a new composite membrane for dehydrating azeotropes has been reported which may impact this application in the short term (Ellinghorst et al., 1996).

For situations in which organics are to be permeated at the expense of water, a considerable number of environmentally oriented uses could develop. In particular, the separation of a wide range of relatively hydrophobic organics—boiling up to and above 200°C—should be separable from water because of the fact that activity coefficients of dilute organics in water can be very high (Hwang et al., 1992a; Hwang et al., 1992b); therefore, there will be a substantial selectivity for them. Pervaporation should begin to be used on relatively small streams—a few gpm. As stream flows increase 10- to 100-fold, steam stripping and air stripping will become formidable competitors from an investment standpoint. Steam stripping will feature a lower investment compared to pervaporation, but pervaporation should have an energy-use advantage. Air stripping will be a possibility in situations in which cheap incineration is available.

References

Baker, R. W., *Membrane and Module Preparation,* in *Membrane Separation Systems,* DOE Final Report No. DE-AC01-88ER30133, 1990.

Baker, R. W. et al., *Journ. Memb. Sci.,* **31,** 259, 1987.

Cheryan, M., *Ultrafiltration Handbook,* Technometric Publishing Co., Lancaster, PA, 1986.

Ellinghorst, G. et al., "Dehydration of Organics to Very Low Water Content by Pervaporation," presented at the 8th Annual Meeting of the North American Membrane Society, Ottawa, Ontario, Canada, May 1996.

Eykamp, W. and J. Steen, "Ultrafiltration and Reverse Osmosis," in Rousseau, R. W., ed., *Handbook of Separation Process Technology,* Wiley-Interscience, New York, 1987.

Fain, D. E., "Inorganic Membranes: The New Industrial Revolution," presented at the 3rd International Conference on Inorganic Membranes, Worcester, MA, July, 1995.

Fain, D. E. and G. E. Roettger, "Hydrogen Production Using Inorganic Membranes," Conference 94051-3, ORNL/FMP-94/1, 51, NTIS, Springfield, VA, 1994.

Ho, W. S. W. and K. K. Sirkar, eds., *Membrane Handbook*, Van Nostrand Reinhold, New York, 1992.

Ho, W. S. W. et al., "Membrane Separation of Aromatics from Fuels," presented at the Engineering Foundation Conference on Separation Technology, Snowbird, UT, 1995.

Humphrey, J. L., et al., "Membranes," in *Separation Technologies—Advances and Priorities,* 133, DOE/ID/12920-1, NTIS, Springfield, VA, 1991.

Hwang, S.-T. and K. Kammermeyer, "Membranes in Separations," in *Techniques of Chemistry,* Vol. 7, Wiley, New York, 1975.

Hwang, Y.-L. et al., *Ind. Eng. Chem. Res.,* **31,** 1753, 1992a.

Hwang, Y.-L. et al., *Ind. Eng. Chem. Res.,* **31,** 1760, 1992b.

Kesting, R. E. and A. K. Fritzsche, *Polymeric Gas Separation Membranes,* Wiley-Interscience, New York, 1993.

Klein, E. et al., "Membrane Processes—Dialysis and Electrodialysis," in Rousseau, ed., *Handbook of Separation Process Technology,* Wiley-Interscience, New York, 1987.

Koros, W. J., *Chem. Eng. Prog.,* **91,** October 1995, p. 68.

Lee, E. K., "Membranes, Synthetic, Applications," in *Encyclopedia of Science and Technology,* Vol. 8, Academic Press, New York, 1987, p. 20.

"Membrane News Column," *Membrane Quarterly,* **9** (2), 33, 1994.

Membrane Separation Systems: A Research Needs Assessment Final Report, Vol. 2, DOE/ER/30133-H1, NTIS, Springfield, VA, 1990.

Noble, R. D., and S. A. Stern, eds., *Membrane Separation Technology: Principles and Applications,* Elsevier, New York, 1995.

Parekh, B. S., *Reverse Osmosis Technology,* Marcel Dekker, New York, 1988.

Porter, M. C., ed., *Handbook of Industrial Membrane Technology,* Noyes Publications, Park Ridge, NJ, 1990.

Rezac, M. E. et al., *Ind. Eng. Chem. Res.* **34,** 862, 1995.

Rosenzweig, M. D., *Chemical Engineering,* **88,** No. 24, 62, 1981.

Sirkar, K. K., "Membrane Separation Technologies: Current Developments and Future Opportunities," paper given at AIChE National Meeting, Miami, November 1995.

Sirkar, K. K. and D. R. Lloyd, eds., *New Membrane Materials and Processes for Separation,* AIChE Symposium Series 261, Vol. 84, New York, 1988.

Sourirajan, S., *Ind. Eng. Chem. (Fund.),* **2**(1), 51, 1963.

Sourirajan, S., *Reverse Osmosis,* Academic Press, New York, 1970.

Spillman, R. W., *Chem. Eng. Prog.,* **41** (1), 1989.

Stanford, P., and P. Vail, *Industrial Water Treatment,* **11,** 52 1992.

Turner, M. K., ed., *Effective Industrial Membrane Processes: Benefits and Opportunities,* Elsevier Applied Science, New York, 1991.

6

Energy Considerations

Introduction

The amount of energy consumed is an important part of the total operating costs of separation processes. In distillation, energy consumption is high and tends to be the weakness of the process. On the other hand, membrane processes tend to be energy-efficient. Indeed we shall show that the energy required for the separation of a gas mixture by a membrane process approaches the thermodynamic minimum as the pressure drop across the membrane approaches zero. The energy efficiency of other processes tend to fall between distillation and membrane processes.

In this chapter we present fundamentals of energy consumption in distillation, and provide guideline values of the amount of reboiler energy required for the purification of specific products. We also provide a comparison of energy consumption in distillation versus extraction processes.

Relationships for thermodynamic efficiency of various separation processes are developed. Whereas energy consumption is useful to determine energy costs for specific applications, thermodynamic efficiency is useful for comparison of performance of processes.

Energy Consumption in Distillation

Fundamentals

Total energy consumption by the chemical process industries (CPI) in the United States is over 5 quads/yr (1 quad = 10^{15} Btu, equivalent to 170 million barrels of oil). Separation processes account for about 41% of this total. Distillation alone consumes about 2.4 quads/yr which is the energy equivalent of over 100,000 barrels of oil per day. This amount of

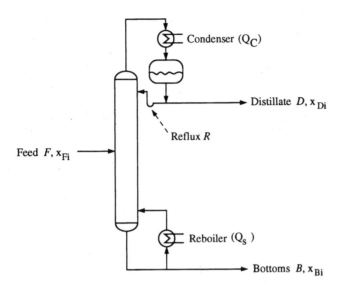

Figure 6.1   Distillation system.

energy is about 40% of the total amount of energy used in the chemical and petroleum refining industries (Humphrey et al., 1991).

An overall heat balance around a distillation column (Figure 6.1) yields:

$$Q_s - Q_c = Dh_D + Bh_B - Fh_F \qquad (6.1)$$

where Q_s = energy required at reboiler, energy/time
$\quad\quad\;\; Q_c$ = energy removed at condenser, energy/time
$\quad\quad\;\; D, B, F$ = rates of overhead product, bottom product, and feed, respectively; mass/time
$\quad\quad\;\; h_D, h_B, h_F$ = enthalpies of overhead product, bottom product, and feed, respectively; energy/mass

In many distillations, the enthalpies of feed and products are approximately equal, and the right-hand side of Equation (6.1) approaches zero. Thus, the energy provided at the reboiler is approximately equal to that lost at the condenser. Equation (6.1) reduces to

$$Q_s = Q_c \qquad (6.2)$$

Equation (6.2) shows that for conventional distillation, the amount of energy supplied at the reboiler is removed from the process at the condenser. Steam is typically used as the source of energy at the reboiler. The relationship between the amount of boiler fuel needed to supply reboiler steam, amount of steam required at the reboiler, and boiler efficiency is given by Equation (6.3).

$$Q_{\text{fuel}} = \frac{Q_s}{E_{\text{boiler}}} \qquad (6.3)$$

where Q_{fuel} = amount of steam boiler fuel needed, energy/time
Q_s = steam required at reboiler, energy/time
E_{boiler} = energy efficiency of steam boiler, %

Equation (6.4), obtained from a heat balance around the overhead condenser, may be used to estimate the amount of reboiler energy needed for distillation as a function of reflux ratio and overhead product rate.

$$Q_s = \Delta H_v (R + 1)D \qquad (6.4)$$

where Q_s = energy required at the reboiler, energy/time
ΔH_v = heat of vaporization of overhead product, energy/mass
R = reflux ratio (L_0/D)
D = overhead product, mass/time

As shown by Equation (6.4), at a given heat of vaporization for overhead product (ΔH_v) and rate of overhead product (D), the amount of reboiler energy (Q_s) is proportional to the reflux ratio. Indeed, most energy saving projects in industry are centered around ways to reduce reflux ratio while maintaining product purity.

Practical energy considerations

Despite poor energy efficiency, the practical side of generating steam sometimes favors distillation, a reality that must be taken into consideration. The generation and use of steam in an industrial complex is highly integrated. Steam is generated at high pressure in a central boiler facility and expanded through turbines to drive pumps and compressors, and is used as process heat. For example, steam at a generated pressure of 600 psig may be expanded through turbines to a lower pressure of 150–200 psig. In a subsequent step, 150–200 psig steam may again be expanded through turbines to a lower level of 20–40 psig.

Steam at all three pressure levels is distributed throughout the plant to drive pumps and compressors and for process heat. Since plant operators generally find lower-pressure steam less useful than high-pressure steam, an oversupply of low-pressure steam develops. The excess low-pressure steam must be vented to the atmosphere and the plant is said to be "out of steam balance." The amount of low-pressure steam that is vented varies from plant to plant. Newer plants do not generally vent low-pressure steam at all.

Low-pressure steam can frequently be used as reboiler heat for distillation columns. Thus, if the plant is out of balance, saving low-pressure steam simply results in more low-pressure steam being vented.

Under such circumstances, no boiler fuel is saved. In recent years, programs have been initiated to correct steam balance problems in older plants. One approach involves installation of condensing turbines to reduce the amount of low-pressure steam. However, many older plants have steam balance problems. In discussions with plant operators, some suggest that only 70% of energy saved in distillation can actually be realized as true fuel savings. However, a more general industrial approach is that if distillation energy can be saved, then ways can ultimately be found to translate these savings into true fuel savings. Before assuming energy can be saved by reducing distillation energy, information on the plant steam balance is needed, as well as on any plant programs that may change this balance.

Energy consumption for specific products

Table 6.1 gives estimated total energy consumption for key distillation separations, while Tables 6.2 to 6.5 provide estimates of energy consumption per unit of product (Btu/lb) for several products (Humphrey, 1991; Mix et al., 1981).

TABLE 6.1   Total Distillation Energy Consumption by Category

Feed	Typical components light/heavy key	Estimated reboiler energy, quads/yr
Petroleum fuel fractions	Gasoline/naphtha	0.493
Crude oil	Light naphtha/heavy naphtha/light distillate	0.423
Liquefied petroleum gas (LPG)	Ethane/propane/butane	0.217
Olefins	Ethylene/ethane, propylene/propane	0.118
Miscellaneous hydrocarbons	Cumene/phenol, acetone/acrylonitrile	0.101
Water—oxygenated hydrocarbons	Methanol/water, water/acetic acid	0.100
Aromatics	Ethylbenzene/styrene, benzene/toluene	0.082
Water—inorganics	Ammonia/water	0.057
Air	Nitrogen/oxygen	0.017
Water—hydrocarbons	p-Xylene/water	0.007
Other		0.302
Total reboiler energy, quads/yr		1.9
Total fuel requirement, quads/yr*		2.4

*Fuel requirement was estimated based on an average boiler efficiency of 80%.

TABLE 6.2   Distillation Energy: Water-Oxygenated Hydrocarbons

Primary product(s)	Components, light key/heavy key	Estimated reboiler energy (Btu/lb)
Ethanol (direct hydration of ethylene)	Ethanol azeotrope/water	5365
Acetic acid	Water/acetic acid	2769
Ethylene glycol	Water/ethylene glycol	1650
Vinyl acetate	Vinyl acetate—water azeotrope/acetic acid water azeotrope	3589
Methanol	Methanol/water	1236
Ethylene oxide	Ethylene oxide/water	1321
Ethylene glycol	Water/ethylene glycol	966
Methanol	Methanol/water	1182
sec-Butanol	sec-Butanol/water	3977
Isopropanol	Isopropanol/water	1104
Isopropanol	Isopropanol-water/mesityl oxide-water	1439
sec-Butanol	sec-Butanol/water	2436

TABLE 6.3   Distillation Energy: Olefins

Primary product(s)	Components, light key/heavy key	Estimated reboiler energy (Btu/lb)
Propylene	Propylene/propane	515
Propylene	Ethane/propylene	1764
Ethylene	Ethylene/ethane	346
Propylene	Propylene/butane	101
Butadiene	1,3-Butadiene/vinyl acetylene	1836
Ethylene	Methane/ethylene	136
Butadiene	Cis-2-butene/1,3-butadiene	1284

TABLE 6.4 Distillation Energy: Aromatics

Primary product(s)	Components, light key/heavy key	Estimated reboiler energy (Btu/lb)
Styrene	Ethylbenzene/styrene	2281
Ethylbenzene	Benzene/ethylbenzene	1025
Benzene, toluene	Benzene/toluene	287
Benzene	Heptane/benzene	254
Ethylbenzene	Ethylbenzene/p-xylene	13,579
o-Xylene	m-Xylene/o-xylene	6670
Ethylbenzene	Ethylbenzene/diethylbenzene	657
Cumene	Benzene/cumene	1229
Toluene, xylene	Toluene/xylene	240
Xylene	Xylene/C9 aromatics	165
Benzene	Benzene/diethylbenzene	292
Benzene, ethylbenzene	Benzene/ethylbenzene	281

TABLE 6.5 Distillation Energy: Miscellaneous Hydrocarbons

Primary product(s)	Components, light key/heavy key	Estimated reboiler energy (Btu/lb)
1,2-Dichloroethane	1,2-Dichloroethane/1,1,2-trichloroethane	1379
Phenol	Cumene/phenol	2351
Formaldehyde	Formaldehyde/methanol	774
Acrylonitrile	Acetone/acrylonitrile	1851
Acrylonitrile	Acrylonitrile/acetonitrile	1643
C8 Aromatics	Sulfolane/C8 aromatics	148
Acrylonitrile	Acylonitrile/cyanohydrin	1528
Phenol	Phenol/acetophenone	968
Ethylene glycol	Ethylene glycol/diethylene glycol	594
Phenol	Phenol/acetophenone	936
Vinyl acetate	Vinyl acetate/ethyl acetate	1417
Vinyl chloride monomer	Vinyl chloride monomer/1,2-dichloroethane	286
1,2-Dichloroethane	1,2-dichloroethane/1,1,2-trichloroethane	280
Propylene oxide	Propylene oxide/propylene dichloride	795
Isopropanol	Benzene/isopropanol	1575
Propylene oxide	Propylene oxide/propylene dichloride	620
Adiponitrile	3-Pentenitrile/adiponitrile	2283
Butadiene	1,3-Butadiene/acetonitrile/water	415
Isopropanol	Diisopropyl ether/isopropanol-water	872
2-Methyl, 3-butenenitrile	2-Methyl, 3-butenenitrile/3- and 4-pentenitrile	1594

Energy Consumption—Distillation Versus Extraction

Comparisons of energy requirements for distillation and extraction have been made by Null (1980). The results are presented in Figures 6.2 and 6.3.

In the extraction process, the major consumption of energy is in the reboiler of the accompanying distillation column required to recover solvent from the extract stream. A comparison of the energy requirements of distillation versus extraction is illustrated in Figure 6.2. In constructing this figure, it was assumed that the extraction solvent is nonvolatile and does not contaminate the raffinate, and requires only a simple flash step for recovery from the extract. For the case when 60% of the feed is taken overhead in a distillation process ($D/F = 0.6$), and the required heating-medium temperature for distillation is 300°F, extraction would consume less energy than distillation providing the distillation reflux is 2.0 or greater.

Figure 6.3 covers the opposite extreme in which two accompanying distillation columns are needed. One column is needed for the solvent-extract separation and one for the solvent-raffinate separation. If each of these columns were to require a reflux ratio, of 2.0 at $D/F = 0.60$, extraction would merit consideration if the required reflux ratio for distillation were 4.0 or greater. In Figure 6.3, the tempera-

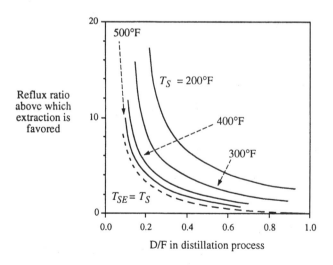

T_S = Required heating-medium temperature for distillation column

T_{SE} = Required heating-medium temperature for extraction solvent recovery column = 600 °F

Figure 6.2 Energy required—distillation versus extraction: case of nonvolatile solvent separation. (*Adapted from Null, 1980, with permission of the American Institute of Chemical Engineers. Copyright © 1980 AIChE. All rights reserved.*)

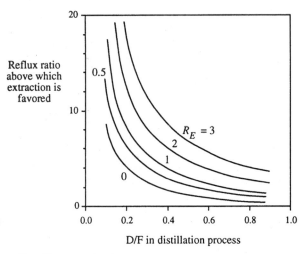

$T_S = T_{SE}$

R_E = Required reflux ratio for solvent-raffinate

Figure 6.3 Energy required—distillation versus extraction: case of difficult solvent separation. (*Adapted from Null, 1980, with permission of the American Institute of Chemical Engineers. Copyright © 1980 AIChE. All rights reserved.*)

ture of the heating medium for solvent stripping is assumed the same as that of the heating medium for distillation—an extreme situation.

Maximum Thermodynamic Efficiency

Fundamentals

Maximum thermodynamic efficiency is useful in comparing performance of different processes. We shall develop thermodynamic efficiency relationships for distillation, extraction, adsorption, and membrane processes and present some example calculations.

The *minimum thermodynamic work* required to separate a homogeneous mixture into pure products at constant temperature and pressure is given by Equation (6.5) (King, 1980).

$$W_{min} = -RT \sum_{i=1}^{n} x_{Fi} \ln \gamma_{Fi} x_{Fi} \tag{6.5}$$

where W_{min} = minimum work consumed per mol of feed, energy/mol
R = gas constant, energy/mol-temperature
T = temperature, °C or °K
x_{Fi} = mol fraction of component i in feed, dimensionless
γ_{Fi} = activity coefficient of component i in feed
n = number of components in feed, dimensionless

The summation in Equation (6.5) is over all components in the feed. Equation (6.5) applies to gas, liquid, and solid mixtures and γ_{F_i} represents the degree of departure from ideal solution behavior. For an ideal solution, $\gamma_F = 1$.

W_{min} is independent of any particular process. Actual processes operate with finite driving forces; they are irreversible, and consume more energy than the thermodynamic minimum given by Equation (6.5).

A sample calculation of W_{min} is presented below for the separation of a liquid mixture containing 53.5 mol% ethylbenzene and 46.5% styrene into pure components at 110°C (Keller, 1982). The mixture is assumed to be ideal (i.e., activity coefficients equal to 1). Under these conditions, for a binary mixture, Equation (6.5) reduces to:

$$W_{min} = -RT(x_{F1} \ln x_{F1} + x_{F2} \ln x_{F2}) \tag{6.6}$$

Substituting the gas constant ($R = 1.986$ Btu/lb mol-°K), temperature, and composition into Equation (6.5) gives:

$$W_{min} = -(1.986)(383)[0.535 \ln (0.5325) + 0.465 \ln (0.465)] \tag{6.7}$$
$$= 525 \text{ Btu/lb-mol feed}$$

On a Btu/lb styrene basis, the minimum work is:

$$\left(\frac{525 \text{ Btu}}{\text{lb mol feed}}\right)\left(\frac{1 \text{ lb-mol feed}}{0.465 \text{ lb-mol styrene}}\right)\left(\frac{1 \text{ lb-mol styrene}}{104 \text{ lb styrene}}\right) \tag{6.8}$$
$$= 10.9 \text{ Btu/lb styrene}$$

Maximum thermodynamic efficiency is a useful parameter to compare performance of competing processes. It is defined as the minimum thermodynamic energy (W_{min}) divided by minimum amount of energy consumption required for the process.

$$E_{max} = W_{min}/Q_{min} \tag{6.9}$$

where E_{max} = maximum thermodynamic efficiency, dimensionless
W_{min} = minimum thermodynamic work for separation, energy/mol
Q_{min} = minimum energy for separation process, energy/mol

Distillation

Minimum energy. For distillation, the minimum amount of energy (Q_{min}) required for separation is the minimum amount of heat required at the reboiler. Minimum energy consumption occurs at the minimum reflux ratio which corresponds to an infinite number of equilibrium stages.

The derivation of minimum energy (Q_{min}) for distillation follows. This derivation is based on the following assumptions:

- Single reboiler and condenser
- Constant relative volatility
- Equal molar overflow
- Minimum reflux ratio
- Complete separation
- Feed at its bubble point

The following relationships are based on Figure 6.1:

Overall material balance:

$$F = D + B \tag{6.10}$$

Material balances for the more volatile component are:

$$Fx_{F1} = D \tag{6.11}$$

$$F(1 - x_{F1}) = B \tag{6.12}$$

where F = feed rate, mass/time
D = overhead product rate, mass/time
B = bottom product rate, mass/time
x_{F1} = concentrations of component 1 in feed, wt fraction or mol fraction

Based on component 1, vapor and liquid compositions are related by:

$$y_1 = \frac{\alpha_{12} x_1}{1 + (\alpha_{12} - 1) x_1} \tag{6.13}$$

where α_{12} = relative volatility for component 1 relative to component 2

From the McCabe-Thiele diagram shown in Figure 6.4, the slope of operating line for the stripper section is L'/V' which is also equal to y/x_F.

Using Equation (6.13) the following relationship may be developed:

$$\frac{L'}{V'} = \frac{\alpha_{12}}{1 + (\alpha_{12} - 1)x_{F1}} = \frac{V' + B}{V'} = \frac{V' + F(1 - x_{F1})}{V'} \tag{6.14}$$

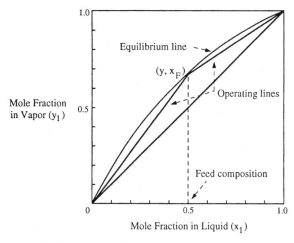

Figure 6.4 McCabe-Thiele diagram for complete separation by distillation.

where L' = liquid rate in stripper section of column, mol/time
V' = vapor rate in stripper section of column, mol/time

Rearranging Equation (6.14) gives:

$$\frac{V'}{F} = \frac{1}{\alpha_{12}-1} + x_{F1} \tag{6.15}$$

Though not used here, the following form of Equation (6.15) can be very useful:

$$R_{min} = \frac{1}{x_{F1}(\alpha_{12}-1)} \tag{6.15a}$$

Based on Equation (6.15), the minimum heat (Q_{min}) required for complete separation by distillation is:

$$Q_{min} = \Delta H_V \left(\frac{V'}{F}\right) = \Delta H_B \left[\frac{1}{\alpha_{12}-1} + x_{F1}\right] \tag{6.16}$$

where Q_{min} = energy required per mol of feed, energy/mass
ΔH_B = molar heat of vaporization of the bottom product

The effects of relative volatility (α_{12}) and feed composition (x_{F1}) on minimum energy required for distillation (Q_{min}) are shown plotted in Figures 6.5 and 6.6, respectively.

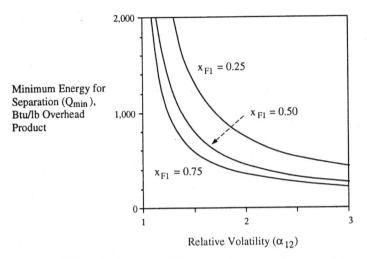

Figure 6.5 Effect of relative volatility on minimum energy required for distillation/heat of vaporization = 8000 cal/g-mol.

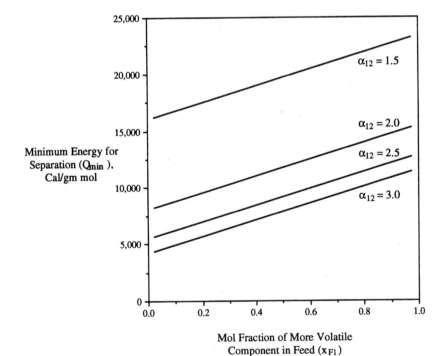

Figure 6.6 Effect of composition on minimum energy required for distillation/heat of vaporization = 8000 cal/g-mol.

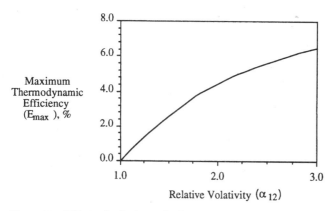

Figure 6.7 Effect of relative volatility on maximum thermodynamic efficiency for equimolar mixture.

Maximum thermodynamic efficiency. Maximum thermodynamic efficiency for distillation, based on an ideal binary mixture, is given in Equation (6.17). Equation (6.17) is the result of dividing Equation (6.6) by Equation (6.16):

$$E_{max} = \frac{W_{min}}{Q_{min}} = -\frac{RT\,(x_{F1}\ln x_{F1} + x_{F2}\ln x_{F2})}{\Delta H_B\left(\dfrac{1}{\alpha_{12}-1} + x_{F1}\right)} \tag{6.17}$$

Based on an equimolar feed ($x_{F1} = 0.5$) and assuming Trouton's rule is valid (i.e., $\Delta H_B \approx 21T_B$, where T_B = normal boiling point of styrene, °K), E_{max} may be plotted as a function of relative volatility. The results are given in Figure 6.7.

Results show that maximum thermodynamic efficiency for distillation (E_{max}) increases as relative volatility (α_{12}) increases. However, even for very easy distillation separations, when α_{12} is 3.0 and above, E_{max} is only about 6%.

The effects of feed composition and relative volatility on maximum thermodynamic efficiency are illustrated in Figure 6.8.

Example problem. An example of how thermodynamic efficiency may be used to evaluate performance of an industrial distillation column follows.

Feed = 53.5 mol% ethylbenzene (EB), 46.5% styrene (S)

$$\text{Average relative volatility} = \frac{\alpha_{\text{Top}} + \alpha_{\text{Bot}}}{2} = \frac{1.43 + 1.32}{2} = 1.375$$

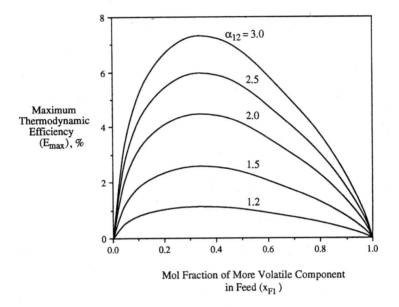

Figure 6.8 Effects of feed composition and relative volatility on maximum thermodynamic efficiency for equimolar mixture.

ΔH_v of styrene (heavy component) = 162 Btu/lb (9350 cal/g-mol)

ΔH_v of 22 psig steam = 937 Btu/lb (9370 cal/g-mol)

$W_{\min, T, P}$ = 526 cal/g-mol feed (see earlier calculation)

Actual steam usage = 0.56 lb/lb feed (3.27 g-mol/g-mol feed)

Minimum heat requirement for complete separation:

$$Q_{\min} = \Delta H_v \left(\frac{1}{\alpha_{12} - 1} + x_{F1} \right)$$

$$= 9350 \left(\frac{1}{1.375 - 1} + 0.535 \right) = 29,900 \text{ cal/g-mol feed}$$

Maximum thermodynamic efficiency:

$$E_{\max} = 100(W_{\min}/Q_{\min}) = \frac{(100)(526)}{29,900} = 1.76\%$$

Actual energy consumption is determined by steam consumption (from plant data)

$$\text{Heat added/g-mol feed} = \left(3.27 \; \frac{\text{g-mol steam}}{\text{g-mol feed}}\right)(9370 \; \text{cal/g-mol steam})$$

$$= 30{,}600 \; \text{cal/g-mol feed}$$

$$E_{act} = 100(W_{min}/Q_{act}) = \frac{(100)(526)}{30{,}600} = 1.72\%$$

Comparison of maximum (1.76%) versus actual thermodynamic efficiency (1.72%) shows the distillation column to be operating efficiently. Although a minor reduction in steam usage might be possible, a major reduction cannot be expected.

Extraction

Minimum energy. Similar to distillation, minimum energy (Q_{min}) may be determined for extraction and other processes (Bravo et al., 1984). In the simplest case, the minimum heat required for liquid/liquid extraction is determined by the amount of energy required to recover solvent from the extract stream by distillation. When the solvent is the higher-boiling component, the minimum energy can be derived as:

$$Q_{min} = \frac{x_{F1}}{x_{S1}} \Delta H_{sol} \left[\frac{1}{\alpha_{12} - 1} + x_{S1} \right] \qquad (6.18)$$

where Q_{min} = minimum heat required, energy/lb mol feed
ΔH_{sol} = heat of vaporization of solvent, energy/lb mol
x_{F1} = mol fraction of solute in feed, dimensionless
x_{S1} = mol fraction of solute in solvent (extract) at equilibrium, dimensionless
α_{12} = relative volatility of solute relative to solvent

The maximum thermodynamic efficiency for liquid/liquid extraction can then be obtained by dividing Equation (6.5) by Equation (6.18).

Adsorption

Minimum energy. For adsorption, the minimum energy requirement, using a thermal swing regeneration cycle, is (Bravo et al., 1984):

$$Q_{min} = x_{F1} \left[C_p (T_d - T_a) \left(MW + \frac{1}{m P x_{F1}} \right) + \Delta H_{ads} \right] \qquad (6.19)$$

where Q_{min} = minimum energy required to thermally regenerate
adsorption bed, energy/mol

x_{F1} = mol fraction adsorbate in feed, dimensionless

C_p = heat capacity of adsorbent, energy/mol-temperature

T_d = desorption temperature, °C or °F

T_a = adsorption temperature, °C or °F

MW = molecular weight of adsorbate

m = slope of isotherm (for gases: loading versus partial pressure), 1/pressure

P = feed pressure, energy/length^3

ΔH_{ads} = heat of adsorption, energy/mol

Maximum thermodynamic efficiency for adsorption, based on thermal swing regeneration cycle, can be obtained by dividing Equation (6.5) by Equation (6.19).

Membranes

For a membrane separation, the minimum amount of energy required is determined by the amount of mechanical work required to compress the feed. For a binary mixture, the amount of compression work may be developed based on permeation of components through two selective membranes. To determine the amount of ideal work required, it is assumed the components in the permeate streams are compressed reversibly and isothermally from permeate pressure up to the pressure of the feed (Figure 6.9).

Based on thermodynamic fundamentals, the amount of reversible isothermal work to compress 1 mol of ideal gas is:

$$W = RT \ln \frac{P}{P_{pi}} \tag{6.20}$$

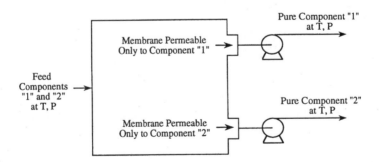

Figure 6.9   Ideal membrane separation process.

where P_{pi} = pressure of permeate of component i, force/length^2
P = feed pressure, force/length^2
T = temperature of feed, $^\circ$R or $^\circ$K
R = gas constant

Therefore, for 1 mol of feed, the work of compression is:

$$W = RT \left[x_{F1} \ln \frac{P}{P_{p1}} + (1 - x_{F1}) \ln \frac{P}{P_{p2}} \right] \qquad (6.21)$$

For a multicomponent mixture, the above expression becomes:

$$W = RT \sum_{i=1}^{n} x_{Fi} \ln \frac{P}{P_{pi}} \qquad (6.22)$$

The maximum thermodynamic efficiency for a membrane process to separate a multicomponent ideal gas mixture may be obtained by dividing Equation (6.5) (with $\gamma_{Fi} = 1$) by Equation (6.22):

$$E_{max} = \frac{-RT \sum_{i=1}^{n} x_{Fi} \ln x_{Fi}}{} \qquad (6.23)$$

When the pressure drop across the membrane is very small, P_{pi} approaches the value of Px_{Fi} and the equation in the denominator of Equation (6.23) becomes identical to the one in the numerator. Thus, as pressure drop across the membrane approaches zero, the work required to make the separation approaches the thermodynamic minimum and maximum thermodynamic efficiency approaches 100%.

References

Bravo, J. L. et al., "Assessment of Potential Energy Savings in Fluid Separation Technologies: Technology Review and Recommended Research Areas (Suppl.)," U.S. Dept. of Energy Report, Contract No. DOE/ID/12473-1, 1984.

Humphrey, J. L. et al., "Separation Technologies—Advances and Priorities," U.S. Dept. of Energy Final Report, DOE/ID/12920-1, February 1991.

Keller, G. E. II, *Adsorption, Gas Absorption, and Liquid-Liquid Extraction: Selecting a Process and Conserving Energy,* Industrial Energy-Conservation Manual 9, MIT Press, Cambridge, MA, 1982.

King, C. J., *Separation Processes,* 2nd ed., McGraw-Hill, 1980.

Mix, T. W. et al., "Energy Conservation in Distillation," U.S. Dept. of Energy Final Report, DOE/CS/40259, July 1981.

Null, H. R., *Chem. Eng. Progr.,* **76**(8), 42, 1980.

7

Process Selection

Introduction

In this chapter we will deal first with the issue of how to choose the right process—or at least narrow down the number of choices—for a given separation. We will then deal with how a series of processes should be oriented to separate a complex mixture. Finally we will discuss in some detail the separation issues in the environmental area. Our approach will be primarily heuristic in nature, that is, we will deal with simple rules of thumb and direction statements, called heuristics. One early proponent of this approach was King (1980), and many others have developed variations on the heuristic theme. Taking cognizance of the dominance of distillation, we will begin with an analysis of why distillation is so dominant and then proceed to define those separation conditions which favor the selection of other processes. We will then move to define the conditions which favor one process over another. Next we will discuss the multistep separation of complex mixtures.

Environmental separations are different enough from most other separation needs that, first of all, distillation plays a very minor role and, second, several other processes which are seldom considered for the main separations within processes can play important roles here. For these reasons, we will cover environmental separations in detail.

Strengths and Weaknesses of Distillation and Other Vapor-Liquid Separation Processes

In this and the following section, we will discuss those situations for which the processes we have discussed earlier cause one or the other to be economically preferred. Some of this information for the individual processes is also given in the earlier chapters, but we bring it together here so that the strengths and weaknesses of each process can be compared.

Economically, if a stream can be easily vaporized or condensed, distillation or a related vapor-liquid separation process is most often the process of choice for separating the components of that stream. There are four very powerful reasons for this dominance, which are enumerated below.

- Distillation can be staged very easily. It is possible to have 100 or more theoretical stages in a distillation column. This means that very low relative volatilities—down to perhaps 1.2 or so—can be accommodated. In addition, mass-transfer rates between phases are generally high enough, and phase contacting and disengaging are easily enough accomplished, that equipment sizes to effect high degrees of separation are usually not excessive. The net result is that equipment costs are often quite reasonable to produce large numbers of stages with the concomitant high degrees of separation.

- Distillation's equipment requirements are small compared to virtually all other processes. Distillation is the epitome of a simple process: usually just a column, a reboiler, a condenser and some relatively small ancillary equipment. By comparison, processes based on mass separating agents (MSAs) typically involve two or more separate mass transfer zones, or semibatch operation, or both to perform the separation. And there must be additional investment for the wherewithal to circulate the MSA. All of these complications often lead to relatively high investments for alternative processes compared to distillation unless separation by the latter is very difficult to effect. Also, because of the factors mentioned above, the volume of the distillation equipment to process a given stream flow will often be several times smaller than the volumes of alternative processes, i.e., the capacity per unit volume of column will be higher for distillation. Flowsheets for other vapor-liquid processes, such as azeotropic and extractive distillation and absorption, are indeed more complex than a distillation flowsheet, but the mass transfer rate and the phase disengagement ease of distillation are also present in these processes, making them economically favored in a number of low relative volatility situations.

- The economics of scale favor distillation for large-scale separations. Distillation technology scales up by roughly a 0.6 factor on column throughput (see Chapter 2), which means that doubling a column's capacity increases the column investment, not by a factor of 2 but rather by a factor of only about 1.5. As a result, distillation columns processing over 2 billion lb/yr of feed are routinely designed and successfully installed—and are highly economical. On the other hand, process-design information on most other separation processes indicates that they have higher scaleup factors. Thus the economic improvement of going to very large scales of operation with alternative processes is not normally as pronounced as it is with distillation. This lowered potential for cost reduction with increasing scale of operation is caused by the much more complex flowsheets of the alternative processes and the necessity in a number of cases of replicating vessels and placing them in parallel to accept the whole feed stream. Membrane modules are an excellent example of this problem. To accept large flows, many modules aligned in parallel must be employed, and some of the savings of going to larger scale are lost.

- Energy costs are not necessarily a dominant consideration. This might seem to be a surprising conclusion at first, because distillation has labored under the stigma for many years of being such a high-energy-usage process that there is a major incentive to replace it in as many applications as possible. But as was pointed out in Chapter 2, energy costs for distillation virtually always run second in impact to capital costs. Therefore the substitution of a more complex, higher-investment but lower-energy-usage process for an existing distillation process has truly an uphill fight to be economically viable.

  In addition, in some plants high-pressure steam is generated primarily to run turbines to produce electrical power, and the resulting low-pressure steam from the turbine exhaust is of little or no value. In such cases, distillation-column reboilers can serve as condensers for this low-value steam, making energy costs very low. Heat integration of distillation columns can also be practiced on very large scales (two major examples are olefin plant cryogenic separation systems and cryogenic air separation) to help reduce energy costs.

In spite of this impressive array of factors which cause distillation to be favored, other considerations do lead us to develop and install separation processes other than distillation in many applications. Below are listed the situations which lead to distillation separation costs which are higher than those of alternative processes, or in which other factors militate against the use of distillation. Table 7.1 gives commercial process examples which illustrate these situations.

- Distillation becomes very expensive when the relative volatility between key components is too low. Seldom will a relative volatility

TABLE 7.1   Examples of Separations Difficult
for Distillation to Accomplish

- More saturated from less saturated

  - Ethylene-ethane*

  - Ethylene-acetylene

  - Propylene-propane*

  - Butadiene-butenes

  - Styrene-ethylbenzene*

  - Detergent-range olefins from paraffins with similar carbon number

- Mixtures of isomers

  - p-Xylene-mixed xylenes

  - n-Paraffins from isoparaffins with similar carbon number

- Mixtures of water and polar organics and inorganics, and especially those mixtures dilute in the nonaqueous component

  - Water-acids such as formic, acetic, hydrochloric, etc.

  - Water-low-carbon-number alcohols such as ethanol, isopropanol, etc.

  - Water-polyols such as glycerin, ethylene glycol, etc.

- Mixtures of compounds with overlapping boiling ranges, such as aromatics–nonaromatics

SOURCE: Keller, 1977; Mix et al., 1978.

*These separations are in fact carried out by distillation but are some of the highest-investment and highest-operating-cost separations made on large scale.

**These separations must be carried out by multi-effect evaporation to give reasonable energy usages.

over part or all of the distillation curve of less than about 1.2 give acceptable economics for distillation. This rapid upswing in cost is discussed in Chapter 2. Relative volatilities less than 1.2 obviously include azeotrope-forming situations, for which the relative volatility will actually become less than 1.0 as the azeotrope is crossed. In such cases, azeotropic or extractive distillation, pervaporation, solvent extraction, and adsorption should be investigated for providing reasonable solutions.

- Feed composition and product purity requirements can present unusual problems. Three problems dominate this category.

  - If the product of interest is a high boiler in low concentration (perhaps 10–20 wt% or less) in the feed, then energy costs per unit of desired product to boil up all of the low-boiling material, as well as column capital costs, can become excessive.

- If a small concentration (only a few percent) of high-boiling contaminants must be removed from a desired product, then in
some cases energy and capital costs for distillation can also
become excessive, especially if the energy cost is based on the
cost per unit of high boiler removed. This situation is virtually
the same as the one above.

- If the boiling range of one set of components overlaps the boiling
range of another set of components from which it must be separated, then many distillation columns would be required to
make a complete separation. Other processes might be able to
separate one set from another in one step if there are chemical
similarities within one set which differ from the chemical similarities of the other set.

- Extremes of temperature or pressure are required. If distillation
temperatures are less than about $-40°C$ (233 °K) or greater than
250°C (523 °K), then the cost of refrigeration to condense column
overheads in the first case and heat to vaporize in the second
escalate rapidly from typical cooling water and steam costs. If
required column pressures are less than about 15 mm Hg (2
KPa), then column-size and vacuum-pump costs can escalate
rapidly. If column pressures of greater than about 750 psi (5
MPa) are required, column investment can also escalate as a
result of the cost of the extra metal required to meet the pressure
requirements.

- Production rates are only a few tons a day or less. Earlier it was
mentioned that distillation has a relatively low scaleup factor—
about 0.6. Unfortunately if a process such as distillation scales up
well economically, that same process does not scale down well, that
is, its investment does not reduce, as throughput is reduced, as
much as another with a higher scaleup factor. Thus alternative
processes which do not compete for economically separating a
stream for large production rates may compete well at smaller production rates. An additional problem for distillation for very small
production rates is that often these situations can occur remotely
from steam plants, and only electricity or another relatively high-
cost energy source may be available for heating. Use of these alternative sources will dramatically increase the energy cost for separation per unit of product.

- Various types of undesirable reactions can sometimes occur at column temperatures.

  - Thermally labile components in the feed can undergo reactions
  which result either in significant product loss or in the forma-

tion of byproducts which are difficult to separate. In the first case, valuable product is lost, while in the second, product quality is adversely affected. In addition, in quite a number of distillations with organic liquids, the intrusion of oxygen from small leaks under high vacuum can cause severe product purity problems through unwanted side reactions.

- Column fouling rates can be unacceptable. Even though the product loss rate might be completely acceptable and there are no product quality concerns, if a very small fraction of the feed material reacts to form a solid precipitate or polymer, then column fouling may occur too rapidly to be manageable. Certainly a column which cannot be operated for only a few days before becoming so fouled that it must be shut down and cleaned will hardly be economical unless alternative processes are incredibly expensive.

- Corrosion can be a problem. With certain feed streams, corrosion can be so severe that even stainless steels can have unacceptably short lifetimes. In such cases, investment costs can rise to unacceptable levels through the need for relatively exotic metals for construction.

- Explosive conditions can be encountered in the column. With some feeds which contain unstable materials such as acetylenic compounds and the like, the large vapor volumes found in many columns can become zones in which free-radical reactions can proceed relatively uninhibited by the presence of radical-quenching surfaces. The result can be deflagrations and detonations.

Pluses and Minuses of Alternative Processes

Despite the dominance of distillation, we have seen that it is not always the right answer for a given separation. The other processes we have discussed obviously have their places in the scheme of things, and in fact, new separation opportunities, since they will often involve more complex and less volatile mixtures than the present-day spectrum of separations, will be served by these alternative processes more often on average than before. In Figure 7.1 we summarize the types of mixtures, the conditions and other factors which favor the use of the various processes which are discussed in this book. In a number of cases this figure will not give a single answer, but it will at least lead the reader to the proper questions to ask of each process. This figure thus serves as a type of heuristic for selecting a process to effect a given separation.

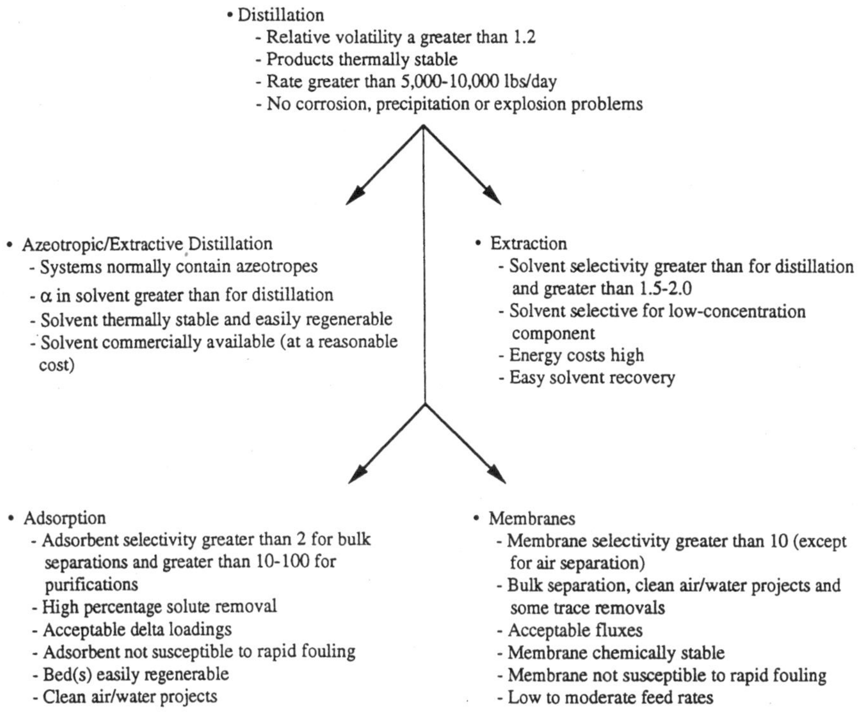

- Distillation
  - Relative volatility a greater than 1.2
  - Products thermally stable
  - Rate greater than 5,000-10,000 lbs/day
  - No corrosion, precipitation or explosion problems

- Azeotropic/Extractive Distillation
  - Systems normally contain azeotropes
  - α in solvent greater than for distillation
  - Solvent thermally stable and easily regenerable
  - Solvent commercially available (at a reasonable cost)

- Extraction
  - Solvent selectivity greater than for distillation and greater than 1.5-2.0
  - Solvent selective for low-concentration component
  - Energy costs high
  - Easy solvent recovery

- Adsorption
  - Adsorbent selectivity greater than 2 for bulk separations and greater than 10-100 for purifications
  - High percentage solute removal
  - Acceptable delta loadings
  - Adsorbent not susceptible to rapid fouling
  - Bed(s) easily regenerable
  - Clean air/water projects

- Membranes
  - Membrane selectivity greater than 10 (except for air separation)
  - Bulk separation, clean air/water projects and some trace removals
  - Acceptable fluxes
  - Membrane chemically stable
  - Membrane not susceptible to rapid fouling
  - Low to moderate feed rates

Figure 7.1   Factors which favor separation processes for liquid mixtures.

Separation of Complex Mixtures and Heuristic Guidelines for Process Selection

Seldom do mixtures to be separated consist of just two components, and in many cases more than two components or groups of components must be separated. In such situations more than one separation step will almost always be needed. If all of the separation steps are simple distillations making sharp (almost complete) separations, then the minimum number of column sequences, S_R, to separate N products is determined by the following equation (Thompson and King, 1972):

$$S_R = [2(N-1)]!/N!(N-1)! \qquad (7.1)$$

The number of column sequences predicted by Equation (7.1) for separating a given number of components is given in Table 7.2, and the five sequences, or flowsheets, for separating four components are shown in Figure 7.2. As can be seen from the table, the number of sequences can get very large with only a few components, and to ana-

TABLE 7.2   Number of Column Sequences Predicted by
Equation (7.1)

Number of products	Number of column sequences
2	1
3	2
4	5
5	14
6	42
7	132
8	429
9	1430
10	4862

lyze each one economically for every separation could be both arduous and expensive. And, of course, to the extent that separation processes other than distillation are considered, and to the extent that partial separations are made in a given step, the number of flowsheets can quickly increase to truly distressingly large values. To simplify the task of analysis, relatively simple methods involving heuristics have been developed. One of the earliest and most popular of the heuristic methods was developed by Heaven (1969) and King (1980). Their method deals for the most part with distillation. They present seven rules, shown in Table 7.3, which can be used to reduce a large number of column sequencing possibilities down to a very small number and sometimes down to just one. King warns, however, that often the four heuristics can conflict with one another, and in such cases more than one sequence will have to be investigated. Nadgir and Liu (1983) have developed a somewhat similar series of heuristics, given in Table 7.4, which can be used to confirm or modify the conclusions from the heuristics of Table 7.3.

These and other sets of heuristics originate from investigating the issues of both investment (as reflected primarily in column sizing) and energy costs. Other issues such as chemical stability, corrosion, etc. may also be involved which can further reduce the number of the most likely possibilities. It must be admitted, however, that the heuristic methods, though they help in rank-ordering various flowsheets, do not help much in giving information about how much more economical one flowsheet might be compared to another. Such information can only be generated by fairly detailed economic-estimation exercises, as will be discussed later.

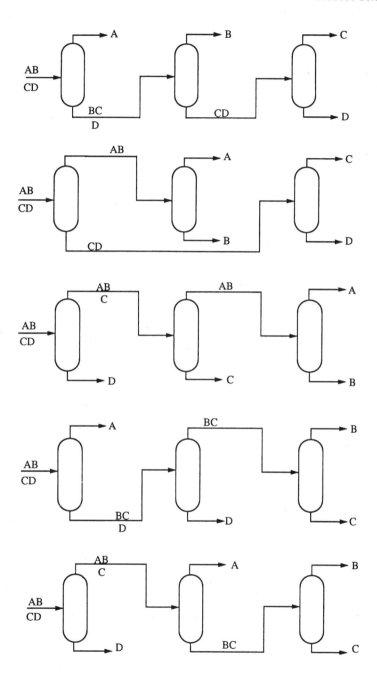

Figure 7.2 Distillation flowsheets for separating a four-component mixture.

TABLE 7.3   Heuristic Rules for Sequencing Simple Distillation Columns for Separating Multicomponent Mixtures

- Separations where the relative volatility of the key components is close to unity should be performed in the absence of nonkey components.

- Sequences which remove the components one by one in column overheads should be favored.

- A product composing a large fraction of the feed should be removed first, or more generally, sequences which give a more nearly equimolal division of the feed between the distillate and bottoms product should be favored.

- Separations involving very high specified recovery fractions should be reserved until late in a sequence.

- Favor sequences which yield the minimum number of products.

- When alternative separation methods are available for the same product split, (1) discourage consideration of any method giving a separation factor close to unity, e.g., less than 1.05, and (2) compare the separation factors attainable with the alternative methods in the light of previous experience with those separation methods (Seader and Westerberg, 1977).

- When a mass separating agent is used, favor recovering it in the next separation step unless it improves separation factors for candidate subsequent separations.

SOURCE: King, 1980.

TABLE 7.4   Alternative Set of Heuristic Rules for Sequencing Simple Distillation Columns for Separating Multicomponent Mixtures

- Favor ordinary distillation unless the relative volatility is too low, and when using a mass separating agent, remove it immediately in the separator following the one in which it is used.

- Avoid vacuum distillation and refrigeration if possible.

- Favor sequences which yield the minimum necessary number of products. Do not favor sequences which separate components which ultimately will be combined into a single product.

- Remove corrosive and hazardous materials first.

- Perform the difficult separations last, after other materials have been removed.

- Remove the most plentiful component first.

- If component compositions do not vary widely, favor sequences which give nearly equimolal splits between distillate and bottoms flows unless relative volatilities are too low.

SOURCE: Nadgir and Liu, 1983.

As we have already seen, simple distillation is not the most economical solution for every separation, and indeed in a number of cases it isn't even practical. So for separating many complex mixtures, more than one type of separation process may be required. To examine the suitability of other processes, one can use the list of situations for which distillation costs escalate rapidly. These situations can supply a type of additional heuristic which leads one away from distillation and toward other processes which may be more economical. In turn, each separation process has its own particular strong points and weak points, and Figure 7.1, given earlier, lists some of the main points for the other processes discussed in this book. From this figure, one can make some rough judgments about which process or processes might be economically the best for a given separation step in the resolution of a complex mixture.

The second level of sophistication for determining the best sequence of separations as well as the processes to use for each one is a combination of heuristics plus additional techniques, such as expert-systems insights. One such technique is given by Barnicki and Fair (1990). Another line of increased sophistication has been the development and use of so-called residue curve maps, which provide a simple means of visualizing ternary mixtures and the concentration paths followed during a distillation operation. These maps are especially enlightening in cases in which one or more azeotropes exist. An example of a residue curve map for a highly complex mixture is shown in Figure 7.3. Residue curve maps for separating more than three components cannot be easily visualized and must be dealt with mathe-

Lines show the trajectories of the residue compositions with time of
one-stage batch distillations.

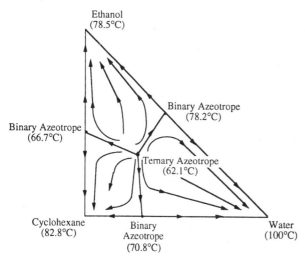

Figure 7.3 Example of a complex ternary system.

matically. An excellent review of heuristic synthesis using residue curve maps is given by Fien and Liu (1994). One of the more intriguing of these second-level methods has been named EXSEP, which is an expert system for flowsheet selection (see, for example, Brunet and Liu, 1993).

Some of the second-level methods can still be done with hand calculators, while in other cases only a minimum of advanced computation is needed. The third level of sophistication involves the use of considerable computational power. These methods have been developed primarily at Carnegie-Mellon University and the University of Massachusetts. Both of these methods are still in the process of evolution, are still mainly concerned with vapor-liquid separations and involve highly complex mathematical and artificial-intelligence approaches. The Carnegie-Mellon studies have been reported by Wahnschafft et al. (1991, 1992, 1993a,b). The University of Massachusetts approach, called Mayflower, is being developed on campus and is funded by a consortium of several large chemical and oil companies. In both of these systems the vast fraction of the information contained in them deals with vapor-liquid separations. There is also a substantial amount of thermodynamic and economic detail present. Furthermore, these techniques can deal with recycle streams, "sloppy" (less than complete separation) cuts and other complexities which are generally outside the domains of the simpler methods. Two companies in North America—Aspentech (Cambridge, MA) and Hyprotech (Calgary, Canada)—offer advanced software for process selection as part of overall process-design packages.

Procedures for Process Selection

The question obviously arises as to what is the proper level of complexity to use in analyzing a given problem and what is the proper procedure for developing the most economical flowsheet. Although no categorical procedure can be given, we recommend the following:

- First, carefully lay out the nature of the separation task, listing the feed rate and its composition and how many different product streams are required and their purities. Specify recycle streams which, when properly accounted for, could in turn alter the presumed feed flow rate and composition, and modify the feed parameters accordingly.

- Begin the analysis using the first-level methods, plus the heuristics given here for selecting the type of separation process. Inspect the predictions for the best flowsheet as well as the next best. Do they seem to be realistic? Are there other factors which might cause you to favor one process over another?

- If some problems arise which cause you to question the realism of the predictions, then advance to the second or third level and redo the analysis.

- Once some confidence has been established that the best one or a few flowsheets have been identified, then proceed to more detailed economic analyses.

Frequently one will find that most methods will identify a best flow-sheet (based on the lowest overall capital plus operating costs) but will also identify one or more others which might be nearly as good. At that point, identifying unambiguously the economically best solution will usually require a detailed economic analysis. Even when one uses major equipment cost as an indicator of overall cost, the former cannot always be trusted to provide a close estimate of the latter. A major reason for this is that the major equipment cost—reactors, heat exchangers, separation equipment, and other large pieces of equipment of any given process or series of steps in a process—constitutes a distressingly small part of the overall investment cost of that process. Typical percentages of the total cost attributable to four major categories of costs are given in Table 7.5. It is very important to note that the major equipment cost, which is often used as a measure of the total investment, must actually be multiplied by a factor of about 4–6 to get the overall cost or total capital investment. (If, instead of constructing a process from the ground up, one buys an assembled overall process component such as a skid-mounted membrane unit which contains pumps, controls, valves, and piping, the installed investment of the overall process drops to only about two to

TABLE 7.5   Total Project Cost Breakdown

Category	Percent of total
• Major equipment, including reactors, separators, heat exchangers, etc.	15–20
• Bulk materials, including site-preparation equipment, concrete, steel, buildings, piping, control systems, electrical equipment, insulation and painting, plus subcontracts for various subassemblies	28–33
• Construction costs	25–32
• Engineering, including process design and detail design	20–25

three times the cost of the skid.) There are so many factors involved in the overall project cost that small differences in major equipment costs between two flowsheets can be more than counterbalanced by higher process-control or piping or other costs associated with one flowsheet compared to another. Added to this complexity are issues such as operability, controllability, and the like, as well as whether existing equipment might be used or modified more easily for one flowsheet compared to another. Another complexity can be added when capital is constrained but utilities are present in abundance, for example.

What, then, are the benefits to be gained from the highly complex approaches discussed above? These benefits seem to lie, first of all, in the analysis of cases which present truly complex separations involving, for example, several azeotropes, the need to determine the optimal solvent for a difficult separation, etc. And second, these approaches involve process design modules which can be used to provide major steps toward an overall process package.

Process Selection for Liquid and Gas Waste Treatment Applications

The strategy recommended by the Chemical Manufacturers Association (1993) for dealing with pollution problems is as follows:

- First, source reduction. Avoid generating the waste or release in the first place.

- Second, recycling. If a waste or release cannot be reduced at the source, then recycle, reuse, or reclaim it. In order of preference: (1) recycle within the process, (2) recycle within the plant, and (3) recycle off-site.

- Third, treatment. If a waste or release cannot be reduced at the source or recycled, reused, or reclaimed, treat it to minimize environmental impacts.

Typical strategies for at-source reduction include:

- Improved separation-process arrangements to minimize the release of various products to waste streams.

- Improved process-control strategies to minimize upsets which cause extra releases to waste streams.

- Improved catalysts and processes which create reduced amounts of by-products.

- Raw-material changes which create reduced amounts of by-products and wastes.

- In a few cases, substitution of one final product for another which has lower pollution-producing tendencies during manufacture.

Once source-reduction strategies have minimized the wastes being generated, the job of cleanup of the remaining waste liquid and gas streams begins. In the solution of these problems, and especially in so-called end-of-pipe situations, distillation plays a nearly negligible role. The number of separation processes for solving these problems is surprisingly large, and the ability of standard process-selection procedures to analyze these situations is not very good. For these reasons, we will cover this area in some detail, indicating which processes are now used as well as which processes are likely to be used in the future. We will not deal here with remediation of soil and groundwater or with the disposal of solid wastes, but rather only with the remediation of liquid and gaseous process effluents.

Liquid waste streams

The vast fraction of liquid waste streams from chemical and petroleum refining plants are primarily aqueous. Contaminants can include dissolved organic monomers, polymers and other materials, ions, particulate matter (including biomass created in the process or in a biotreatment facility), and finely suspended droplets and emulsions. In general, streams containing relatively high concentrations (10–50 wt% or more, for example) of organic species are either separated to recover useful products or intermediates, or, if they are too complex in composition to recover products economically, they are incinerated.

The general end-of-pipe method of cleanup of high-water-composition streams containing organic species is a biotreatment facility. There is an excellent reason for the wide use of biotreatment. This one process destroys a wide spectrum of the organic species in waste streams. In a sense, biotreatment can be thought of as a separation process in which contaminants are destructively separated from the water, thus purifying it.

Table 7.6 gives the acceptable effluent concentrations for organic pollutants, expressed in terms of biological oxygen demand (BOD). Such levels can, for the most part, be met by biotreatment. Very approximately, half of the incoming organics, as measured by the chemical oxygen demand (COD), are not metabolized by the biomass. Priority pollutants, examples of which are given in Table 7.7, must be removed down to about 3 orders-of-magnitude lower concentrations

TABLE 7.6   Acceptable Concentrations for Organic Pollutants in Aqueous Waste Streams

- For commodity organic chemicals, the maximum monthly average = 30 ppm BOD and 46 ppm total suspended solids (TSS).

- For specialty organic chemicals, the maximum monthly average = 45 ppm BOD and 57 ppm TSS.

- For 56 organics and derivatized organics, typical monthly average concentrations, except for a few cases, are less than *50 ppb,* or approximately three orders of magnitude lower concentration than non-priority pollutants.

Notes:

- BOD is related to the oxygen required to destroy those organic species which are susceptible to biodegradation. COD is related to the oxygen required for total destruction of all organic species in the stream. The COD-to-BOD ratio, though quite variable from stream to stream, probably averages about 2 for inlet streams to a biodegradation facility, and about 10 or more for outlet streams.

- In addition to the above restrictions, state and federal agencies can add further restrictions which cover specific chemicals, and especially priority pollutants.

SOURCE: Williams, 1995.

TABLE 7.7   Examples of Priority Pollutants

Acenaphthene	Various phthalate esters
Acrolein	Various polynuclear aromatic hydrocarbons
Acrylonitrile	
Benzene	2,3,7,8-Tetrachlorodibenzodioxin
Benzidine	Various pesticides, including aldrin, dieldrin, chlordane, DDT, endosulfan, endrin, heptachlor, hexachlorocyclohexanes, PCBs and their metabolites
Various chlorinated benzenes, dienes, alkanes, ethylenes, and phenols	
2-Chloronaphthalene	Toxaphene
Dinitrotoluene1,2-diphenylhydrazine	Inorganic species
Ethylbenzene	Metals, including antimony, arsenic, beryllium, cadmium, chromium, copper, lead, mercury, etc.
Fluoranthene	
Other haloethers	Cyanide
Other halomethanes	Asbestos
Hexachlorocyclopentadiene	
Isophorone	
Naphthalene	
Nitrobenzene	
Various nitrophenols	
Various nitrosamines	
Phenol	

Abbreviations: DDT, dichlorodiphenyltrichloroethane; PCB, polychlorinated biphenyls.

than those of the sum of the regular pollutants, and on certain occasions these levels cannot be met by biotreatment alone. In such cases, either at-source removal or removal by a step downstream of the biotreatment facility must by accomplished. Federal and state regulations can also restrict the outlet concentrations of other individual species in streams.

The economics of biotreatment are notoriously difficult to estimate and generalize about. Different feed contaminants and concentrations can change the size of the biotreatment unit for a given feed rate. Different companies calculate investments and operating costs in different ways, allocating these costs sometimes to processing units and sometimes as a stand-alone facility. And, in addition, various add-on steps for preremoval of highly volatile pollutants, specialized pretreatment of the influent, removal of priority pollutants, etc. may be required. In Tables 7.8 and 7.9 are given estimates of the investment and operating cost for a more-or-less standard 1 million gal/day (3.8 million L/day) biotreatment facility. For this flow rate, we estimate an installed investment of approximately $5 million. Dealing with the situations mentioned above can add significantly to this number, perhaps doubling both the investment and the operating cost in some cases.

Recent regulations make it necessary to preremove highly volatile organic compounds from a waste stream prior to its introduction to a biotreatment facility, unless this facility is covered by a roof. This is because, when oxygen is dissolved into the water through the action of the aerators, highly volatile materials can be vaporized into the air passing through the water, fortuitously solving part of the water pollution problem but creating an air pollution problem.

TABLE 7.8   Investment Estimate for a Biotreatment Waste Facility

Basis: 1 million gal/day (3.8 million L/day) of influent wastewater, greenfield site

Estimated Investment

- $5 million (range: $3–8 million). Most expensive items: aeration tanks and aeration system.

- Does not include:

  – Upstream pretreatment, e.g., steam stripping or extensive prefiltration, if necessary

  – Covering of ponds, if necessary

  – Activated-sludge-disposal facility, if necessary

  – Tertiary-treatment facility for handling priority pollutants, if necessary

  – Land cost

SOURCE: Williams, 1995.

TABLE 7.9   Operating Cost Estimate for a Biotreatment Waste Facility

Basis: 1 million gal/day (3.8 million L/day) of influent wastewater, greenfield site

Estimated Operating Cost

- Total influent flow basis: $1–4/1000 gal or $360 thousand–$1.5 million/year. Assume an average value of $1.0 million/yr.

- Total organic carbon (TOC) basis: $0.50–2.00/lb TOC. Although some facilities use TOC for estimating operating costs, others use influent flow. We will use influent flow.

- If tertiary treatment is required for priority pollutants, operating costs increase by 50–100%. Also, if extensive problems exist in removal of fine particulate biomass from the effluent, costs of several hundred thousand dollars per year can result from the need to add polymers to coagulate and settle these solids.

- Largest source of operating cost: energy. Sources of energy cost:

  - Aeration—the more important

  - Pumping

SOURCE: Williams, 1995.

The primary means of removal of volatile components from water are air stripping and steam stripping. Design data have been presented by Hwang, et al. (1992a–c), which allow one to determine which materials can be air stripped or steam stripped down to no more than one-thousandth of the initial concentration. Even though steam stripping requires more energy than does air stripping, the latter can simply transfer a water pollution problem into an air pollution problem and require a second process such as adsorption or membrane permeation to remove pollutants from the air stream.

Table 7.10 gives the conditions for removal of volatile pollutants by steam or air stripping down to no more than one-thousandth of their initial concentrations. It is also possible in some cases to use natural-gas stripping (Humphrey, 1989) and simply burn the pollutants along with the natural gas in a boiler. The boiler must be rated for combustion of the pollutants, however. It is interesting that the vast majority of the organic priority pollutants listed in Table 7.7 (exceptions are phenols, derivatized phenols, acrolein, acrylonitrile, and a few others) can be steam-stripped or, somewhat less often, air-stripped. The thermodynamic analysis and data tables presented by Hwang et al. (1992b, 1992c) show that almost invariably the degree of separation improves with increasing temperature, and hence steam stripping will nearly always be more effective than air or natural gas stripping, given the same molar-flow-rate ratios of gas to liquid in the columns for the two processes.

TABLE 7.10   Design Criteria for Steam and Air Stripping

Design criteria

- Reduction in pollutant concentration by 99.9%
- Stripper contains 10 theoretical stages
- Steam-to-water mole ratio = 0.02, or air-to-water mole ratio = 0.02
- $K^\infty = y/x \geq 100$

   where K^∞ = mole ratio of pollutant in the vapor phase to that in the
   liquid phase at the limit of infinite dilution

SOURCE: Hwang, 1992a.

TABLE 7.11   Other Materials Removable by Steam or Air Stripping

- All aliphatic and cyclic paraffins from C_1 to at least C_{20}.
- All olefins and diolefins to at least C_9.
- All monocyclic aromatics to at least C_{11} by steam stripping. Air stripping is problematical for some of these materials.
- All C_1 and C_2 halogenated hydrocarbons except for hexahalogenated species by steam stripping. Air stripping is problematical for a number of these.
- All C_2 and C_3 mono-, di-, and trihalogenated olefins.
- Mono- and dihalogenated monocyclic aromatics by steam stripping only.
- Possibly some polychlorinated biphenyls (PCBs) by steam stripping only.
- Aliphatic and some cyclic ethers up to at least C_6 by steam stripping only.
- Various sulfur-containing materials such as carbon disulfide and monosulfides containing up to at least four carbons by steam stripping only.

SOURCE: Hwang et al., 1992c.

The materials which can be removed by steam and air stripping are the relatively hydrophobic species, that is, materials which have infinite-dilution activity coefficients in water of well above one. Many of these have limited solubilities in water, making it possible to recover them in very high concentrations, even with steam stripping. Various materials removable by steam or air stripping are listed in Table 7.11. The number of compounds which can be thus removed is quite surprising, and the conclusion is easily made that steam and air stripping are generally underutilized processes for aqueous-stream cleanup.

Pervaporation is another possibility for removal of relatively hydrophobic species from wastewater. The flow rates often encountered in end-of-pipe streams are for now considerably larger than those normally treated by pervaporation, however, so that this process may only be applicable to streams smaller than normally found in plants with a common treatment system for essentially all waste streams.

The question now arises as to what separation processes might possibly replace biotreatment and achieve reduced costs and/or improved removals. One conclusion must be stated at the outset: The capability of biotreatment to remove a very wide spectrum of organic pollutants cannot be matched by one single separation process. Rather, several processes will be required to effect the same removal.

A second conclusion is also important: Since biotreatment destroys pollutants and alternative processes largely recover these pollutants, this fact can present additional problems in some cases and opportunities in others. If the pollutant mix is too complex to be reseparated and recovered as useful products, then incineration or an equivalent process must be employed. On the other hand, there may be opportunities for additional income if the pollutant mix can be separated economically and the products sold. Both of these conclusions make the problem of economic comparison of biotreatment facilities with alternatives difficult to attack and to arrive at general conclusions.

One means of analyzing the question of which processes could replace biotreatment is to characterize the various pollutants commonly encountered. Figure 7.4 shows a means of characterizing pollutants and the processes for their removal. This figure uses an infinite dilution, dimensionless Henry's law coefficient, K^∞, for the pollutant in water as one measure and molecular weight as the other. This coefficient is defined as (Hwang et al., 1992a):

$$K^\infty = y/x \text{ at infinite dilution} = f^\circ \gamma^\infty / P \qquad (7.2)$$

where f° = fugacity, which can be approximated by the vapor pressure
γ^∞ = infinite dilution activity coefficient
P = total pressure

The larger the value of K^∞, the more hydrophobic the material. The figure shows three regions in which different processes can come to the fore. The most difficult of the regions to deal with is that of highly hydrophilic, relatively low-molecular-weight pollutants, called Region I. These materials cannot be steam- or air-stripped to high degrees and are difficult to remove by activated carbon in many cases and by RO membranes. Reverse osmosis processes have an additional problem in that the osmotic pressures associated with low-molecular-weight compounds can be immense, and the degree of concentration which can be reached is seldom more than 10–20 wt%. Activated car-

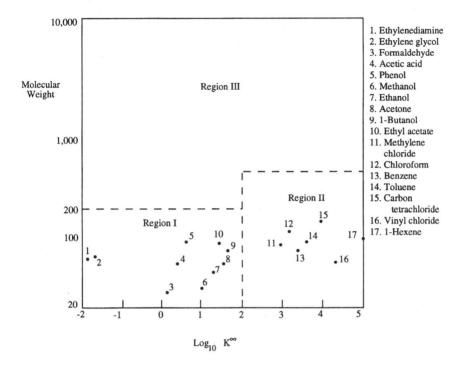

Figure 7.4 Categorization of wastewater organic pollutants.

bon and silicalite are also often of limited value, since water will compete for the adsorption sites and reduce pollutant loadings substantially. Some of these pollutants can be removed by solvent extraction (e.g., phenol), but the processes can be costly.

Generally as one moves from left to right through Region I, the separation problem becomes easier to deal with, and if removals of only about 90% are satisfactory, then adsorption, RO, and solvent extraction all become more reasonable. The second problem with RO is that fluxes are typically low [10 gal/ft²/day (GFD) or less], which, when treating high-flow-rate waste streams, can result in unacceptably high investments. Future improvements in RO performance in terms of investment per unit of flux will certainly help in making this process viable.

If pollutants in Region I are generally the most difficult to remove from water, then those in Region II are perhaps the easiest. Because these materials have very high infinite-dilution activity coefficients, they can be removed easily by steam or air stripping as well as by adsorption on activated carbon or silicalite. Solvent extraction is also a possibility, although its investment implications may be too great in most cases. For relatively small stream flows, pervaporation is also a possibility for pollutants in this region.

Region III also represents a region in which several processes can compete. Vapor-liquid processes are not applicable in this region because these pollutants have very low vapor pressures. However, they can be removed and concentrated by nanofiltration or ultrafiltration membranes. Because of a much reduced osmotic pressure barrier, higher concentrations in the retentate stream can be attained than can be in Region I. Solvent extraction is also a possibility for pollutants on the right-hand side of this region.

One final class of "separation process" must be mentioned: the so-called advanced oxidation processes (see Figure 7.5). These processes use either ozone or hydrogen peroxide in the presence of ultraviolet light to destroy—oxidize to carbon dioxide, water, and inorganic species from any heteroatoms—the last remaining traces of organic pollutants. Removals down to parts per billion levels can be achieved. These processes are too expensive—several dollars per pound of pollutant removed—to be used as a first removal step for most waste streams. On the other hand, advanced oxidation can be used as a polishing step after other processes have removed most of the pollutants. And the very low levels of pollutants which can be reached make possible the desirable concept of total water recycle to a process, that is, the recycle of all water except that which might be formed in reactions inside the process.

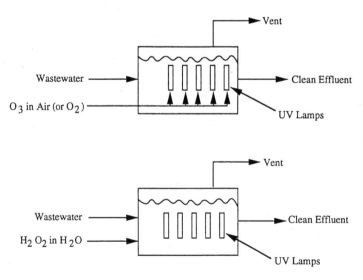

Figure 7.5 Schematic diagrams of advanced oxidation processes.

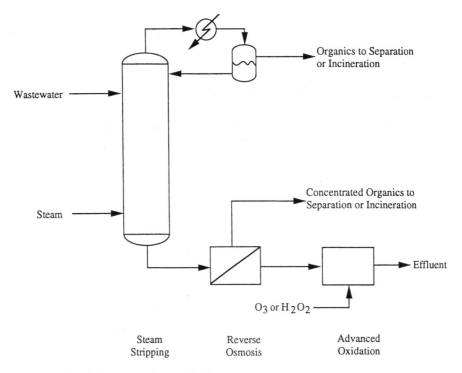

Figure 7.6 Possible wastewater purification process.

From the above discussion we can begin to synthesize flowsheets which can provide alternatives to biotreatment and perhaps improve on some of the shortcomings of this process. In Figure 7.6 is given one such flowsheet, which consists of steam stripping, RO, and advanced oxidation. Organic species are removed in all three steps, with the first two steps removing the bulk of these species and the advanced oxidation unit removing the last traces. The economics of this flowsheet, compared to those of a biotreatment facility, are dependent on several factors:

- If the feed requires steam or air stripping prior to introduction to the biotreatment facility, the cost for this step will be relatively the same for both this flowsheet and a biotreatment facility. This need should therefore not be a deciding factor between biotreatment and its alternatives. If steam is priced at $4.00/1000 lb, and its usage is about 3 lb/100 lb of wastewater, then the daily cost is about $960, and the yearly cost is roughly $350,000.

■ The key item from an investment standpoint will be the RO unit. If we assume that at most $3–4 million can be allocated for the installed cost of this unit, that the membrane flux is 10 GFD, and that nearly all of the stream is permeated, then the total allowable installed cost per square foot of membrane is $30–40/sq ft (about $300–400/sq m). Such an installed cost is on the borderline but may in some cases be realistic. Perhaps the most critical factor in the analysis is the assumption of the flux rate. Certainly a value of one GFD will cause the membrane unit to be out of the question economically.

■ The final disposition of the retentate will be important economically. If valuable materials can be recovered, then the value of these materials could help pay for any further separation equipment. If, however, this stream must be incinerated, an additional cost may be incurred.

■ The cost of the advanced oxidation unit should be less than $1 million. Its operating cost will be directly proportional to the weight of organic species oxidized to carbon dioxide and water.

These rough calculations show that alternatives to biotreatment may in some cases be economically on a par with or superior to biotreatment. At the very least, before deciding to install a biotreatment facility, one should investigate the use of separation processes to perform the cleanup task. Operating labor may be a deciding factor in some cases.

Gaseous waste streams

At present there is not a single separation process, like biotreatment for liquid streams, which dominates in the recovery of volatile organic compounds (VOCs) from gas streams. Rather, recovery of VOCs is an application for which several separation processes compete, including membranes, absorption, pressure-swing adsorption, fixed-bed adsorption, rotary-wheel adsorption, condensation, and freezing. And in many cases, if an emissions problem exists, the stream is simply incinerated. Except for rotary-wheel adsorption and freezing, economic case studies were completed for each of these processes (Humphrey et al., 1996) to assist in placing priorities on the various processes. Results have been presented based on removal of acetone from air at differing air flow rates, acetone concentrations in the feed, and at two levels of acetone recovery. It was assumed that the VOC removal units were installed in a chemical plant on the U. S. Gulf Coast.

The capital and operating costs were determined using the following bases:

1. Air feed streams (fresh feed)
   a. Rates
      (1) Low value: 100 standard cubic feet per minute (scfm)
      (2) High value: 10,000 scfm
   b. Compositions (one organic component: acetone)
      (1) Low value: 0.3 vol% acetone in feed
      (2) High value: 2.0 vol% acetone in feed
      (3) No moisture or solids
   c. Percent recovery of organic
      (1) Low value: 50% of acetone in feed
      (2) High value: 95% of acetone in feed
   d. Pressure of feed stream: atmospheric
   e. Temperature of feed stream: 16°C (60°F)

2. Capital costs
   a. Processes provided as new process units
   b. Location: U.S. Gulf Coast chemical plant
   c. No tearout or rework required in existing plant
   d. Explosion-proof equipment required
   e. Existing plot available
   f. Offsites available (utilities and fire protection)
   g. On-stream factor: 90%

3. Operating costs
   a. Organic by-product acetone credit: $2/gal
   b. Natural gas: $2.50/1000 standard cubic feet (scf)
   c. Steam: $4.00/1000 lb
   d. Electricity: 4.5¢/kWh
   e. Cooling tower water: 7¢/1000 gal (16°C)
   f. Chilled water: 65¢/1000 gal (7°C)
   g. Depreciation schedule: 10%/yr
   h. Salaries including fringe benefits: operators = $70,000/yr; engineers = $120,000/yr
   i. Water treatment cost: $1/1000 gal

Results of comparing the capital and operating costs of processes for 95% acetone recovery are given in Figure 7.7. Lower capital and operating costs were the criteria used to determine the regions in which each process is favored.

In clean-air applications, where acetone concentrations must be reduced to ppm levels, the membrane surface area required becomes

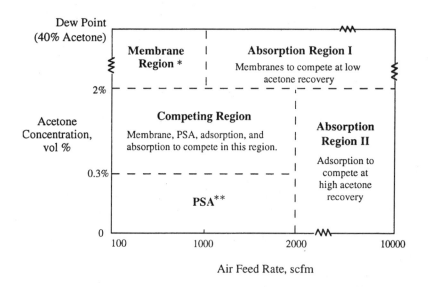

Figure 7.7   Processes for recovery of acetone from air.

excessive as would the number of trays required in an absorption column. Under these conditions, adsorption processes become the methods of choice. They can be used to achieve 99.99% removals. PSA is favored for higher concentrations and lower rates, TSA is favored at lower concentrations and higher rates.

As can be seen in Figure 7.7, there are five regions which can be distinguished. The first is called the Membrane Region. Because the membrane has lower capital costs than other processes in this region, its best opportunity for commercialization occurs at acetone concentrations above 2% and for gas feed rates less than about 1000 scfm. For clean-air applications (removals down to the parts per million level), membrane surface area becomes excessive, and pressure-swing adsorption (PSA) would likely become the favored process.

The second region is called Absorption Region I. Absorption is the favored process for acetone concentrations greater than 2% and for gas feed rates greater than 1000 scfm. Because absorption is a tower-type process, capital costs increase with rates according to the six-tenths power rule. At high rates, capital costs of absorption are lower than for other processes. Though attractive for the acetone/air application, absorption quickly loses favor if the VOC forms a homogeneous azeotrope with water because simple distillation can no longer be used for water recovery. Though acetone does not form an azeotrope with water, a number of organics including many containing oxygen do form azeotropes. The membrane process may compare favorably with absorption at rates higher than 1000 scfm, but applicability is limited to lower acetone recoveries of 50% or so.

The third region is termed the Competing Region. Membrane, adsorption and absorption processes compete for acetone concentrations in the range of 0.3–2%, and for air feed rates less than 2000 scfm. Condensation is not included here because it does not compete economically with the other processes. The process selected will depend on acetone concentration, air feed rate, and acetone recovery. Within this region, the factors favoring each process are

- Membranes are favored by high acetone concentrations and low rates
- PSA by low rates and high acetone recovery
- Adsorption by low acetone concentrations and high acetone recovery
- Absorption by high rates and low acetone recovery

The fourth region is called Absorption Region II. Absorption is favored for acetone concentrations of up to 2% and for gas feed rates above 2000 scfm. However, when organic concentrations must be reduced to low levels, such as those required for clean air applications, adsorption will likely become the favored process.

The final region is the PSA Region, and this process is favored with acetone concentrations below 0.2% and gas feed rates less than 2000 scfm. For clean air applications and rates less than 2000 scfm, PSA may compete favorably over all ranges of acetone concentration in the feed. There is a small area of opportunity—at rates less than 100-200 scfm—for membranes within this region.

Descriptions of processes included in the above analysis are given below. Two processes—wheel-based adsorption and freezing—are not included because economic and process information is not complete enough to render a judgment on them at this time.

Membrane. The membrane unit, a flowsheet for which is given in Figure 7.8, is comprised of a blower to increase feed pressure to 2 psig. Feed to the blower is a mixture of fresh feed and a gas and acetone mixture which is recycled from the acetone recovery system. When the resulting mixture is fed to the membrane module and encounters the membrane, acetone diffuses through the membrane faster than nitrogen or oxygen or other fixed gas, thereby increasing the concentration of acetone in the permeate.

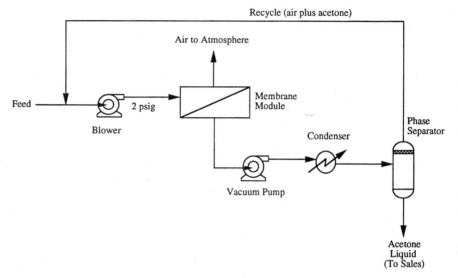

Figure 7.8 Membrane process for recovery of acetone from air.

The retentate from the membrane module is vented to the atmosphere or, as appropriate, processed in a downstream unit. In order to maintain the driving force across the membrane, a vacuum is maintained on the permeate side. The discharge from the vacuum pump, which is an enriched mixture of acetone in gas, is fed to a condenser where acetone is condensed, recovered, and sold as a by-product.

The membrane process was simulated on the computer by using the combination of a membrane model and the Design II Process Simulator. The membrane model is based on a log mean partial pressure driving force requiring permeance, feed rate and composition, feed pressure, permeate pressure, and desired key component recovery as input data. This simulation determines the required membrane surface area, component partial pressure, driving force, and component flux.

Absorption. In the absorption process (Figure 7.9), fresh gas feed is blown into the bottom of the absorber column where it is contacted in a countercurrent fashion with water. The acetone is absorbed by the water so that the effluent air stream has a substantially lower concentration of acetone. The water/acetone mixture from the absorber is fed to a distillation column which separates the acetone as an overhead stream and water as the bottom product. The acetone is sold as by-product, and the bottom water is recycled to the absorber.

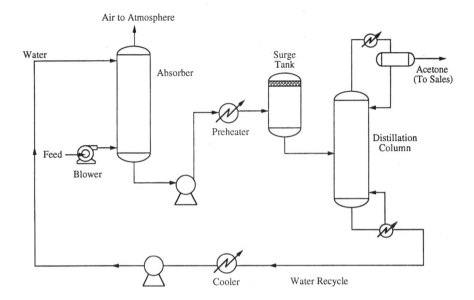

Figure 7.9   Absorption process for recovery of acetone from air.

PSA. The PSA process, shown schematically in Figure 7.10, uses a styrene/divinyl benzene copolymer as the adsorbent. So that the feed stream can be flowed continuously to the process, two adsorbers in parallel are used. Most of the energy and operating costs are incurred in the regeneration step. The feed is supplied at slightly above atmospheric pressure, and the regeneration step occurs under vacuum. In the parlance used earlier, this process might be called a vacuum-swing adsorption (VSA) process.

The adsorption step is continued for 5–15 min, while the regeneration step takes 3–5 min. During regeneration the process uses the combination of vacuum, heat accumulated in the bed during adsorption, and a backpurge of clean gas to regenerate the adsorbent. The discharge stream from the vacuum pump is condensed to recover the acetone. The gas mixture leaving the condenser carries an equilibrium concentration of acetone and is recycled to the feed. The unit requires only electricity to drive a vacuum pump and a cooling fluid for operation. The process was developed by the Dow Chemical Company, first for its own applications and then later for licensing. A similar process is reported to be available from Silica-Verfahrenstechnik GmbH (Berlin, Germany). Since 1988 over 20 commercial units have been placed in operation for removal of hydrocarbons, chlorofluorocarbons (CFCs), chlorinated solvents, aromatics, and monomers from various gas streams.

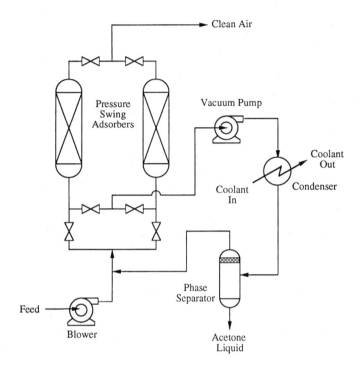

Figure 7.10   Pressure-swing adsorption for recovery of acetone from air.

PSA units can routinely achieve 99.99% organic removal from vent streams which range from 50 to 2000 scfm and feed concentrations of organics from 0.1 to 16%. Because vacuum pumps are expensive at large flow rates, PSA units have an upper economic capacity of about 2000 scfm. Continuous cycling between the adsorption and desorption cycles is required. The feed stream must be kept free of solids and contaminants which would plug or foul the bed.

TSA.   In the TSA process (Figure 7.11), the gas feed is blown into an adsorber where activated carbon selectively adsorbs the acetone. To permit a continuous feed, two (or four) fixed beds are used in parallel; feed is introduced to one (or two beds) while regeneration is occurring in the other one (or two). Two beds were used for the 100 scfm case and four for the 10,000 scfm case. Regeneration is accomplished by feeding steam to the bed. The resulting steam/acetone mixture is condensed and distilled, and the acetone is then sold.

Condensation (Figure 7.12) is conceptually the simplest of the recovery processes. The feed is compressed and acetone is subsequently condensed. Glycol is used as the cooling fluid in the condenser. The amount of acetone which can be condensed is limited by a

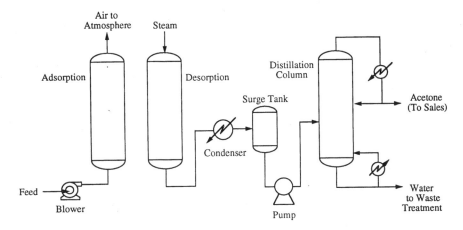

Figure 7.11 Fixed-bed adsorption process for recovery of acetone from air.

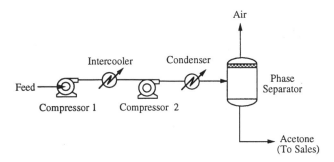

Figure 7.12 Condensation process for recovery of acetone from air.

minimum allowable gas-side temperature of 6°C. It is essential to adhere to this minimum to prevent the formation of ice crystals. Following condensation and phase separation, the gas is either vented or sent downstream for further processing. Recovered acetone is sold. It was found that the cost of compression significantly increases capital and operating costs of this process, rendering it less attractive than competing processes for recovery of acetone from air.

Comparative costs. The capital and operating costs for the various processes investigated are illustrated in Figures 7.13 and 7.14. Results show that the capital costs tend to vary more than operating costs, and conclusions are weighted accordingly. Of the processes considered, it was found that only condensation could not compete economically over a wide range of conditions. Because acetone is highly volatile, it is necessary to compress the feed before condensation can

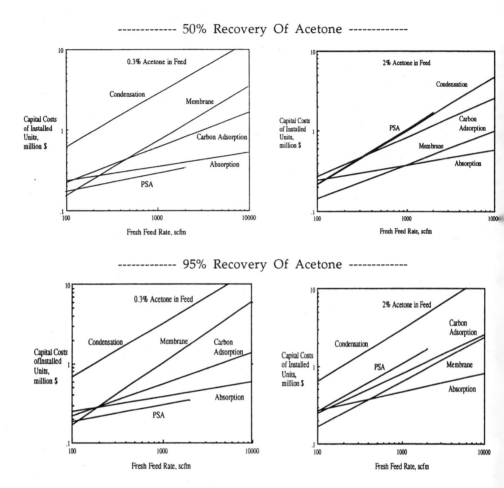

Figure 7.13   Capital costs of installed skid unit plants for recovery of acetone from air.

occur, and especially in cases in which high percentage removals are required, the compression costs become far too high.

The question can be raised as to how the boundaries and conclusions from Figure 7.7 will change with different solvents. Some qualitative directions are given below.

■ As an organic's hydrophilicity decreases, that is, as it becomes less and less water-soluble, water as an absorbent for it becomes less and less an option. In place of absorption (in Absorption Regions I and II), it could well be that wheel-based adsorption will compete for recoveries of such organics for feed concentrations of about 0.1–0.2% and below, while TSA will compete for feed concentrations up to 2–5%. Higher organic feed concentrations may require

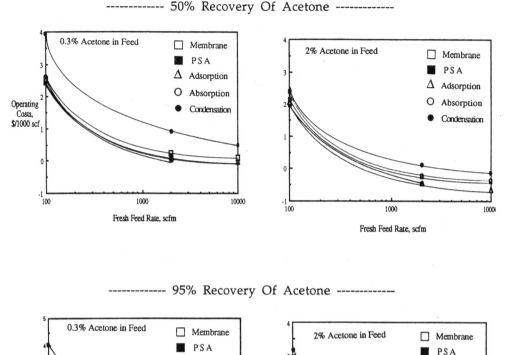

Figure 7.14 Operating costs of installed skid unit plants for recovery of acetone from air, $/1000 SCF.

more than one process in series to reach satisfactory removals. The use of a heavy hydrocarbon as an absorbent is also a possibility in some cases.

■ As an organic's vapor pressure is reduced below that of acetone at a given temperature, then condensation may compete for streams which approach their saturation points and for relatively low percentage removals. Absorption's domain may also grow, particularly at high rates, if easy separation from the solvent can be effected.

References

Barnicki, S. D., and J. R. Fair, *Ind. Eng. Chem. Res.,* **29,** 421, 1990.

Brunet, J. C. and Y. A. Liu, *Ind. Eng. Chem. Res.,* **32,** 315, 1993.

Chemical Manufacturers Association (CMA), *Designing Pollution Prevention into the Process,* copyright CMA, 1993.

Fien, G.-J. A. F. and Y. A. Liu, *Ind. Eng. Chem. Res.,* **33,** 2505, 1994.

Heaven, D. L., M. S., thesis in chemical engineering, University of California, Berkeley, 1969.

Humphrey, J. L., *Utilization of Natural Gas in Large Scale Separation Processes,* Final Report, Gas Research Institute, ID No. 89/0005, February 1989.

Humphrey, J. L., et al., *Membranes Versus Competing Processes for Recovery of VOCs from Air,* presented at the 8th Annual Meeting of the North American Membrane Society, Ottawa, Canada, May 1996.

Hwang, Y.-L. et al., *Ind. Eng. Chem. Res.,* **31,** 1753, 1992a.

Hwang, Y.-L. et al., *Ind. Eng. Chem. Res.,* **31,** 1760, 1992b.

Hwang, Y.-L. et al., supplementary material to Hwang, Y.-L. et al., *Ind. Eng. Chem. Res.,* **31,** 1760, 1992c. Available from the American Chemical Society. This material contains a vapor-liquid-equilibrium databank for 404 organic pollutants commonly found in process water and wastewater streams of chemical plants and refineries.

Keller, G. E., personal communication to C. J. King, 1977, quoted in King, C. J., *Separation Processes,* 2nd ed., McGraw-Hill Book Co., New York, 1980.

King, C. J., *Separation Processes,* 2nd ed., McGraw-Hill Book Co., New York, 1980.

Mix, T. W. et al., *Chem. Eng. Prog.,* **74,** April 1978, p. 49.

Nadgir, V. M. and Y. A. Liu, *AIChE Journal,* **29,** 926, 1983.

Seader, J. D. and A. W. Westerberg, *AIChE Journal,* **23,** 951, 1977.

Thompson, R. W. and C. J. King, *AIChE Journal,* **18,** 942, 1972.

Wahnschafft, O. M. et al., *Computers Chem. Engineering,* **15,** 565, 1991.

Wahnschafft, O. M. et al., *Ind. Eng. Chem. Res.,* **31,** 2345, 1992.

Wahnschafft, O. M. et al., *Ind. Eng. Chem. Res.,* **31,** 1121, 1993a.

Wahnschafft, O. M. and A. W. Westerberg, *Ind. Eng. Chem. Res.,* **32,** 1108, 1993.

Williams, J. B., Union Carbide Corp., personal communication, March 21, 1995.

A

Distillation Column Design Procedures

Design Versus Rating

The following procedures describe the design of a new column and the rating (retrofit) of an existing column. Procedures for columns equipped with sieve trays, random packings, and structured packings are given. To provide stand-alone procedures, there is some duplication of the information and equations presented in Chapter 2.

The two basic approaches used in distillation column design are *design* and *rating*. The design approach involves the design of a new column. In this approach, the separation is specified and the problem is to determine the column height and diameter required to make the separation. The fundamentals of the design approach, based on stage-to-stage calculations, were first published by Lewis and Matheson (1932). The design approach is often called the Lewis-Matheson approach. It is the rigorous analog of the Fenske/Underwood/Gilliland *shortcut* method for determining of the number of equilibrium stages.

The rating approach involves the retrofit of an existing column. In this approach, the column diameter and height are fixed and the capacity and separation are to be determined. Because this approach has convergence advantages, it is the one which is most often used in computer algorithms which do stage-to-stage calculations. The rating approach is often referred to as the Thiele-Geddes method, based on the original work of Thiele and Geddes (1933).

For the design of a new column or rating of an existing one, the design engineer must obtain final designs and quotes from equipment suppliers. In order to evaluate such quotes a preliminary design is needed. In the preliminary design, the following questions must be answered:

- Should trays or packings be used?
- If trays are used, what kind of trays?
- If a packing is used, should it be a random or structured packing?
- What correlations and models should be used to predict performance?
- What is the expected performance?

Design of New Distillation Column

Figure A.1 shows the steps required for designing a new column. Although auxiliary equipment is needed, the focus is on the performance of the column internals and the required column diameter and height.

Step I. Establish bases

Establishing the bases defines the objectives of the project. The bases should include feed and product compositions, feed rate, and physical properties. Any special conditions, such as maximum temperature or pressure drop, should also be specified.

Step II. Operating conditions

First, the column pressure must be determined. Column pressure is used to determine the top and bottom column temperatures by dew point and bubble point calculations, respectively. In establishing column pressure, the following factors should be considered:

- *Undesirable side effects.* Chemical degradation, polymerization, and fouling may occur at the highest (bottom) temperature. By lowering pressure and temperature, it may be possible to avoid side effects.
- *Cost of cooling medium.* If possible, water or air should be used as the cooling medium. The selection of the cooling medium determines the overhead process temperature. For cooling tower water, the maximum overhead process temperature is probably around 140°F.
- *Cost of heating medium.* The higher the column pressure and temperature at the bottom of the column, the more expensive the heating medium will be.
- *Relative volatility.* A lower column pressure and temperature generally results in a higher relative volatility and a lower reflux ratio (or a lower number of equilibrium stages).
- *Cost of column shell.* Higher operating pressures trigger higher column shell costs.

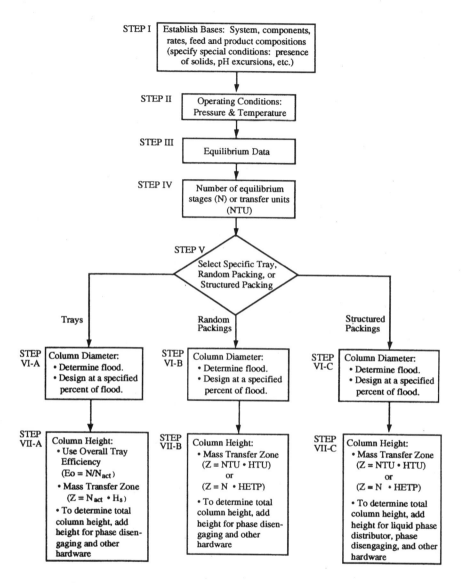

STEP I — Establish Bases: System, components, rates, feed and product compositions (specify special conditions: presence of solids, pH excursions, etc.)

STEP II — Operating Conditions: Pressure & Temperature

STEP III — Equilibrium Data

STEP IV — Number of equilibrium stages (N) or transfer units (NTU)

STEP V — Select Specific Tray, Random Packing, or Structured Packing

Trays

Random Packings

Structured Packings

STEP VI-A — Column Diameter:
• Determine flood.
• Design at a specified percent of flood.

STEP VI-B — Column Diameter:
• Determine flood.
• Design at a specified percent of flood.

STEP VI-C — Column Diameter:
• Determine flood.
• Design at a specified percent of flood.

STEP VII-A — Column Height:
• Use Overall Tray Efficiency $(E_o = N/N_{act})$
• Mass Transfer Zone $(Z = N_{act} \cdot H_s)$
• To determine total column height, add height for phase disengaging and other hardware

STEP VII-B — Column Height:
• Mass Transfer Zone $(Z = NTU \cdot HTU)$ or $(Z = N \cdot HETP)$
• To determine total column height, add height for phase disengaging and other hardware

STEP VII-C — Column Height:
• Mass Transfer Zone $(Z = NTU \cdot HTU)$ or $(Z = N \cdot HETP)$
• To determine total column height, add height for liquid phase distributor, phase disengaging, and other hardware

Figure A.1 Procedures for design of a new distillation column.

- *Operating costs.* Significant deviations in pressures and temperatures away from atmospheric conditions increase operating costs.

- *Physical property effects.* At higher pressures and temperatures, vapor pressures increase and viscosities decrease. The result is higher efficiency and vapor capacity, and lower column diameter. The downside is that at higher pressures (and temperatures), relative volatility is likely to decline, and more stages may be required to make the separation.

Step III. Equilibrium data

Once the bases and operating conditions are established, the phase equilibria must be determined. Either experimental or predicted phase equilibria can be used. Most of the large simulators contain thermodynamic data bases to predict phase equilibria. Sources of experimental vapor-liquid equilibria are presented in Table A.1.

TABLE A.1   Key Sources of Vapor-Liquid Equilibrium (VLE) Data

Source	Description	Reference(s)
Design Institute for Physical Property Data (DIPPR), American Institute of Chemical Engineers	Extensive data compilation project	Gess et al., 1991
Gmehling (Dechema series)	Includes basic data plus parameters for popular models for predicting liquid phase activity coefficients. Also includes equilibria for liquid/liquid systems.	Gmehling et al., 1979, 1994
Handbook of Applied Thermodynamics	Provides guidance on approaches to VLE measurement and prediction.	Palmer, 1987
S. M. Walas	One of the most extensive references on phase equilibria.	Walas, 1985
Hirata et al.	Includes basic data plus parameter for models to predict liquid phase activity coefficients. Easy to use.	Hirata et al., 1975
Hala et al.	Includes references of source publications which contain basic data.	Hala et al., 1967, 1968; Wichterle et al., 1975
Chu et al.	Data for 476 systems. Vapor and liquid compositions are reported at system conditions.	Chu et al., 1950 (two references)
Hadden and Grayson	K values for hydrocarbon systems plus hydrogen presented in the form of nomographs	Hadden and Grayson, 1961
Natural Gas Processors Association	Comprehensive and reliable displays of hydrocarbon K values	Natural Gas Processors Suppliers Assn., 1972
Horsley	Azeotropic data	Horsley, 1973
American Petroleum Institute Research Project No. 44	Thermodynamic data on 935 hydrocarbons including vapor pressure data	American Petroleum Institute, 1953 and supplements; Chao and Seader, 1961; Gallant, 1965–1970

TABLE A.2   Models for Predicting Vapor-Liquid Equilibria

Model	Description	Reference(s)
Van Laar	Fits extremely nonideal systems well such as alcohols and hydrocarbons. Cannot represent maxima or minima in liquid phase activity coefficients.	Abbott and Prausnitz, 1987
Wilson	Best for homogeneous mixtures exhibiting large positive deviations from ideal behavior such as alcohols-hydrocarbons	Wilson, 1964
NRTL	Unlike Wilson model, it can predict interior extremes in activity coefficients and can thus predict formation of two liquid phases.	Renon and Prausnitz, 1968
UNIQUAC	Flexible and powerful model. Can predict multicomponent mixture behavior from binary data.	Abrams and Prausnitz, 1975; Anderson and Prausnitz, 1978; Maurer Prausnitz, 1978; Prausnitz et al., 1980
UNIFAC	Used when no data are available. Functional groups are used to predict mixture behavior and liquid phase activity coefficients. Procedure is described in Fredenslund et al. (1977). Functional group parameter values are published in Gmehling et al. (1982) and Macedo et al. (1983). Special parameters are available for liquid-liquid equilibria in Magnussen et al. (1981).	Fredenslund et al., 1975, 1977; Gmehling et al., 1982, 1994; Macedo et al., 1983; Magnussen et al., 1981; Skjold-Jorgensen et al., 1979

The industrial mixture is normally a multicomponent mixture. However, most experimental vapor-liquid equilibria are for binary systems. The popular approach to predict phase equilibria for multicomponent mixtures is to use the data for the binary pairs, in combination with a model to predict multicomponent behavior. The Wilson, NRTL, and UNIQUAC models described in Table A.2 may be used for this purpose. These models are used to predict liquid phase activity coefficients in the multicomponent mixture as a function of composition. Once activity coefficients are known, vapor and liquid equilibrium compositions can be determined. When binary data are not available, the UNIFAC model, described in Table A.2, is used.

Step IV. Equilibrium stages

Shortcut methods.   Sometimes shortcut methods are used to determine the number of equilibrium stages. Though these methods are available via computer software, they are presented here to provide fundamentals of the methods. Of the shortcut methods, the Fenske-Underwood-Gilliland (FUG) Method is perhaps the most popular and is given below. Kister (1992) presents the strengths and weaknesses of both shortcut and comprehensive stage-to-stage methods for determining the number of equilibrium stages.

To determine the number of equilibrium stages (N), the minimum number of stages (at total reflux) and the minimum reflux (infinite number of equilibrium stages) must first be determined; the steps required are given below.

Minimum number of equilibrium stages. The minimum number of equilibrium stages (N_{min}) is determined by Fenske's method (Fenske, 1932).

$$N_{min} = \frac{\ln\left[\left(\frac{x_{LK}}{x_{HK}}\right)_D \left(\frac{x_{HK}}{x_{LK}}\right)_B\right]}{\ln(\alpha_{LK/HK})_{av}} \tag{1}$$

where N_{min} = minimum number of equilibrium stages, dimensionless

x_{LK} = mol fraction of light key, dimensionless

x_{HK} = mol fraction of heavy key, dimensionless

D   = when used as subscript signifies overhead product

B   = when used as subscript signifies bottoms product

$(\alpha_{LK/HK})_{av}$ = average value of relative volatility of light key relative to heavy key, dimensionless

The geometric average value of relative volatility is calculated from the top column temperature (dew point of overhead vapor) and the bottoms temperature (bubble point of bottom liquid product).

$$(\alpha_{LK/HK})_{av} = \sqrt{(\alpha_{LK/HK})_D (\alpha_{LK/HK})_B} \tag{2}$$

Determinations of the dew and bubble points are partially by trial and error, since the distributions of the nonkey components in the distillate and bottoms are not known.

Minimum reflux ratio. Underwood (1948) proposed the following two equations for calculation of minimum reflux ratio:

$$\sum_{i=1}^{n} \frac{\alpha_i x_{Fi}}{\alpha_i - \theta} = 1 - q \tag{3}$$

where α_i = average relative volatility for component i relative to heavy key [use Equation (2) to determine average]

x_{Fi} = mol fraction of component i in feed, dimensionless

q = number of moles of saturated liquid produced on the feed tray per mol of feed

In Equation (3), the value of θ must be determined by trial and error. The correct value of θ will lie between the relative volatility of the two key components. Once the value of θ is known, Equation (4) may be used to determine minimum reflux ratio.

$$R_{min} + 1 = \sum_{i=1}^{n} \frac{\alpha_i x_{Di}}{\alpha_i - \theta} \tag{4}$$

where R_{min} = minimum reflux ratio, dimensionless

n = number of components, dimensionless

x_{Di} = mol fraction of component i in distillate (D), dimensionless

Number of equilibrium stages. The number of equilibrium stages is calculated as function of the minimum number of equilibrium stages and minimum reflux ratio. Several methods have been proposed, including that proposed by Gilliland (1940). The Gilliland method was first developed as a plot, but was later transformed by Eduljee (1975) to the following equation:

$$\frac{N - N_{min}}{N + 1} = 0.75 \left(1 - \left[\frac{R - R_{min}}{R + 1} \right]^{0.566} \right) \tag{5}$$

where N = number of equilibrium stages, dimensionless

R = operating reflux ratio, dimensionless

The optimum reflux ratio may be determined by optimizing capital versus operating costs. Low reflux ratios require a large number of equilibrium stages and a higher capital cost. On the other hand, high reflux ratios translate to high energy costs. The optimum value of the operating reflux ratio (R) usually lies between 1.1 and 1.5 times the minimum reflux ratio (R_{min}) and must be specified by the designer. Once the operating reflux ratio is specified, the number of equilibrium stages may be calculated using Equation (5).

Distributed components by Fenske equation. Values for the distributed or nonkey components may be determined after the number of minimum stages has been calculated. The Fenske equation may be modified and used to determine concentrations of distributed components in the distillate and bottoms products. The modified form of the Fenske equation can be written, for any component i, as follows:

$$\frac{x_{Di}}{x_{Bi}} = (\alpha_i)_{av}^{N_{min}} \times \frac{(x_{HK})_D}{(x_{HK})_B} \tag{6}$$

where x_{Bi} = mol fraction of component i in bottoms product (B), dimensionless

$(\alpha_i)_{av}$ = average relative volatility of component i relative to heavy key. Use Equation (2) to determine averages

Location of feed tray. Kirkbride's method may be used to determine the ratio of trays above and below the feed point (Kirkbride, 1944):

$$\log \frac{M}{P} = 0.206 \log \left\{ \frac{B}{D} \left(\frac{x_{HK}}{x_{LK}} \right)_F \left[\frac{(x_{LK})_B}{(x_{HK})_D} \right]^2 \right\} \tag{7}$$

where B = bottom product rate, mass/time

D = overhead product rate, mass/time

F = feed rate to distillation column, mass/time

M = number of equilibrium stages above the feed tray

P = number of equilibrium stages below the feed tray

F, B, D = where used as subscripts refer to feed, bottom, and overhead streams, respectively

Use of simulators. In industry, a simulator containing stage-to-stage calculation algorithms is almost always used to determine the number of equilibrium stages. Such simulators also include thermodynamic data bases which also allow determinations of the vapor/liquid equilibria. And these simulators usually offer options, which not only allow the user to select a particular algorithm for stage calculations, but also the model/approach to determine vapor/liquid equilibria. Some simulators also allow the user to incorporate tray efficiencies, as a function of composition, so the number of actual trays can be determined as well. Users of such simulators must have a good understanding of fundamentals in order to appreciate limitations of the methods and models, and to recognize errors when they occur.

There are a number of simulators which are commercially available which will determine distillation stage requirements. Some examples of these simulators, their suppliers, and brief descriptions of their capabilities are given in Table A.3.

TABLE A.3   **Examples of Commercially Available Simulators Which Will Determine Stage Requirements for Distillation Processes**

Name of simulator	Supplier/ address	Software description/for determination of distillation stage requirements
ASPEN PLUS	Aspen Technology, Inc., Ten Canal Park, Cambridge, MA 02141	In addition to conventional distillation, this simulator addresses absorption, stripping, extraction, and azeotropic distillation, including nonideal mixtures. The models are based on an inside-out two-tier algorithm. A description of the options follows. RADFRAC can handle a second liquid phase and a solid phase anywhere in the column, as well as any combination of rate-controlled processes.
		PETROFRAC is specifically designed for petroleum refinery applications such as preflash towers, crude units, vacuum units, main fractionators, and delayed coker fractionators.
		MULTIFRAC solves systems of interlinked columns. It is applicable to absorber/stripper columns, air separation columns, and ethylene fractionators.
		BATCHFRAC™ simulates the dynamic behavior of batch columns. RATE-FRAC™ is a rate-based non-equilibrium model that simulates tray and packed columns based on actual trays and packings. It calculates tray efficiencies and HETPs, allowing more accurate results than with the conventional equilibrium stage approach.
PRO/II with PROVISION	Simulation Sciences, Inc., 601 S. Valencia Ave., Brea, CA 92621	PRO/II provides five types of robust solutions for distillation, each based on rigorous stage-to-stage solutions for the MESH equations.
		PRO/II provides I/O and Sure for general purpose solutions, Chemdist for nonideal chemical systems and reactive distillation, Eldist for electrolyte systems, and LLE for liquid/liquid extractors.
		PRO/II offers numerous methods for calculating thermodynamic properties such as K values, enthalpies, entropies, densities, gas and solid solubilities in liquids, and vapor fugacities.
		These methods include: generalized correlations, equations of state, liquid activity coefficients, vapor fugacity methods, and special methods for calculating the properties of specific systems of components such as alcohols, amines, glycols, sour water, and electrolyte systems.

Name of simulator	Supplier/ address	Software description/for determination of distillation stage requirements
DESIGN II	WinSim, Inc., 9800 Centre Parkway, Ste. 430, Houston, TX 77251	DESIGN II performs complete heat and material balance calculations for a wide variety of petrochemical, chemical, and refinery processes. The DIST module is used for rigorous simulation of absorbers, fractionators, strippers and other types of single-column configurations. The DIST module automatically runs the shortcut distillation first if the column has a partial or total condenser. The shortcut fractionator uses standard textbook techniques in calculating feed tray location (Kirkbride), reflux ratio (Underwood), theoretical stages (Fenske; Erbar-Maddox; Gilliland).
		DESIGN II includes fundamental data such as K-values, enthalpies, densities, transport properties, and surface tension correlations.
		With ChemTran, VLE/LLE data can be regressed to user-defined correlations [Renon (NRTL), Wilson, UNIFAC, UNIQUAC].
HYSYS Conceptual Design	Hyprotech Ltd., 300 Hyprotech Centre, 1110 Centre St. North, Calgary, Alberta, Canada, T2E 2R2	HYSYS allows users to capture and visualize the system thermodynamics of ideal, nonideal, and azeotropic mixtures to assist in distillation design. Options include: (1) choosing the property method which best predicts the known behavior of the component mixture; (2) fitting interaction parameters for specific operating conditions; (3) predicting azeotropes, based on design and property method(s); (4) examining the separation space; distillation boundaries and regions, azeotropes and liquid/liquid regions; and (5) designing a binary or ternary distillation tower to achieve the desired separation. The package developed in C++, uses the HYSYS graphical interface, and makes extensive use of PXY, TXY, and McCabe-Thiele diagrams and runs on a 32-bit Windows operating system.
		HYSYS Conceptual Design is the product of a technical alliance with Drs. Doherty and Malone of the University of Massachusetts—the developers of Mayflower; the Thermodynamic Research Center of Texas A&M University; and Hyprotech, Ltd.

*Abbreviations:*NRTL, nonrandom two liquid; VLE/LLE, vapor liquid equilibria/liquid liquid equilibria.

Step V. Trays versus packings

Performance comparisons. For any application, one must compare performance and costs to determine whether a tray, random packing, or structured packing is best. For high-pressure distillations, one should consider using one of the newer trays such as the Nye™, Max-Frac™, Ultra-Frac™, ECMD, or EEMD trays. These new trays offer a whole new "toolbox" of options.

A comparative analysis of efficiency and capacity of trays and packings, both random and structured, has been made by Kister et al., 1994 and 1993. This comparison is based on differences between the efficiency and capacity of an optimally designed tray and an optimally designed packed tower. Highlights of the details follow.

Efficiency comparisons. Efficiency comparisons of trays and packings were expressed as a function of the height equivalent to a theoretical plate (HETP), and the flow parameter (FP). The flow parameter is given by:

$$FP = \frac{L}{V} \sqrt{\frac{\rho_G}{\rho_L}} \tag{8}$$

where FP = flow parameter
L = liquid flow rate, lb/h ft^2 of cross-sectional area
V = vapor flow rate, lb/h ft^2 of cross-sectional area
ρ_G = gas density, lb/ft^3
ρ_L = liquid density, lb/ft^3

Low flow parameters, less than about 0.1, are typical of vacuum distillations. High flow parameters, greater than 0.3, are typical of high pressure or high liquid rate operation. Table A.4 shows the ranges of values of the flow parameter for each type of distillation.

Figure A.2 shows the relationship between HETP and FP for optimized trays (at 24-in. tray spacing), random packing (Nutter Rings™), and structured packing (Norton Intalox 2T). In this figure, adjustments were made for the vertical column height consumed by distributors, redistributors, and the end tray.

TABLE A.4   Values of Flow Parameter vs. Distillation Pressure Ranges

Values of flow parameter (FP)	Pressure distillation
0.02–0.1	Vacuum
0.1–0.3	Low to medium pressure
0.3–0.5	High pressure

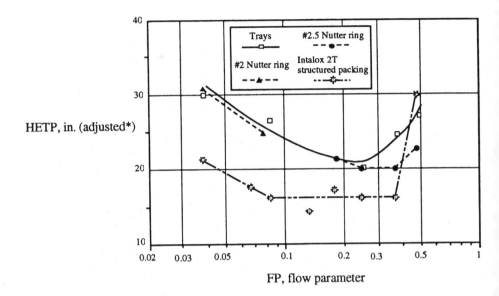

* Adjusted for vertical height consumed by distributor, redistributor and end tray

Figure A.2 Trays vs. packings—comparison of efficiency. (*Adapted from* Chemical Engineering Progress, *February 1994 with permission of the American Institute of Chemical Engineers. Copyright © 1994 AIChE. All rights reserved.*)

Figure A.2 shows minima in the plots of HETP versus FP. At FPs up to 0.2, HETP declines with a rise in FP. In this region, the HETP of optimized trays is about the same as random packing, and both are about 50% higher than that of the structured packing.

At flow parameters beyond 0.3, HETP rapidly rises with FP. The rate of rise is the smallest with random packings and the largest with structured packings. At 400 psia and an FP of about 0.5, the HETP of Intalox 2T structured packing is slightly higher than the optimized trays and about 20% higher than that of Nutter Rings™.

Capacity comparisons. In comparing capacity of trays versus random and structured packings, results are expressed as a function of the flood capacity factor, or C-factor, and the flow parameter (FP). The C-factor at flood is a measure of the vapor load at flood and is defined as:

$$C_s = U_G \sqrt{\frac{\rho_G}{\rho_L - \rho_G}} \qquad (9)$$

where C_s = C-factor, ft/s, based on column superficial area
U_G = gas velocity, ft/s, based on column superficial area

Figure A.3 shows that at FPs of 0.05, and less, the capacity of structured packing is 30–40% greater than that of either trays or random

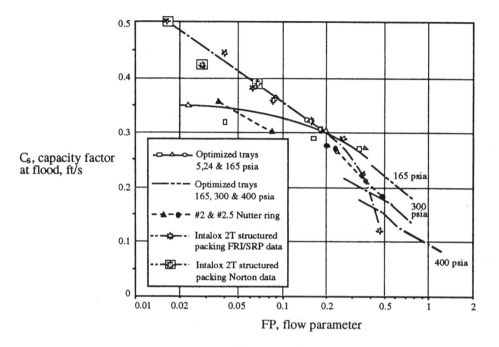

Figure A.3 Trays vs. packings—comparison of capacity. (*Adapted from* Chemical Engineering Progress, *February 1994 with permission of the American Institute of Chemical Engineers. Copyright © 1994 AIChE. All rights reserved.*)

packing. As FP increases, however, the capacity of the structured packing declines much faster than the capacity of the optimized tray or random packing. When the FP reaches 0.2, the capacity of structured packing approaches that of the optimized tray.

At FPs beyond 0.2–0.3, results show that the capacity of the structured packing declines rapidly with a rise in FP. The capacity of the optimized trays declines more slowly, and that of random packing declines at the slowest rate. At an FP of 0.5 and pressure of 400 psia, structured packing has the lowest capacity, while the capacity of optimized trays at 24-in. tray spacing is about 20% higher, and the capacity of random packing is 20% higher.

Summary.   When trays (24-in. tray spacing) were compared with state-of-the-art random and structured packings, all optimally designed, it was found that:

At FPs of 0.02–0.1 (corresponding to vacuum distillation):

■ Structured packing efficiency is about 50% higher than either trays or random packings.

As FP increases from 0.02 to 0.1:

- The capacity advantage of structured packing declines from 30 to 40% to zero
- Trays and random packing have about the same efficiency and capacity

As FP increases from 0.1 to 0.3:

- Efficiency and capacity of trays, random packing, and structured packing decline with an increase in flow parameter.
- The rate of decline in the capacity and efficiency is the most significant with structured packing, and least significant in random packings.

At an FP of 0.5 and 400 psia (corresponding to high-pressure distillation):

- Random packing appears to have the highest capacity and efficiency, and the structured packing the lowest.

The above comparisons of efficiency and capacity are based on data obtained for optimized designs and under ideal test conditions. To translate results to actual columns, one must consider liquid and vapor maldistribution. Liquid and vapor maldistribution are far more detrimental to the efficiency of packings than to the efficiency of trays.

Factors favoring trays versus packings.   In addition to the performance comparisons presented in the previous section, Chuang et al. (1992) compared trays and packings and presented the following guidelines:

- Trays are favored when:
  Operating pressure and liquid rate are high (high-efficiency random packing can also be attractive—see previous section)
  Column diameter is large

- Random packings are favored when:
  Column diameter is small
  Corrosion and foaming are present
  Batch columns are involved

- Structured packings are favored when:
  Column operates at vacuum and low pressures
  Low pressure drop is needed
  Low liquid holdup is desired
  Low column temperature is required

In the final analysis, one must consider relative costs versus benefits for each application. For example, the installed cost of structured packings is about three times that of traditional trays for the same diameter column (Yeoman, 1994).

Step VI-A. Diameter of sieve tray column

Flood velocity for distillation sieve trays can be calculated using the Fair Correlation given in Figure A.4 (Fair, 1961). In conjunction with Figure A.4, the flood velocity is determined from the following equation.

$$U_{nf} = C_{sb}\left(\frac{\sigma}{20}\right)^{0.2}\sqrt{\frac{\rho_L - \rho_G}{\rho_G}} \qquad (10)$$

where U_{nf} = net velocity of vapor at flood, ft/s
C_{sb} = Souders and Brown Factor at flood
σ = surface tension, dynes/cm

The Souders and Brown factor (C_{sb}) must be determined from Figure A.4. The flow parameter (FP) is the abscissa and the correlating parameter is tray spacing, and both must be determined in order to obtain the value of C_{sb}. After C_{sb} is determined (by reading the appropriate ordinate value from Figure A.4) the net velocity of vapor at flood (U_{nf}) may be calculated using Equation (10). Standard tray spacings in dis-

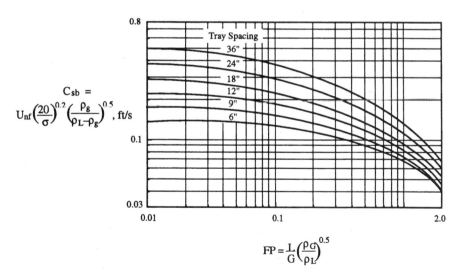

Figure A.4 Fair flood correlation for tray distillation columns. (*Adapted from Fair and Matthews, Petro. Refin., April 1958 with permission of Gulf Publishing Co. Copyright © 1958 Gulf Publishing Co, Houston, TX. All rights reserved.*)

tillation are 18 and 24-in. for large-diameter columns, although 12 and 36 in. are also used.

Gas velocity (U_n) is then determined by assuming it is 50–90% of the net velocity at flood (U_{nf}). After U_n is determined, net column cross-sectional area (A_n) and total column cross-sectional area (A_c) are calculated by Equations (11) and (12).

$$A_n = \frac{Q_G}{U_n} \tag{11}$$

and

$$A_c = A_n + A_d \tag{12}$$

where A_n = net column area, length^2
   $\quad Q_G$ = volumetric flow rate of the vapor, length^3/time
   $\quad U_n$ = net vapor velocity, length/time
   $\quad A_c$ = total column cross-sectional area, length^2
   $\quad A_d$ = downcomer area, length^2

Column diameter is then calculated by: $D = \sqrt{\dfrac{4A_c}{\pi}}$

Step VII-A. Height of sieve tray column

To convert equilibrium stages to actual trays, one must take into account that equilibrium is not reached with actual trays. Equilibrium stages are converted to actual trays using overall tray efficiency.

$$E_o = \frac{N}{N_{act}} \tag{13}$$

where E_o = overall tray efficiency, dimensionless
   $\quad N_{act}$ = number of actual trays

There are several ways to determine overall tray efficiency (E_o). The most rigorous method is to begin with point efficiency, and then convert to tray efficiency. For multicomponent systems, this approach is not practical because of the uncertainties in the mechanisms and models involved. For preliminary design, the O'Connell Correlation (1946) is often used to estimate E_o. The original correlation was in graphical form, the equivalent analytical expression is given below (Lockett, 1986).

$$E_o = 0.492(\mu_L \alpha)^{-0.245} \tag{14}$$

where μ_L = liquid viscosity, centipoise
   α = relative volatility

Based on the value of the overall tray efficiency and a specified tray spacing, the mass transfer zone (the part of the column containing the trays) can be determined. The height of the column shell must be tall enough to include the trays and provide extra height at the top and bottom of the column for phase disengagement and hardware. Thus, the total height of the column shell is:

$$H_c = (N_{act} - 1)H_s + DH \qquad (15)$$

where H_c = height of column shell, length
   H_s = tray spacing, length
   DH = total height of column required for phase disengagement and hardware (in a commercial column several feet may be required at both the top and bottom of the column), length

Step VI-B. Diameter of column with random packings

The pressure drop at the flood point may be estimated using the Kister and Gill correlation (Kister and Gill, 1991).

$$\Delta P_{flood} = 0.115 F_p^{0.7} \qquad (16)$$

where ΔP_{flood} = pressure drop at flooding, inches water/ft of column height
   F_p = packing factor, characteristic of each packing (normally provided by packing supplier) ft^{-1}

Based on the operating reflux, liquid and vapor rates can be determined. These rates together with the values of liquid and gas densities allow determination of the flow parameter (FP).

   Column diameter may then be determined using the following steps:

- With the values of the pressure drop at flood (ΔP_{flood}), FP, and the Eckert correlation given in Figure A.5, one can determine the C-factor (C_s).

- Once the C-factor is known, Equation (18) may be used to determine the superficial velocity at flood.

- The operating velocity, which is normally chosen as 50–90% of the flood velocity, can then be determined.

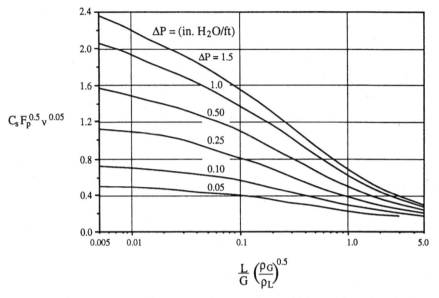

Figure A.5 Eckert generalized pressure drop correlation. (*Adapted from Strigle, R. F. Jr., Random Packings and Packed Towers, 1987 with permission of Gulf Publishing Co. Copyright © 1987 Gulf Publishing Co., Houston, TX. All rights reserved.*)

- The cross-sectional area of the column (A_c) may then be determined by dividing the vapor volumetric flow rate (Q_G) by gas velocity (U_G) at operating conditions.

- Column diameter is determined by: $D = \sqrt{\dfrac{4A_c}{\pi}}$

- Based on operating gas velocity, the pressure drop at operating conditions may then be estimated by using Figure A.5.

The ordinate in Figure A.5 is called the capacity parameter, which is defined as:

$$\text{Capacity parameter} = C_s F_p^{0.5} \nu^{0.05} \qquad (17)$$

and

$$C_s = U_G \sqrt{\frac{\rho_G}{\rho_L - \rho_G}} \qquad (18)$$

where ν = kinematic viscosity, centistokes

The guidelines presented in Table A.5 may be used to cross-check values of the maximum operating pressure drop (Kister, 1992).

TABLE A.5 Maximum Operating Pressure Drop for Packed Distillation Columns. (*Adapted from Kister, 1992.*)

Column pressure	Flow parameter (FP)	Maximum pressure drop (in. water/ft of packing)
Vacuum	0.02–0.10	0.01–0.6
Low to medium pressure	0.10–0.30	0.5–1.0
High pressure	>0.5	$0.15\,F_p^{0.7}\dfrac{\rho_L}{\rho_{H_2O}}$

Step VII-B. Height of column equipped with random packings

The height of a transfer unit (HTU) and number of transfer units (NTU) are defined as:

$$\text{HTU} = \frac{G}{(K_G a_e)A_c} \tag{19}$$

$$\text{NTU} = \int_{y_2}^{y_1} \frac{dy}{y-y^*} \tag{20}$$

where

G = molar flow of vapor, mass/time
K_G = overall mass transfer coefficient, mass/time-length^2
a_e = area of interfacial contact between liquid and vapor phase per unit volume of contactor, length^2/length^3
A_c = total column cross-sectional area, length^2
y = mol fraction in vapor phase, dimensionless
y^* = mol fraction of component in vapor phase in equilibrium with liquid phase, dimensionless
HTU = height of a transfer unit for an individual phase, length
NTU = number of transfer units, dimensionless

and

$$Z = \text{HTU} \cdot \text{NTU} \tag{21}$$

where Z = height of mass transfer zone (height of column containing pacing), length

The number of transfer units (NTU) depends on the concentration at the top and bottom of the column, and on the equilibria. Height of a transfer unit (HTU) depends on vapor flow rate (G), overall mass transfer coefficient (K_G), and on the interfacial contact area between the vapor and liquid phases (a_e). The relationship between HTU and HETP is:

$$\text{HETP} = \text{HTU}\,\frac{\ln \lambda}{\lambda - 1} \tag{22}$$

where HETP = height equivalent of a theoretical plate, length
$\quad\quad\quad\lambda$ = ratio of slope of equilibrium line (m) to the slope of the operating line (L/G), dimensionless
$\quad\quad\quad\text{HTU}_G$ = contribution of vapor phase to overall HTU, length
$\quad\quad\quad\text{HTU}_L$ = contribution of liquid phase to overall HTU, length

and

$$Z = \text{HETP} \cdot N \tag{22a}$$

where N = number of equilibrium stages, dimensionless

Although HTU is more rigorous, HETP is used more frequently. It allows the designer to compare the efficiency of tray and packed columns.

Rules of thumb for predicting HETP for a column equipped with random packings are presented in Table A.6 (Kister, 1992).

A global transfer unit incorporates the contribution of both the gas and liquid phase resistances (Treybal, 1980):

$$\text{HTU} = (\text{HTU})_G + \lambda(\text{HTU})_L \tag{23}$$

where

$$(\text{HTU})_G = \frac{V_G}{k_G a_i} \tag{23a}$$

$$(\text{HTU})_L = \frac{L}{k_L a_i} \tag{23b}$$

TABLE A.6   Rule-of-Thumb for Predicting HETP of Columns Equipped with Random Packings

HETP	Column diameter
HETP is equal to or greater than the diameter of the column (in.)	For column diameters less than 2 ft
HETP = 1.5×packing diameter (in.)	For column diameters greater than 2 ft

$$\lambda = \frac{m}{\left(\dfrac{L}{G}\right)} = \frac{\text{slope of equilibrium line}}{\text{slope of operating line}} \qquad (23c)$$

where V_G = total flow of vapor phase, mol/time
$\quad k_G$ = individual mass transfer coefficient, ft/s
$\quad k_L$ = mass transfer coefficient for liquid phase, mol/time-length^2
$\quad a_i$ = interfacial area between phases per unit volume of packed column, length^2/length^3
$\quad m$ = slope of equilibrium line, dimensionless

Considering that $G/A_c = U_G\,\rho_M$, and dividing the mass transfer coefficient in Equation (19) by the density of the vapor, we arrive at Equation (24) where the mass transfer coefficients (k_G and k_L) are expressed in units of velocity:

$$\text{HTU} = \text{HTU}_G + \lambda \text{HTU}_L = \frac{U_G}{k_G a_e} + \lambda \frac{U_L}{k_L a_e} \qquad (24)$$

where U_L = superficial velocity for liquid, length/time
$\quad a_e$ = "effective" area for mass transfer, length^2

There are a number of ways to determine HTU or HETP. The preferred route is to obtain large-scale plant data. Kister (1992) presents efficiency data for a number of packings and systems. Cornell et al. (1960, two references) have proposed a model to predict HTU_G and HTU_L. A design procedure may also be found in Henley and Seader (1981).

In another approach, developed by Bravo and Fair (1982), one determines HETP by first determining HTU. This model involves estimating mass transfer coefficients and an "effective" interfacial area. In this approach, mass transfer coefficients are calculated by:

$$\frac{k_G}{a_p D_G} = 5.23\,\text{Re}_G^{0.7}\,\text{Sc}_G^{0.333}(a_p D_p)^{-2} \qquad (25)$$

where a_p = packing surface area, ft^2/ft^3
$\quad D_G$ = diffusion coefficient for vapor phase, ft^2/s
$\quad \text{Re}_G$ = Reynolds number for vapor phase, dimensionless
$\quad \text{Sc}_G$ = Schmidt number for vapor phase, dimensionless
$\quad D_P$ = packing diameter, in.

and

$$k_L\left(\frac{\rho_L}{0.000672g\mu_L}\right)^{0.333} = 0.0051\left(\text{Re}_L\,\frac{a_p}{a_w}\right)^{0.667}\text{Sc}_L^{-0.5}(a_p D_p)^{0.4} \qquad (26)$$

where g = gravity, 32 ft/s²
       Re_L = Reynolds number for liquid phase, dimensionless
       Sc_L = Schmidt number for liquid phase, dimensionless
       a_w = area of wetted packing, ft²

The ratio of wetted to packing area is:

$$\frac{a_w}{a_P} = 1 - \exp\left[-1.45 \, Re_L^{0.1} \, Fr_L^{-0.05} \, We_L^{0.2} \left(\frac{\sigma_c}{\sigma}\right)^{0.75}\right] \tag{27}$$

where Fr_L = Froude number of liquid, dimensionless
       We_L = Weber number of liquid, dimensionless
       σ_c = critical surface tension, dynes/cm (61 dynes/cm for ceramic packings; 75 dynes/cm for structured packings; 33 dynes/cm for polyethylbenzene packings)

$$Re_L = \frac{\rho_L U_L}{0.000672 \, a_P \mu_L} \tag{28}$$

where

$$Re_G = \frac{\rho_G U_G}{0.000672 \, a_P \mu_G} \tag{29}$$

where μ_G = viscosity of vapor phase, centipoise

$$Fr_L = \frac{a_P U_L}{g} \tag{30}$$

$$We_L = \frac{\rho_L U_L^2}{a_P(\sigma/453.23)} \tag{31}$$

$$Sc_L = \frac{0.000672 \mu_L}{\rho_L D_L} \tag{32}$$

where D_L = diffusion coefficient for liquid, ft²/s

$$Sc_G = \frac{0.000672 \mu_G}{\rho_G D_G} \tag{33}$$

The effective area for mass transfer is:

$$a_e = a_P \sigma^{0.5} \, HTU^{0.4} \, (Ca_L \, Re_G)^{0.392} \tag{34a}$$

and

$$Ca_L = \frac{0.304 \, U_L \mu_L}{\sigma} \tag{34b}$$

where Ca_L = capillary number for liquid phase, dimensionless

Finally, the individual height of transfer units (HTU_G and HTU_L) and global height of a transfer unit (HTU) are calculated with Equation

(24). HETP may then be determined with Equation (22). The mass transfer zone (Z) may be determined with Equation (22a) and column height with Equation (35).

$$H_C = Z + \text{DH} \tag{35}$$

As a rule-of-thumb, in packed columns, liquid should be collected and redistributed every 15–20 ft.

Step VI-C. Diameter of column equipped with structured packings

There are two types of metallic structured packings: (1) wire mesh and (2) corrugated sheet metal. Examples of the wire mesh type are Goodloe®, Hyperfil®, and Sulzer BX®. Examples of the structured sheet metal type are Mellapak®, Norton 2T, Flexipac®, Gempak®, Intalox®, and Montz™.

One procedure used to calculate the cross-area of a column with structured packing is similar to that used for random packing. These steps follow:

- Equation (16) is used to calculate pressure drop at flood.
- The Kister and Gill Correlation given in Figure A.6, which is similar in format to the Eckert Correlation, is used to determine flood velocity (Kister and Gill, 1991).
- Operating velocity is determined as a percent approach to flood velocity.
- Figure A.6 may then be used to determine pressure drop at operating conditions.

The cross-sectional area of the packed column (A_c) is then obtained by dividing the volumetric flow rate (Q_G) by the vapor velocity at operating conditions (U_G). Column diameter is determined by:

$$D = \sqrt{\frac{4 A_c}{\pi}} \tag{36}$$

Rocha et al. Model (SI units). A more recent model that may be used to determine column diameter in columns equipped with structured packings has been presented by Rocha et al. (1993). Liquid holdup, as well as pressure drop, are functions of liquid and vapor superficial velocities. The details follow. Liquid holdup is given by Equation (37).

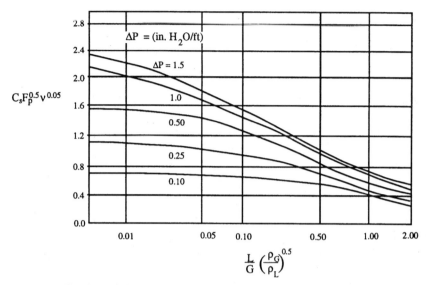

Figure A.6 Kister and Gill—generalized pressure drop correlation for structured packings. (*Adapted from Kister, H. Z.,* Distillation Design, *with permission of McGraw-Hill, Inc. Copyright © 1992 McGraw-Hill. All rights reserved.*)

$$h_t = \left[\frac{4F_t}{S}\right]^{2/3} \left[\frac{3\mu_L U_L}{\rho_L \, \varepsilon \sin\theta g \, \frac{\rho_L - \rho_G}{\rho_L}\left(1 - \frac{\Delta P/\Delta Z}{(\Delta P/\Delta Z)_{\text{flood}}}\right)}\right]^{1/3} \tag{37}$$

where h_t = liquid holdup, dimensionless
$\quad\ \ F_t$ = parameter to determine liquid holdup, dimensionless
$\quad\ \ S$ = length of a corrugation in structured packing, m
$\quad\ \ \Delta P$ = pressure drop, Pascals
$\quad\ \ \Delta Z$ = incremental height, m

$$\frac{\Delta P}{\Delta Z} = \frac{(\Delta P/\Delta Z)_{\text{dry}}}{[1-(0.614 + 71.35 \cdot S)h_t]^5} \tag{38}$$

$$F_t = \frac{29.12(\text{We}_L \text{Fr}_L)^{0.15} S^{0.359}}{\text{Re}_L^{0.2}\varepsilon^{0.6}(1-0.93\cos\gamma)(\sin\theta)^{0.3}} \tag{39}$$

Dimensionless numbers are calculated based on the following:

$$\text{We}_L = \frac{U_L^2\rho_L S}{\sigma} \tag{40}$$

$$\mathrm{Fr_L} = \frac{U_L^2}{Sg} \qquad (41)$$

where

$$\mathrm{Re_L} = \frac{U_L S \rho_L}{\mu_L} \qquad (42)$$

For metallic structured packings:

$$\cos \gamma = 0.9 \quad \text{for} \quad \sigma < 0.055 \text{ N/m} \qquad (43)$$

where γ = wetting angle between liquid and solid surface, degrees (°)

$$\cos \gamma = 5.211 \times 10^{-16.835\sigma} \quad \text{for} \quad \sigma \geq 0.055 \text{ N/m} \qquad (44)$$

Combining Equations (37) with (38) and using Equations (39)–(42) gives:

$$1 - \left[\frac{\dfrac{0.1775\rho_G}{S\,\varepsilon^2(\sin\theta)^2} U_G^2 + \dfrac{88.774\,\mu_G}{S^2\,\varepsilon\sin\theta} U_G}{\Delta P / \Delta Z} \right]^{0.2}$$

$$- (0.614 + 71.35 \cdot S) \left[\frac{82.71\mu_L^{0.2}\,(R_1 U_G)^{0.4}}{S^{0.841}\sigma^{0.15}\rho_L^{0.05}\varepsilon^{0.6}(1-0.93\cos\gamma)(\sin\theta)^{0.3}} \right]^{\frac{2}{3}}$$

$$\times \left[\frac{3\mu_L R_1 U_G}{\rho_L\,\varepsilon\sin\theta\,g\,\dfrac{\rho_L-\rho_G}{\rho_L}\left(1 - \dfrac{\Delta P/\Delta Z}{(\Delta P/\Delta Z)_{\text{flood}}}\right)} \right]^{\frac{1}{3}} = 0 \qquad (45)$$

where ε = void fraction of packing, dimensionless

$$R_1 = \frac{U_L}{U_G} \qquad (46)$$

Because $\Delta P/\Delta Z$ is known, U_G can be determined by trial and error. The cross-sectional area of the column may be determined by dividing volumetric gas flow rates (Q_G) by the gas velocity (U_G).

Step VII-C. Height of column equipped with structured packings

To determine the total height of the packing required to make the separation, one must first determine HTU or HETP. Three optional methods are presented below.

Harrison and France (1989) have presented the following rule of thumb for the quick estimation of HETP for structured packing.

$$\text{HETP} = \frac{1200}{a_p} + 4 \tag{47}$$

where HETP = height equivalent to a theoretical plate, in.
      a_p = packing surface area, ft²/ft³

Alternately, a more accurate approach would be to use the interpolation-extrapolation of packing efficiency data presented by Kister (1992). A newer model for structured packings is also available (Rocha et al., 1996). This model is based on first calculating HTU and then using Equation (22) to determine HETP. In Step VII-C, as well as in Step VI-C, System International Units (SI) are used (m, s, kg).

First the mass transfer coefficients which appear in Equation (24) are obtained with the following equations:

$$k_G = 0.054 \frac{D_G}{S} \left[\frac{(U_{Ge} + U_{Le})\rho_G S}{\mu_G} \right]^{0.8} \left[\frac{\mu_G}{D_G \rho_G} \right]^{0.33} \tag{48}$$

where U_{Ge} = effective velocity for vapor phase, m/s
      U_{Le} = effective velocity for liquid phase, m/s

and

$$k_L = 2 \sqrt{\frac{D_L U_{Le}}{\pi S}} \tag{49}$$

where D_L = diffusion coefficient for liquid, m²/s

Effective velocities are obtained from:

$$U_{Ge} = \frac{U_G}{\varepsilon(1-h_t)\sin \theta} \tag{50}$$

$$U_{Le} = \frac{U_L}{\varepsilon\, h_t \sin \theta} \tag{51}$$

The effective interfacial area is calculated by:

$$\frac{a_e}{a_p} = F_{SE} \frac{29.12\,(\text{We}_L \text{Fr}_L)^{0.15} S^{0.359}}{\text{Re}_L^{0.2} \varepsilon^{0.6}(1-0.93\cos\gamma)(\sin\theta)^{0.3}} \tag{52}$$

where F_{SE} = surface enhancement factor, dimensionless

TABLE A.7   Structured Packings—Surface
Enhancement Factors

Type packing	Surface enhancement factor (F_{SE})
Gempak®	0.344
Flexipac®	0.350
Intalox®	0.415

For the specific case of gauze structured packings

$$\frac{a_e}{a_p} = 0.65 \tag{53}$$

The surface enhancement factor is a function of the type of treatment on the surface of the packing. Surface enhancement factors (F_{SE}) for several structured packings are given in Table A.7.

Superficial velocity (U_G) was calculated in the previous section in determining column diameter (Step VI-C). With Equations (48)–(52), mass transfer coefficients and the effective area (a_e) can be computed. With these results, Equation (24) may be used to calculate HTU. HETP may then be determined with Equation (22).

Equation (22a) may be used to calculate the required height of packing, but extra column height (DH) must be included for hardware for distribution of the liquid and vapor phases and redistribution of the liquid phase. Equation (54) is used to determine total height of the column shell.

$$H_C = Z + DH \tag{54}$$

Retrofit of Existing Column

In a rating (retrofit) project, column height and diameter are fixed. In a retrofit project, involving a distillation column operating at high pressure (i.e., greater than 50–100 psig), a typical project might involve replacing conventional trays with higher capacity trays like the Nye™, Max-Frac™, Ultra-Frac™, ECMD, or EEMD trays.

When the objective is to increase capacity in vacuum distillation columns, case studies show that replacing sieve trays with structured packings can often be economically attractive (Humphrey and Seibert, 1992). The steps involved are shown in Figure A.7. The following case is based on replacing sieve trays with structured packing.

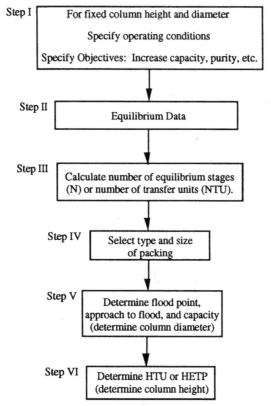

Figure A.7 Retrofit procedures for replacement of sieve trays with structured packing.

Step I. Bases

The objective of the retrofit must be clearly stated. For example, is the objective to increase capacity, increase purity, or save energy? From flowsheets, one must obtain precise information about the height and diameter of the column and its existing internal hardware. Accurate information must also be obtained about feed composition, rates, and physical properties.

Step II. Equilibrium data

Use the same procedures and sources of equilibrium data described in New Column Design—Step III.

Step III. Calculate number of equilibrium stages

Use the same procedures described in New Column Design—Step IV.

Step IV. Select type and size of packing

If increased capacity is the objective, structured packings with larger spacings between packing elements should be considered. If increased purity is needed, packings with smaller spacings and larger surface area, which provide the equivalent of more equilibrium stages, would probably be needed. And it is best to use packings for which large-scale data are available.

Step V. Determine flood point, approach to flood, and capacity

One must determine if the diameter of the column is large enough to accommodate desired gas and liquid rates. One must compare specified rates versus the flood rates. We suggest using the procedures and equations presented in Step VI-C for the design of a new column.

If the Rocha et al. method is used, Equation (45) must be solved by trial and error to determine $\Delta P/\Delta Z$. Then the liquid holdup (h_t) may be determined with Equation (37). Note that SI units are used in these calculations.

Step VI. Determine HTU or HETP

The value of HETP for structured packing must be determined. The rule of thumb method of France (presented earlier) may be used to get a quick estimate of the HETP. For more precise estimates, follow one of these procedures:

1. Use interpolation-extrapolation of packing efficiency data (Kister, 1992).

2. Rocha et al. method. These procedures are the same as those used for Step VII-C for the design for a new column.

If the Rocha et al. method is used, the value of $\Delta P/\Delta Z$ is calculated and compared with the value at flooding. The resulting value of HETP is compared with that required for the separation.

References

Abbott, M. M. and J. M. Prausnitz, "Phase Equilibria," in R. W. Rousseau, ed., *Handbook of Separation Process Technology,* Wiley, 1987.
Abrams, D. S. and J. M. Prausnitz, *AIChE Journal,* **21,** 116 (1975).
American Petroleum Institute, *Selected Values of Physical and Thermodynamic Properties of Hydrocarbons and Related Compounds,* Project 44, Carnegie Press, Pittsburgh, PA, 1953 and supplements.
Anderson, T. F. and J. M. Prausnitz, *Ind. Eng. Chem. Process Des. Dev.,* **17,** 552, 1978.
Bravo, J. L. and J. R. Fair, *Ind. Eng. Chem. Proc. Des. Dev.,* **21,** 162, 1982.
Chao, K. C. and J. D. Seader, *AIChE Journal,* **7,** 598, 1961.
Chu, J. C. et al., *Distillation Equilibrium Data,* Reinhold, New York, 1950.

Chu, J. C. et al., *Vapor-Liquid Equilibrium Data,* Edwards, Ann Arbor, MI, 1950.

Chuang, K. T. et al., *Can. J. of Chem. Eng.,* **70,** 794–799, August 1992.

Cornell, D. et al., *Chem. Eng. Prog.,* **56**(8), 48–53, 1960.

Cornell, D. et al., *Chem. Eng. Prog.,* **56**(7), 68–74, 1960.

Eduljee, H. E., *Hydrocarbon Proc.,* **54**(9), 120, 1975.

Fair, J. R., *Petro/Chem. Engr.,* **33**(10), 45, 1961.

Fair, J. R. and F. Matthews, *Petro. Refin.,* Gulf Publishing, Houston, TX, April 1958.

Fenske, M. R., *Ind. Eng. Chem.,* **24,** 482, 1932.

Fredenslund, A. et al., *Vapor-Liquid Equilibria Using UNIFAC,* Elsevier, Amsterdam, 1977.

Fredenslund, A. et al., *AIChE Journal,* **21,** 1086, 1975.

Gallant, R. W., *Physical Properties of Hydrocarbons, Hydrocarbon Process,* 44–49, July 1965–January 1970.

Gess, M. A. et al., "Thermodynamic Analysis of Vapor-Liquid Equilibria: Recommended Models and a Standard Data Base," in *Design Inst. for Phys. Prop. Data,* American Institute of Chemical Engineering, New York, 1991.

Gilliland, E. R., *Ind. Eng. Chem.,* **32,** 1220, 1940.

Gmehling, J. et al., *Azeotropic Data,* DM 598, ISBN 3-527-28671-3, Germany, 1994.

Gmehling, J. et al., *Ind. Eng. Chem. Process Des. Dev.,* **21,** 118, 1982.

Gmehling, J. et al., *Vapor-Liquid Equilibrium Collection* (continuing series), DECHEMA, Frankfurt, 1979.

Hadden, S. T. and H. G. Grayson, *Petrol. Refiner.,* **40**(9), 207, 1961.

Hala, E. et al., *Vapor-Liquid Equilibrium at Normal Pressures,* Pergamon, Oxford, 1968.

Hala, E. et al., *Vapor-Liquid Equilibrium,* 2nd ed., Pergamon, Oxford, 1967.

Harrison, M. E. and J. J. France, *Chem. Eng.,* 121, April 1989.

Henley, E. J. and J. D. Seader, *Equilibrium-Stage Separation Operations in Chemical Engineering,* Wiley, 1981.

Hirata, M. et al., *Computer Aided Data Book of Vapor-Liquid Equilibria,* Elsevier, Amsterdam, 1975.

Horsley, L. H., *Azeotropic Data-III,* American Chemical Society Advances in Chemistry Series 116, Washington, DC, 1973.

Humphrey, J. L. and A. F. Seibert, *Chem. Eng. Prog.,* March 1992.

Kirkbride, C. G., *Petrol. Refiner.,* **23**(32), 1944.

Kister, H. Z. and D. R. Gill, *Chem. Eng. Prog.,* **87**(2), 32, 1991.

Kister, H. Z., *Distillation Design,* McGraw-Hill, New York, 1992.

Kister, H. Z., *Distillation Operation,* McGraw-Hill, New York, 1990.

Kister, H. Z. et al., "Capacity and Efficiency: How Trays and Packings Compare," presented at the AIChE Spring Meeting, Houston, TX, March–April 1993.

Kister, H. Z. et al., *Chem. Eng. Prog.,* 23–32, February 1994.

Lewis, W. K. and G. L. Matheson, *Ind. Eng. Chem.,* **24,** 494, 1932.

Lockett, M. J., *Distillation Design Fundamentals,* Cambridge University Press, 1986.

Macedo, E. A. et al., *Ind. Eng. Chem. Process Des. Dev.,* **22,** 676, 1983.

Magnussen, T. et al., *Ind. Eng. Chem. Process Des. Dev.,* **20,** 331, 1981.

Maurer, G. and J. M. Prausnitz, *Fluid Phase Equilibria,* **2,** 91, 1978.

Natural Gas Processors Suppliers Assn., *Engineering Data Book,* 9th ed., Tulsa, OK, 1972.

O'Connell, H. E., *Trans. AIChE,* **42,** 741, 1946.

Palmer, D. A., *Handbook of Applied Thermodynamics,* CRC Press, Boca Raton, FL, 1987.

Prausnitz, J. M. et al., *Computer Calculations for Multicomponent Vapor-Liquid and Liquid-Liquid Equilibria,* Prentice-Hall, Englewood Cliffs, NJ, 1980.

Renon, H. and J. M. Prausnitz, *AIChE Journal,* **14,** 135, 1968.

Rocha, J. A. et al., "Distillation Columns Containing Structured Packings: A Comprehensive Model for Their Performance. 2. Mass Transfer Model," *Ind. Eng. Chem. Research,* **35**(5), 1660–1667, 1996.

Rocha , J. A. et al., *Ind. Eng. Chem. Research,* **32**(4), 641–651, 1993.

Skjold-Jorgensen, S. et al., *Ind. Eng. Chem. Process Des. Dev.,* **18,** 714, 1979.

Strigle, R. F., Jr., *Random Packings and Packed Towers,* Gulf Publishing, 1987.

Thiele, E. W. and R. L. Geddes, *Ind. Eng. Chem.,* **25,** 290, 1933.

Treybal, R. E., *Mass Transfer Operations,* 4th ed., McGraw-Hill, New York, 1980.

Underwood, A. J. V., *Chem. Eng. Prog.,* **44,** 603, 1948.

Walas, S. M., *Phase Equilibria in Chemical Engineering,* Butterworth, Boston, 1985.

Wichterle, I. et al., *Vapor-Liquid Equilibrium Data Bibliography,* Elsevier, Amsterdam, 1975.

Wilson, G. M., *Journal of the Am. Chem. Soc.,* **86,** 127, 1964.

Yeoman, N., Koch Engineering, letter communication, June 9, 1994.

B

Equipment Suppliers

The number of companies supplying distillation, extraction, adsorption, and membrane equipment are far too numerous to list them all. More extensive lists can be found in the *Thomas Register* and in the *Chemical Week Buyer's Guide*. The coverage here is more complete for U.S. companies.

TABLE B.1   Some Suppliers of Distillation Equipment

Company	Address	Product description
ACS Industries, Inc.	14208 Industry Rd., Houston, TX 77053	Structured column packing, tower trays, internals, mist eliminators
APV Crepaco, Inc.	395 Fillmore Ave., P.O. Box 366, Tonawanda, NY 14151-0366	Packaged distillation systems
Artisan Industries, Inc.	73 Pond St., Waltham, MA 02154	Tower trays, column packings
Cielcote	The Cielcote Company, 140 Sheldon Rd., Berea, OH 44017	Thermoplastic random tower packings
Chem-Pro Corporation	P.O. Box 1248, Fairfield, NJ 07007	Structured and dumped tower packings, column internals
Croll-Reynolds Co., Inc.	751 Central Ave., Westfield, NJ 07091	Dumped tower packing
Distillation Engineering Co.	105 Dorsa Ave., Livingston, NJ 07039	Distillation columns and pilot plant equipment
Edwards Engineering Corp.	101 Alexander Ave., Pompton Plains, NJ 07444	Hydrocarbon vapor recovery, solvent vapor recovery
Finish Company, Inc.	921 Greengarden Rd., Erie, PA 16501-1591	Distillation equipment
Glitsch, Inc.	P.O. Box 226227, Dallas, TX 75266	Packing, trays, other distillation equipment
Hoechst Celanese Corp.	13800 South Lakes Dr., Charlotte, NC 28273	Membrane phase contactors for absorption/stripping
Hoyt	Hoyt Corporation, Forge Rd., Westport, MA 02790-0217	Distillation systems
Interel Corporation	P.O. Box 4676, Englewood, CO 80155	Distillation systems
Jaeger Products, Inc.	P.O. Box 1563, Spring, TX 77383-9955	Structured and dumped tower packings, column internals
Kimbre, Inc.	P.O. Box 570846, Perrine, FL 33257-0846	Mist eliminators and tower packings

TABLE B.1   Some Suppliers of Distillation Equipment (*Continued*)

Company	Address	Product description
Koch Engineering Co., Inc.	P.O. Box 8127, Wichita, KS 67208	Distillation and extraction equipment including packings and trays
KŪHNI	CH-4123 Allschwil-Basle, Switzerland, Muhlebachweg 9-15	Structured packing
Leybold-Heraeus	UIC, Inc., P.O. Box 863, 1225 Channahon Rd., Joliet, IL 60434	Short-path distillation systems
Norton	P.O. Box 350, Akron, OH 44309	Structured and dumped tower packings, trays, other distillation equipment
Nutter Engineering	P.O. Box 700480, Tulsa OK 74170	Structured and dumped tower packings, trays, other column internals
Otto York Company, Inc.	42 Intervale Rd., P.O. Box 3100, Parsippany, NJ 07054-0918	Tower packings, mist eliminators
Petro Ware Inc.	P.O. Box 220, 713 Keystone St., Crooksville, OH 43731	Ceramic random packing
Pope	N90 W14337 Commerce Dr., P.O. Box 495, Menomonee Falls, WI 53051	Wiped-film stills
Praxair, Inc.	P.O. Box 44, Tonawanda, NY 14151-0044	Trays, other internals for air separation
Sulzer Chemtech Ltd.	P.O. Box 65, CH-8404 Winterthur, Switzerland 052-262 11 22	Packings and trays including new Optiflow packing.
Sulzer Canada, Inc.	60 Worcester Rd., Rexdale (Toronto)/Ontario, Canada M9W 5X2	Packings and trays including new Optiflow packing.
UOP	175 E. Park Dr., Tonawanda, NY 14161-0044	Trays
Vereinigte Fullkorper-Fabriken	P.O. Box 2020, Rheinstr. 176, Ransbach-Baumbach 2, Germany	Tower packings, column internals
Wright-Austin Co.	3250 Franklin St., Detroit, MI 48207	Gas/liquid separators

TABLE B.2   Some Suppliers of Extraction Equipment

Company	Address	Product description
APV Chemical Machinery	1000 Hess St., Saginaw, MI 48601	Centrifugal contactors
CCS Computer Chemical Systems	Rt. 42 & Newark Rd., P.O. Box 683, Avondale, PA 19311	Computerized supercritical fluid chromatograph
CF Technologies, Inc.	One Westinghouse Plaza, Ste. 200, Hyde Park, MA 02136-2059	Supercritical fluid extraction equipment
Chem-Pro	A division of Otto H. York Co., P.O. Box 3100, Parsippany, NJ 07054-0918	Liquid extraction equipment
Glitsch, Inc.	P.O. Box 660053, Dallas, TX 72566-0053	Supercritical fluid extraction equipment
Hoechst Celanese Corporation	13800 South Lakes Dr., Charlotte, NC 28273	Membrane phase contactors for extraction
Koch Engineering Co., Inc.	P.O. Box 8127, Wichita, KS 67208	Structured packings for extraction
KŪHNI	CH-4123 Allschwil-Basle, Muhlebachweg 9-15, Switzerland	Solvent extraction columns
Lee Scientific	4426 South Century Dr., Salt Lake City, UT 84123-2513	Supercritical fluid chromatography
Milton Roy Co.	201 Ivyland Rd., Ivyland, PA 18974-0577	Supercritical fluid extraction equipment
Newport Scientific, Inc.	8246-E Sandy Court, Jessup, MD 20794-0189	Supercritical fluid extraction equipment
Osmonics, Inc.	5951 Clearwater Dr., Minnetonka, MN 55343	Liquid/liquid coalescers
Otto York Co., Inc.	42 Intervale Rd., P.O. Box 3100, Parsippany, NJ 07054-0918	Mechanically aided extractors; other extraction equipment
Phasex Corporation	360 Merrimack St., Lawrence, MA 01843	Supercritical fluid extraction technology
Sepracor Inc.	33 Locke Dr., Marlborough, MA 01752	Membrane solvent extraction

Some Suppliers of Adsorbents and Adsorption Processes

The companies listed in Table B.3 are among those with substantial businesses in the sale of adsorbents and adsorption processes. Much more extensive lists of companies can be found, for example, in (1) the *Thomas Register* under Adsorbents, Adsorbers, Adsorbers: Carbon, Dryers: Adsorptive, Filters: Activated Carbon, Filters: Adsorption, and Filters: Air; and (2) the *Chemical Week Annual Buyers' Guide* under Adsorbents, and Adsorption Systems.

TABLE B.3   Some Suppliers of Adsorbents and Adsorption Processes

Company	Address	Product description
Advanced Separation Technologies, Inc.	5315 Great Oak Drive, Lakeland, FL 33801-3180	ISEP continuous adsorption process
Air Products and Chemicals, Inc.	7201 Hamilton Blvd., Allentown, PA 18195	Carbon molecular sieves, PSA and VSA processes
Alcoa Industrial Chemicals Division	P.O. Box 300, Bauxite, AR 72011	Activated Al_2O_3, other Al_2O_3-based adsorbents
American Norit Co.	420 Agmac Ave., Jacksonville, FL 32205	Activated carbon
Artisan Industries	73 Pond St., Waltham, MA 02254	Activated carbon-based processes
Atochem, Inc.	266 Harristown Rd., Glen Rock, NJ 07452	Activated carbon
Barnebey & Sutcliffe Corp.	835 N. Cassady Ave., Columbus, OH 43216	Activated carbon, activated carbon regeneration, activated carbon-based processes
Bergbau Forschung	Franz-Fischer-Weg, 4300 Essen 13 (Kray), GmbH, Germany	Carbon molecular sieves (CMS) and CMS-based processes
Calgon Carbon Corp.	P.O. Box 717, Pittsburgh, PA 15230	Activated carbon and activated carbon-based processes
Chematur Engineering AB	Box 430, S-691, 27 Karlskoga, Sweden	Moving-bed adsorption processes
C. M. Kemp Mfg. Co.	490 Baltimore-Annapolis Blvd., Glen Burnie, MD 21061	CMS, CMS-based processes, activated carbon and activated carbon-based and other adsorbent-based processes
Culligan International Co.	1 Culligan Parkway, Northbrook, IL 60082	Activated carbon and activated carbon-based processes
Durr Industries, Inc.	14492 Sheldon Rd., Plymouth, MI 48170	Wheel-based processes
Envirogen, Inc.	Princeton Res. Ctr., 4100 Quakerbridge Rd., Lawrenceville, NJ 08648	Biosorption processes
ICI Americas, Inc.	P.O. Box 15391, Wilmington, DE 19850	Activated carbon
ICI Katalco	Two Transom Plaza Dr., Oakbrook Terrace, IL 60181	Irreversible adsorbents
LaRoche Chemical Co.	P.O. Box 1031, Airline Highway, Baton Rouge, LA 70821	Activated Al_2O_3
Munters Zeol	Kalksteenscagen 1, S-22 378 Lund, Sweden	Adsorbent wheel–based processes

TABLE B.3   Some Suppliers of Adsorbents and Adsorption Processes (*Continued*)

Company	Address	Product description
Norton Co.	60 E. 42nd St., New York, NY 10017	Activated Al_2O_3
Progress Water Technologies, Inc.	P.O. Box 33042, St. Petersburg, FL 33733	ISEP continuous adsorption processes
Seibu Giken Co., Ltd.	1043-5 Wada, Sasaguri-Machi, Kasuya-Gun, Fukuoka, Japan/T811-24	Adsorbent wheels and wheel-based processes
Seitetsu Kagaku Co., Ltd.	Sumitomo Bldg, No. 2, 4-7-28, Kitahama, Chuo-ku, Osaka 541, Japan	PSA processes
Tigg Corp.	P.O. Box 11661, Pittsburgh, PA 15228	Activated carbon and activated carbon-based processes
UOP	25 E. Algonquin Rd., Des Plaines, IL 60017	ZMS, silicalite, and ZMS- and silicalite-based processes
U.S. Filter/IWT	4669 Shepherd Trail, Box 560, Rockford, IL 61105	Various ZMS- and activated carbon-based processes
Westates Carbon	2130 Leo Ave., Los Angeles, CA 90040	Carbon, impregnated carbons, aluminas, solvent recovery, and regeneration
Westvaco Corp.	299 Park Ave., New York, NY 10171	Activated carbon
W. R. Grace & Co. (Davidson)	5500 Chemical Blvd., Baltimore, MD 21226	Activated Al_2O_3, SiO_2, ZMS

Abbreviations: ISEP, ZMS (zeolite molecular sieves).

Suppliers of Membranes and Membrane Processes

Membrane and membrane-process-equipment suppliers are legion. Below are listed some representative companies active in this area. The reader should be aware, however, that this list is far from complete. Further listings are given in the book by Ho and Sirkar (1992) in the references section of Chapter 5. Other listings can be found in the *Thomas Register* and in the *Chemical Week Annual Buyers' Guide*. The list in Table B.4 is more nearly complete in its listing of U. S. companies.

TABLE B.4   Some Suppliers of Membranes and Membrane Processes

Company	Address	Product description
Amicon, Inc.	72 Cherry Hill Dr., Beverly, MA 01915	RO, UF modules and systems
Air Products & Chemicals Corp. (Permea)	7201 Hamilton Blvd., Allentown, PA 18195	Gas-separation modules and systems
Aqua-Chem, Inc., Water Technologies Div.	Box 421, Milwaukee, WI 53201	RO systems
Aqualytics (division of The Graver Co.)	7 Powder Horn Dr., P.O. Box 4904, Warren, NJ 07059	ED modules and systems
Bend Research, Inc.	64550 Research Rd., Bend, OR 97701-8599	RO, UF modules and systems
Carbone Lorraine	540 Branch Dr., P.O. Box 1189, Salem, VA 24153	Pervaporation processes
CeraMem Separations	12 Clematis Ave., Waltham, MA 02154	Ceramic membranes for UF and MF
Desalination Systems, Inc.	1238 Simpson Way, Escondido, CA 92025	RO, UF modules
Dow Corp. (Generon)	Box 4641, Houston, TX 77210	Gas-separation modules and systems
E.I. DuPont de Nemours & Co.	Permasep Products, Glasgow Site, Wilmington, DE 19898	RO modules
Fluid Systems Co.	10124 Old Grove Rd., San Diego, CA 92151	RO, UF modules
GFT	400 Myrtle Ave., Boonton, NJ 07005	Vapor permeation modules and systems
GKSS-Forschungs-zentrum Geesthacht GmbH	D-21502 Geesthacht, Germany	RO, UF, gas-separation membranes and modules
Grace Membrane Systems	7125 W. Tidwell, Houston, TX 77092	Gas-separation modules and systems
Hoechst Celanese	13800 South Lakes Dr., Charlotte, NC 28217	Hollow fiber solvent extraction modules
Hydranautics Co. (Nitto Denko)	8444 Miralani Dr., San Diego, CA 92126	RO, UF, MF modules and systems
Ionics, Inc.	65 Grove St., Watertown, MA 02172	ED, MF modules and systems
Koch Membrane Systems	850 Main St., Wilmington, MA 01887	RO, UF modules; UF systems
LCI Corp.	Box 16348, Charlotte, NC 28297	RO, UF systems

TABLE B.4   Some Suppliers of Membranes and Membrane Processes (*Continued*)

Company	Address	Product description
Membrane Products kiryat Weizmann	Rehovot, Israel	RO, UF modules and systems
Millipore Corp.	80 Ashby Rd., Bedford, MA 10730	UF, ED, MF modules; UF, ED, MF systems
MTR Corp.	1360 Willow Rd., Menlo Park, CA 94025	Vapor permeation and pervaporation modules and systems
New Logic, Intl.	1155 Park Ave., Emeryville, CA 94608-3631	Mechanically agitated MF and UF systems
Osmonics, Inc.	5951 Clearwater Drive, Minnetonka, MN 55343	RO, UF, MF modules and systems
Osmonics, Inc. (Japan)	2-2 Nihonbashi-muromachi Chuo-ku, Tokyo 103, Japan	RO, UF, MF modules and systems
UOP	50 East Algonquin Rd., P.O. Box 5016, Des Plaines, IL 60017	Gas-separation systems
U.S. Filter/IWT	4669 Shepherd Trail, Box 560, Rockford, IL 61105	Ceramic membranes for UF and MF
Zenon Environmental, Inc.	13 Estates Drive, Sussex, NJ 07461	RO, UF, pervaporation systems

C

Terminology For Membranes and Membrane Processes[1]

International Union of Pure and Applied Chemistry
Macromolecular Division
Working Party on Membrane Nomenclature

Contents

General Terms

1. asymmetric membrane: membrane (§18) constituted of two or more structural planes of non-identical morphologies.

2. co-current flow: flow pattern through a membrane module in which the fluids on the upstream and the downstream sides of the membrane move parallel to the membrane surface and in the *same* directions. (*Note:* see Figure C.1a.)

3. completely-mixed (perfectly-mixed) flow: flow through a membrane module in which the fluids on both the upstream and downstream sides of the membrane are individually well-mixed. (*Note:* see Figure C.1b.)

4. composite membrane: membrane having chemically or structurally distinct layers.

---

[1]Koros, W. J., Y. H. Ma, and T. Shimidzu, "Terminology for Membranes and Membrane Processes," terminology has been submitted for publication to International Union of Pure and Applied Chemistry (IUPAC), Copyright © 1996.

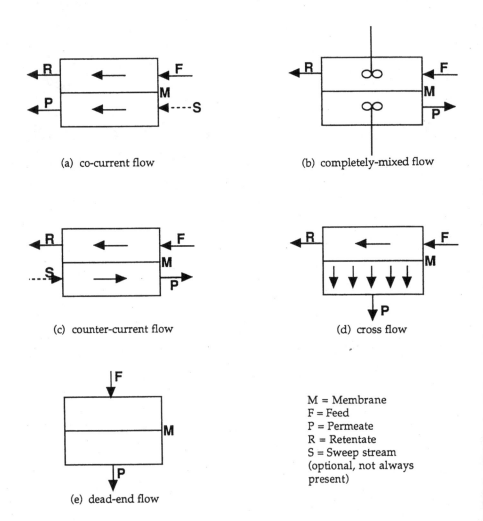

(a) co-current flow

(b) completely-mixed flow

(c) counter-current flow

(d) cross flow

(e) dead-end flow

M = Membrane
F = Feed
P = Permeate
R = Retentate
S = Sweep stream
(optional, not always
present)

Figure C.1 Types of ideal continuous flows used in membrane-based separations.

5. continuous membrane column: membrane module(s) arranged in a manner to allow operation analogous to that of a distillation column, with each module acting as a stage.

6. counter-current flow: flow through a membrane module in which the fluids on the upstream and downstream sides of the membrane move parallel to the membrane surface but in *opposite* directions. (*Note:* see Figure C.1c.)

7. cross flow: flow through a membrane module in which the fluid on the upstream side of the membrane moves parallel to the membrane surface and the fluid on the downstream side of the membrane moves away from the membrane in the direction normal to the membrane surface. (*Note:* see Figure C.1d.)

8. dead-end flow: flow through a membrane module in which the only outlet for upstream fluid is through the membrane. (*Note:* see Figure C.1e.)

9. dense (non-porous) membrane: membrane with no detectable pores.

10. downstream: side of a membrane from which permeate emerges.

11. dry-phase separation membrane formation: process in which a dissolved polymer is precipitated by evaporation of a sufficient amount of solvent to form a membrane structure. (*Note:* Appropriate mixtures of additives are present in solution with the polymer to alter its precipitation tendency during solvent evaporation.)

12. dry-wet phase separation membrane formation: combination of the dry (§11) and the wet-phase formation processes (§45).

13. dynamic membrane formation: process in which an active layer is formed on the membrane surface by the deposition of substances contained in the fluid being treated.

14. flux, J_i, [kmol m^{-2} s^{-1}]: number of moles, volume, or mass of a specified component i passing per unit time through a unit of membrane surface area normal to the thickness direction. {*Note:* other commonly used units for J_i include [m^3 m^{-2} s^{-1}], or [kg m^{-2} s^{-1}], or [m^3 (measured at standard temperature and pressure) m^{-2} s^{-1}].}

15. fouling: process resulting in loss of performance of a membrane due to the deposition of suspended or dissolved substances on its external surfaces, at its pore openings, or within its pores.

16. homogeneous membrane: membrane with essentially the same structural and transport properties throughout its thickness.

17. Langmuir-Blodgett (LB) membrane: synthetic composite membrane formed by sequential depositing of one or more monolayers of surface-active component onto a porous or nonporous support.

18. membrane: structure, having lateral dimensions much greater than its thickness, through which mass transfer may occur under a variety of driving forces.

19. membrane compaction: compression of membrane structure due to a pressure difference across its thickness.

20. membrane conditioning (pretreatment): process carried out on a membrane after the completion of its preparation and prior to its use in a separation application. (*Note 1:* thermal annealing to relieve stresses or pre-equilibriation in a solution similar to the feed stream it will contact are examples of conditioning treatments.) [*Note 2:* conditioning treatments differ from post-treatments (§25) since the latter occur before exposure to feed-type solutions, while conditioning may occur using actual feed solutions.]

21. membrane distillation: distillation process in which the liquid and gas phases are separated by a porous membrane, the pores of which are not wetted by the liquid phase.

22. membrane module (cell): manifold assembly containing a membrane or

(a) Hollow fiber

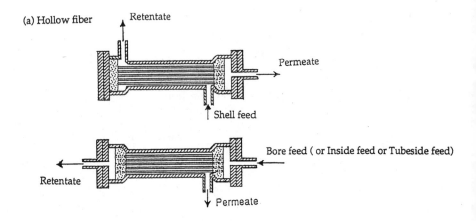

(b) Plate-and-frame

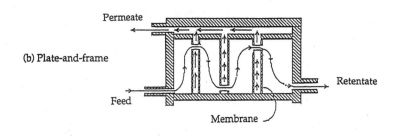

(c) Spiral wound

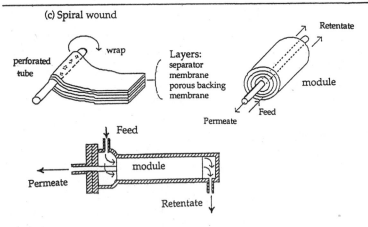

Figure C.2 Types of modules used in membrane-based separation.

membranes to separate the streams of feed, permeate, and retentate. (*Note:* see Figure C.2a–c.)

23. membrane partition (distribution) coefficient: parameter equal to the equilibrium concentration of a component ($c_i^{(m)}$ in a membrane divided by the corresponding equilibrium concentration of the component in the external phase in contact with the membrane surface, $c_i^{(e)}$ (viz., $K = c_i^{(m)}/c_i^{(e)}$).

24. membrane physical aging: change in the transport properties of a membrane over a period of time due to physical chemical structural alterations.

25. membrane post-treatment: process carried out on a membrane after its essential structure has been formed but *prior* to its exposure to an actual feed stream (§20).

26. membrane reactor: device for simultaneously carrying out a reaction and membrane-based separation in the same physical enclosure.

27. penetrant (permeant): entity from a phase in contact with one of the membrane surfaces that passes through the membrane.

28. permeability coefficient, P_i, [kmol m m^{-2} s^{-1} kPa^{-1}]: parameter defined as a transport flux, J_i, per unit transmembrane driving force per unit membrane thickness, ℓ, viz., $P_i = J_i/[(\text{transmembrane driving force of component } i/\ell)$ {*Note:* other commonly used units for P_i include [m^3 m m^{-2} s^{-1} kPa^{-1}], [kg m m^{-2} s^{-1} kPa] or [m^3 (measured at standard temperature and pressure m m^{-2} s^{-1} kPa^{-1})].}

29. permeance (pressure normalized flux), [kmol m^{-2} s^{-1} kPa^{-1}]: transport flux per unit transmembrane driving force, viz., P_i/ℓ (§28). {*Note:* other commonly used units include [m^3 m^{-2} s^{-1} kPa], [kg m^{-2} s^{-1} kPa^{-1}], or [m^3 (measured at standard temperature and pressure) m^{-2} s^{-1} kPa^{-1}].}

30. permeate: stream containing penetrants that leaves a membrane module. (*Note:* see Figures C.1 and C.2.)

31. perstraction: separation process in which membrane permeation and extraction phenomena occur by contacting the downstream with an extracting solvent.

32. relative recovery $\eta_{n,B}$ (substance efficiency): amount-of-substance of a component B collected in a useful product, $\eta_{B,out}$, divided by the amount-of-substance of that component entering the process, $\eta_{B,in}$, viz., $\eta_{\eta,B} = \eta_{B,out}/\eta_{B,in}$. [*Note:* in membrane separations, the useful product may be either the retained material (or retentate) or the permeated material (or permeate).]

33. rejection factor, R: parameter equal to one minus the ratio the concentrations of a component (i) on the downstream and upstream sides of a membrane {*Note 1:* $R = 1-[(c_i)_{downstream}/(c_i)_{upstream}]$.} [*Note 2:* concentrations may be either in the bulk ("Apparent Rejection Factor") or at the membrane surface ("Intrinsic Rejection Factor").] [*Note 3:* rejection factor refers to a local relationship between upstream and downstream concentrations while retention factor (§35) and relative recovery (§32) refer to feed and retentate or permeate leaving the module.]

34. retentate (raffinate): stream that has been depleted of penetrants that leaves the membrane modules without passing through the membrane to the downstream.

35. retention factor, r_F: parameter defined as one minus the ratio of permeate concentration to the retentate (§37) concentration of a component (i) {*Note*

1: $r_F = 1-[(c_i)_p/(c_i)_r]$} [*Note 2:* p and r refer to permeate (§30) and retentate (§34). See Figure C.1.] [*Note 3:* Compare rejection factor (§33).]

36. selective membrane skin: region, often located at the upstream face of an asymmetric membrane, that forms a thin, distinguishable layer primarily responsible for determining the permeability of the asymmetric membrane.

37. separation coefficient, S_C(AB):   ratio of the compositions of component A and B in the downstream relative to the ratio of compositions of these components in the upstream {*Note 1:* For example, if compositions are expressed in mole fractions (X_A and X_B), $S_C(AB) = [X_A/X_B]_{downstream}/[X_A/X_B]_{upstream}$} (*Note 2:* The separation coefficient can also be defined equivalently in terms of concentrations in the downstream and upstream, since only ratios are involved) [*Note 3:* The separation coefficient refers to a local relationship between concentrations on the upstream (§44) and downstream (§10) concentrations while the separation factor (§38) refers to retentate (§34) and permeate (§30) leaving the module.]

38. separation factor, S_F(AB):   ratio of the compositions of components A and B in the permeate relative to the composition ratio of these components in the retentate {*Note 1:* For example, $S_F(AB) = [X_A/X_B]_{Permeate}/[X_A/X_B]_{Retentate}$} (*Note 2:* The separation factor can also be defined in terms of concentrations in the permeate and retentate since only ratios are involved. *Note:* see Figure C.1.) [*Note 3:* Compare separation coefficient (*see* §37).]

39. sol-gel membrane formation: multistep process for making membranes by a reaction between two chemically multifunctional materials, dissolved in a solvent, that results in a network structure with solvent retained in the network followed by heat treatment to achieve a desired pore structure.

40. stage cut: parameter defined as the fractional amount of the total feed entering a membrane module that passes through the membrane as permeate (*Note:* see Figure C.1.)

41. synthetic (artificial) membrane: membrane formed by a process not occurring in nature.

42. thermally-induced phase-separation membrane formation: process in which a dissolved polymer is precipitated or coagulated by controlled cooling to form a membrane structure.

43. track-etch membrane formation: process for forming porous membranes with well-defined pores by exposing a dense film to ion bombardment followed by etching of the damaged region (*Note:* Usually produces pores with a narrow size distribution.)

44. upstream: side of a membrane into which penetrants enter from the feed stream.

45. wet-phase separation membrane formation: process in which a dissolved polymer is precipitated by immersion in a non-solvent bath to form a membrane structure.

Carrier-Mediated (Facilitated) Separations

46. anchored (bound) carrier: distinct species bonded chemically to fixed sites within a membrane for the purpose of increasing the selective sorption and flux of a specific component in a feed stream relative to all other components.

47. carrier complexation coefficient, K_c, [kmol m^{-3}]: parameter defined as the ratio of the rate constants for the second order complexation and first order decomplexation reaction between a carrier and a penetrant; viz., $A + M = AM$ [Note 1: $K_c = k_c/k_d$, where A & M, resp., are a penetrant and a carrier site within a membrane (see note 2).] [Note 2: both anchored carrier sites (§46) and mobile carrier sites (§59) are possible.]

48. carrier complexation: phenomenon in which carrier molecules form a coordinated structure with penetrant molecules.

49. carrier deactivation: chemical transformations involving a carrier entity which render it less capable of undergoing the desired interaction with a penetrant.

50. carrier leaching: loss of carrier due to its partitioning by mass transport into one or both external phases.

51. carrier-mediated (facilitated) transport: process in which chemically distinct carrier species (§46, §59) form complexes with a specific component in the feedstream, thereby increasing the flux (§14) of this component relative to other components.

52. complexation rate constant, k_c, [kmol^{-1} m^3 s^{-1}]: carrier complexation rate divided by the product of the local concentrations of the carrier and the complexable component, viz., $k_c = $ (complexation rate)/$[(c)_{carrier}(c)_{complexable\ component}]$ where concentrations are given in [kmol m^{-3}] and complexation rate is given in [kmol m^{-3} s^{-1}].

53. coupled transport: process in which the flux of one component between the upstream and downstream is linked to the flux of a second component.

54. Damkohler number: dimensionless number equal to the characteristic time (ℓ^2/D_{AM} for diffusion of complexed component across a membrane of thickness, ℓ, divided by the characteristic time (k_d^{-1}) for the decomplexation reaction between a carrier (M) and a complexed penetrant, A, viz., $\ell^2/D_{AM}k_d$) when D_{AM} is the effective diffusion coefficient of the complexed carrier entity in the membrane.

55. decomplexation rate constant, k_d, [s^{-1}]: ratio of the decomplexation rate to the product of the local concentration of the complexed carrier, viz., $k_d = $ (decomplexation rate)/$(c)_{complexed\ carrier}$ [Note: typical units for decomplexation rate are (kmol m^{-3} s^{-1}), and for complexed carrier are (kmol m^{-3}).]

56. enhancement factor, ε: ratio of the flux of a component, i (§14) across a carrier-containing membrane divided by the transmembrane flux of the same component across an otherwise identical membrane without carrier. {Note: $\varepsilon = [(J_i)_{with\ carrier}/(J_i)_{without\ carrier}].$}

57. facilitation factor: parameter equal to the enhancement factor (§56) minus one. (*Note: F* = $\varepsilon - 1$.)

58. liquid membrane: liquid phase existing either in supported or unsupported form that serves as a membrane barrier between two phases.

59. mobile carrier: distinct species moving freely within a membrane for the purpose of increasing the selective sorption and flux of a specific component in a feed stream relative to all other components.

60. uphill transport: process in which diffusion of a component occurs from a less concentrated feed stream to a more concentrated permeate stream.

Dialysis, Nanofiltration, Ultrafiltration, and Microfiltration Separations

61. backflush: temporary reversal of the direction of the permeate flow.

62. bubble point: pressure at which bubbles first appear on one surface of an immersed porous membrane as gas pressure is applied to the other surface.

63. cake layer: layer comprised of rejected particulate materials residing on the upstream face of a membrane.

64. concentration polarization: concentration profile that has a higher level of solute nearest to the upstream membrane surface compared with the more-or-less well-mixed bulk fluid far from the membrane surface.

65. concentration factor: ratio of the concentration of a component i in the retentate to the concentration of the same component in the feed {*Note 1:* $c_F =$ $[(c_i)_{\text{retentate}}/(c_i)_{\text{feed}}]$ (*Note:* see Figure C.1.) [*Note 2:* Compare retention factor (§35).]

66. dialysis: membrane process in which transport is driven primarily by concentration differences, rather than by pressure or electrical-potential differences, across the thickness of a membrane.

67. dialysis permeability coefficient: permeability coefficient (§28) based on a transmembrane driving force expressed in terms of the concentration difference of a given component.

68. gel fouling layer: highly swollen fouling layer comprising a three-dimensional, possibly network, structure residing at the surface of a membrane.

69. hemodialysis: dialysis (§66) in which undesired metabolites and toxic by-products, such as urea and creatine, are removed from blood.

70. hemofiltration: ultrafiltration (§75) in which undesired metabolites and toxic by-products, such as urea and creatine, are removed from blood.

71. hindered transport: combined partition, diffusion, and convection process in which the effective partition, diffusion, and viscous drag coefficients in a restricted environment depend upon the ratio of the effective radius of the penetrant molecule to that of the pore.

72. microfiltration: pressure-driven membrane-based separation process in which particles and dissolved macromolecules larger than 0.1 μm are rejected.

73. molecular-weight cutoff: molecular weight of a solute corresponding to a 90% rejection coefficient (§33) for a given membrane.

74. nanofiltration: pressure-driven membrane-based separation process in which particles and dissolved molecules smaller than about 2 nm are rejected.

75. ultrafiltration: pressure-driven membrane-based separation process in which particles and dissolved macromolecules smaller than 0.1 μm and larger than about 2 nm are rejected.

Electrically Mediated Separations

76. anion-exchange membrane: membrane containing fixed cationic charges and mobile anions that can be exchanged with other anions present in an external fluid in contact with the membrane.

77. bipolar membrane: synthetic membrane containing two oppositely charged ion-exchanging layers in contact with each other.

78. cation-exchange membrane: membrane containing fixed anionic charges and mobile cations which can be exchanged with other cations present in an external fluid in contact with the membrane.

79. charge-mosaic membranes: synthetic membrane composed of two- or three-dimensional alternating cation- and anion-exchange channels throughout the membrane.

80. Donnan exclusion: reduction in concentration of *mobile* ions within an ion exchange membrane due to the presence of *fixed* ions of the same sign as the mobile ions.

81. electro-dialysis: membrane-based separation process in which ions are driven through an ion-selective membrane under the influence of an electric field.

82. electro-osmosis: process by which water is transported across the thickness of an anion-exchange (§76) or cation-exchange membrane (§78) under an applied electric field.

83. limiting current density: current density at which dramatic increases in resistance are observed in an ion-exchange membrane system under the influence of an applied electric field between the upstream and downstream.

Gas, Vapor, and Pervaporation Separations

84. ideal separation factor: parameter defined as the ratio of the permeability coefficient of component A to that of component B and equal to the "separation factor" (§37) where a perfect vacuum exists at the downstream membrane face for gas and vapor permeation systems.

85. pervaporation: membrane-based process in which the feed and retentate streams are both liquid phases while permeant emerges at the downstream face of the membrane as a vapor.

86. solution-diffusion (sorption-diffusion):   molecular-scale process in which penetrant is sorbed into the upstream membrane face from the external phase, moves by molecular diffusion in the membrane to the downstream face and leaves into the external gas, vapor, or liquid phase in contact with the membrane.

87. sweep: nonpermeating stream directed past the downstream membrane face to reduce downstream permeant concentration.

Reverse Osmosis Separations

88. brackish water: water having a total dissolved-solids content that is less than that of sea water but above that of potable water.

89. feed pretreatment: process carried out on a crude feed stream, prior to feeding to a membrane separation system, to eliminate objectionable components such as biological agents and colloids that might impede the stable operation of the membrane.

90. permeate post-treatment: one or more final conditioning steps to improve permeate quality, e.g., contacting with anion-exchange resins to remove trace ions in the permeate of a reverse osmosis product stream.

91. potable water: term used to indicate water having a total dissolved solids content of less than 500 ppm with a sufficiently low level of biological agents, suspended solids, organic odour- and colour-generating components to be safe and palatable for drinking.

92. reverse osmosis: liquid-phase pressure-driven separation process in which applied transmembrane pressure causes selective movement of solvent against its osmotic pressure difference.

References

Audinos, R. and P. Isoard, eds., *Glossaire des termes techniques des procédés à membranes;* France, Société Française de Filtration, 1986.

Glossary of Atmospheric Chemistry Terms, compiled by Jack G. Calvert, Applied Chemistry Division, Commission on Atmospheric Chemistry, IUPAC, 1990.

Porter, Mark, *Handbook of Industrial Membrane Technology,* Park Ridge, NJ: Noyes Publications, 1990.

Quantities, Units and Symbols in Physical Chemistry, I. M. Mills et al., Blackwell Scientific, 1993.

Standard D1129-90, ASTM Committee on Water, Subcommittee on Membrane and Ion Exchange, D19.08, Vol. 11.01, April 1991.

Standard D5090, ASTM Committee on Water, Subcommittee on Membrane and Ion Exchange D19.080, Vol. 11.02, May 1991.

Terminology for Electrodialysis, prepared by Karl Hattenback, European Society of Membrane Science and Technology, issued November 1988.

Terminology for Membrane Distillation, prepared by A. C. M. Franken and S. Ripperger, University of Twente.

Terminology for Pressure Driven Membrane Operations, prepared by Vassilis Gekas, European Society of Membrane Science and Technology, issued June 1986.

Terminology in Pervaporation, prepared by K. W. Boddeker, European Society of Membrane Science and Technology, issued November 1989.

Conversion Factors: English to System International Units

TABLE D.1 Conversion Factors: English to SI Units

English units	Multiply by	To obtain SI units
Acceleration		
feet/(second^2)	0.3048	meter/(second^2)
Mass		
pound mols	453.592374	mols
pound mols	0.453592374	kilogram mols
Area		
sq ft	0.09290304	sq meters
Density		
pounds mass/cu ft	16.01846	kilograms/(cu meter)
pounds mass/gallon	119.83946	kilograms/(cu meter)
Diffusion coefficient		
sq ft/second	0.09290304	sq meters/second
Energy		
Btu	1055.056	joules
foot pounds force	1.35582	joules
Force		
pounds force	4.44822	newtons
Heat flux		
Btu/(sq ft second)	11356.5	watts/(sq meter)
Heat transfer coefficient		
Btu(hr sq ft °R)	5.6782	watts/(sq meter K)
Interfacial tension		
pound force/ft	14.5939	newton/meter

TABLE D.1   Conversion Factors: English to SI Units (*Continued*)

English units	Multiply by	To obtain SI units
	Kinematic viscosity	
sq ft/second	0.09290304	sq meters/second
	Length	
feet	0.3048	meters
	Mass	
pounds mass	0.453592374	kilograms
	Mass concentration	
pounds mass/(cu ft)	16.01846	kilograms/(cu meter)
	Mass flux	
pound mass/(sq ft second)	4.882428	kilogram/(sq meter second)
	Mass transfer coefficient	
pound mass/(sq ft second)	4.8824	kilogram/(sq meter second)
pound force second/cu ft	157.09	kilogram/(sq meter second)
	Molar concentration	
pound mols/(cu ft)	16018.463	kilogram mols/(cu meter)
	Molar flux	
pound mols/(sq ft second)	4882.428	mols/(sq meter second)
	Permeability	
lb mols/(ft second pound force/sq ft)	31.0809	mols/(meter second pascal)
lb mols/(ft second pound force/sq in)	0.2158397	mols/(meter second pascal)
cu ft/(ft second pound force/sq ft)	0.0019403	cu meters/(meter second pascal)
cu ft/(ft second pound force/sq in)	1.35E-05	cu meters/(meter second pascal)
	Power	
(ft pounds force)/second	1.355818	watts
horsepower	745.7	watts
	Pressure	
pounds force/(sq ft)	47.8803	pascals
pounds force/(sq in)	6894.76	pascals
atmospheres	101325	pascals

TABLE D.1   Conversion Factors: English to SI Units (*Continued*)

English units	Multiply by	To obtain SI units
Specific heat		
Btu/(pound mass °R)	4186.801	joule/(kilogram K)
Btu/(cu ft °R)	67066.112	joule (cu meter K)
Btu/(pound mol R)	4.1868	joule/(mol K)
Surface tension		
pound force/ft	14.5939	newton/meter
Temperature		
(°F-32)	5/9	°C
°R	5/9	K
Thermal conductivity		
Btu/(hr ft °R)	1.7307	watts/(meter K)
Velocity		
ft/second	0.3048	meters/second
Viscosity		
pound mass/(ft second)	1.4882	kilogram/(meter second)
pound mass/(ft second)	14.882	poise
(pound force second)/(sq ft)	47.88	kilogram/(meter second)
(pound force second)/(sq ft)	478.8	poise
Volume		
cu ft	0.028316847	cu meters
gallons	0.003785	cu meters
Volumetric flow rate		
cu ft/second	0.028316847	cu meters/second
gallons/second	0.003785	cu meters/second
Ideal gas constant		
—	8314	(cu meters pacals/
—	8.314	(mol K)
J/(mol K)		
(cu ft atmospheres)/	0.7302	—
(pound mol °R)	1.987	—
Btu/(pound mol °R)		

Derived SI units as follows: 1 Newton [=] 1 kilogram meter/(second ^2); 1 Pascal [=] 1 Newton/(sq meter); 1 joule [=] 1 Newton meter; 1 Watt [=] 1 joule/s.

Derived unit conversions for SI-cgs as follows: 1 poise [=] 0.1 kilogram/(meter second); 1 dyne [=] 1 E-5 newton.

Index

ABOUT THE AUTHORS

JIMMY L. HUMPHREY is President of J.L. Humphrey & Associates, a consulting firm which specializes in application of leading-edge separation technologies. He has a Ph.D. in chemical engineering and is an Adjunct Professor at the University of Texas at Austin, where he is former associate head of the Separations Research Program. Dr. Humphrey is also former chair of the Separations Division of the American Institute of Chemical Engineers and a former Production Manager in the chemical industry. He has been elected Fellow of the American Institute of Chemical Engineers.

GEORGE E. KELLER II is Senior Corporate Research Fellow and Manager of the Separations and Process Fundamentals Group at Union Carbide Chemicals and Plastics Co., Inc., in South Charleston, West Virginia. Dr. Keller has received a number of prestigious awards, including an Outstanding Achievement Award from *Chemical Engineering* magazine and the Gerhold Award for contributions in separations. He was named a Chemical Pioneer by the American Institute of Chemists for his work in long-range hydrocarbon technology. He holds a Ph.D. in chemical engineering. He has been elected to the National Academy of Engineering and is a Fellow of the American Institute of Chemical Engineers.